FM 3-05.210

Special Forces Air Operations

February 2009

DISTRIBUTION RESTRICTION: Distribution authorized to U.S. Government agencies and their contractors only to protect technical or operational information from automatic dissemination under the International Exchange Program or by other means. This determination was made on 2 January 2009. Other requests for this document must be referred to Commander, United States Army John F. Kennedy Special Warfare Center and School, ATTN: AOJK-DTD-SF, Fort Bragg, NC 28310-9610.

DESTRUCTION NOTICE: Destroy by any method that will prevent disclosure of contents or reconstruction of the document.

FOREIGN DISCLOSURE RESTRICTION (FD 6): This publication has been reviewed by the product developers in coordination with the United States Army John F. Kennedy Special Warfare Center and School foreign disclosure authority. This product is releasable to students from foreign countries on a case-by-case basis only.

Headquarters, Department of the Army

This publication is available at
Army Knowledge Online (www.us.army.mil) and
General Dennis J. Reimer Training and Doctrine
Digital Library at (www.train.army.mil).

FM 3-05.210, C1

CHANGE NO. 1

Headquarters
Department of the Army
Washington, DC, 14 April 2010

SPECIAL FORCES AIR OPERATIONS

In accordance with United States Army Special Operations Command (USASOC) Regulation 350-2, *Training: Airborne Operations*, dated 4 June 2008, FM 3-05.210, dated 27 February 2009, is changed as follows:

    1. New or changed material is identified by a vertical bar (|) in the margin opposite the changed material.

    2. File this transmittal sheet in front of the publication for reference purpose.

    3. Remove old pages and insert new pages as indicated below:

Remove old pages:	**Insert new pages:**
7-1 and 7-2	7-1 and 7-2

DISTRIBUTION RESTRICTION: Distribution authorized to U.S. Government agencies and their contractors only to protect technical or operational information from automatic dissemination under the International Exchange Program or by other means. This determination was made on 2 January 2009. Other requests for this document must be referred to Commander, United States Army John F. Kennedy Special Warfare Center and School, ATTN: AOJK-DTD-SF, Fort Bragg, NC 28310-9610.

DESTRUCTION NOTICE: Destroy by any method that will prevent disclosure of contents or reconstruction of the document.

FM 3-05.210, C1
14 April 2010

By Order of the Secretary of the Army:

GEORGE W. CASEY, JR.
General, United States Army
Chief of Staff

Official:

[signature]

JOYCE E. MORROW
Administrative Assistant to the
Secretary of the Army
1011805

DISTRIBUTION:

Active Army, Army National Guard, and U. S. Army Reserve: To be distributed in accordance with the initial distribution number (IDN) 111113, requirements for FM 3-05.210.

PIN: 081798-001

*FM 3-05.210

Field Manual
No. 3-05.210

Headquarters
Department of the Army
Washington, DC, 27 February 2009

Special Forces Air Operations

Contents

		Page
	PREFACE	x
Chapter 1	SPECIAL FORCES AIR OPERATIONS	1-1
	Mission	1-1
	Characteristics of Special Forces Air Missions	1-1
	Types of Special Forces Air Operations	1-2
	Airborne Infiltration Techniques	1-3
	Aerial Resupply	1-9
	Combat Considerations	1-13
Chapter 2	PREMISSION PREPARATION	2-1
	Planning Considerations	2-1
	Emergency Procedures	2-5
	En Route Evasion Plan of Action	2-6
	Joint Mission Briefing	2-7
	Rates of Descent	2-7
Chapter 3	DROP ZONES	3-1
	Selection of Drop Zone	3-1
	Types of Drop Zones	3-9
	Point of Impact	3-13
	Drop Zone Reports	3-14
	Drop Zone Formulas for GMRS and VIRS	3-15
	Drop Zone Markings	3-22

DISTRIBUTION RESTRICTION: Distribution authorized to U.S. Government agencies and their contractors only to protect technical or operational information from automatic dissemination under the International Exchange Program or by other means. This determination was made on 2 January 2009. Other requests for this document must be referred to Commander, United States Army John F. Kennedy Special Warfare Center and School, ATTN: AOJK-DTD-SF, Fort Bragg, NC 28310-9410.

DESTRUCTION NOTICE: Destroy by any method that will prevent disclosure of contents or reconstruction of the document.

FOREIGN DISCLOSURE RESTRICTION (FD 6): This publication has been reviewed by the product developers in coordination with the United States Army John F. Kennedy Special Warfare Center and School foreign disclosure authority. This product is releasable to students from foreign countries on a case-by-case basis only.

*This publication supersedes FM 3-05.210, 31 August 2004.

Contents

	Static-Line Drop Zone Marking Patterns 3-24
	Verbally Initiated Release System 3-26
	MFF Drop Zone Marking Pattern 3-29
	Organization and Operations of a Drop Zone 3-31
	Postmission Requirements 3-37
Chapter 4	LANDING ZONES 4-1
	Authorization 4-1
	Fixed-Wing Landing Zone Training Operations 4-1
	Aircraft Classifications 4-1
	Landing Zone Considerations 4-2
	Helicopter Landing Zones 4-3
	Types of Landing Zones 4-8
	Marking Landing Zones 4-9
	Light and Medium STOL and Medium Aircraft Landing Zones 4-12
	Water Landing Zone (Single- and Twin-Engine Aircraft) 4-27
	Snow Landing Zones 4-31
Chapter 5	ARMY SPECIAL OPERATIONS AVIATION UNITS AND AIRCRAFT 5-1
	Unit Organization 5-1
	Mission 5-1
	MH-6M Helicopter 5-1
	AH-6M Helicopter 5-5
	MH-60L Helicopter 5-10
	MH-60K Helicopter 5-16
	MH-47G Helicopter 5-22
	MH-47E Helicopter 5-27
Chapter 6	AIR FORCE SPECIAL OPERATIONS ORGANIZATION AND AIRCRAFT 6-1
	Unit Organization 6-1
	Concept of Operations 6-2
	Environment 6-2
	Capabilities 6-2
	AFSOF Limitations 6-3
	Augmenting USAF Forces 6-3
	Aircraft Capabilities 6-4
	Combat Aviation Advisory Teams 6-10
	Special Tactics Squadron 6-11
	MC-130E Combat Talon I and MC-130H Combat Talon II 6-11
	MC-130W Combat Spear 6-13
	AC-130H/U Gunship 6-14
	MC-130P Combat Shadow 6-18
	EC-130J Commando Solo 6-21
	C-5/C-17 SOLL II 6-22
	CV-22 Osprey 6-23
	HH-60G Pave Hawk 6-25
Chapter 7	NONSTANDARD AIRCRAFT USED DURING AIRBORNE OPERATIONS 7-1
	C-23B/B+ Sherpa 7-1
	C-27A Spartan 7-5

	CASA-212	7-7
	UV-18B Twin Otter and De Havilland DHC-6 Twin Otter	7-13
	U-21A Ute	7-18
	C-208B Caravan	7-21
Chapter 8	**CARGO SLINGS, AIRDROP CONTAINERS, AND PONCHO-EXPEDIENT PARACHUTE**	**8-1**
	A-Series Containers	8-1
	Rigging Procedures	8-1
	A-7A Cargo Sling	8-2
	A-21 Cargo Bag	8-3
	A-22 Cargo Bag	8-4
	Cargo Parachute Rigging on A-Series Containers	8-6
	Poncho-Expedient Parachute	8-6
	Steel Strapping	8-7
	Rigging Knots	8-7
Chapter 9	**SPECIAL PATROL INFILTRATION AND EXFILTRATION SYSTEM**	**9-1**
	Training Objectives	9-1
	Preoperations Briefings and Procedures	9-1
	Key Personnel Qualifications	9-2
	Personnel Duties and Responsibilities	9-3
	SPIES Equipment	9-5
	Inspection of SPIES	9-6
	Operational Requirements	9-7
	Helicopter Rigging	9-8
	SPIES Operations	9-14
	After-Operations Procedures	9-18
Chapter 10	**FAST-ROPE INSERTION AND EXTRACTION SYSTEM**	**10-1**
	Objectives	10-1
	Guidance for Commanders	10-1
	Safety	10-2
	FRIES Equipment	10-5
	FRIES Hardware Kits	10-8
	Operational Requirements and Limitations	10-11
	FRIES Qualification Training	10-12
	FRM Selection and Qualification Training	10-14
	Key Personnel Duties and Responsibilities	10-15
	Rigging of Aircraft	10-19
	Commands and Signals	10-23
	FRIES Procedures	10-23
	Procedures Specific to Each Aircraft	10-25
	Equipment-Lowering Procedures	10-27
	Emergency Actions	10-30
Chapter 11	**AIR-WATER OPERATIONS**	**11-1**
	Safety	11-1
	Air-Water Operations Qualification Training	11-2
	Personnel Qualification Requirements	11-3
	Personnel Duties and Responsibilities	11-4

Contents

            Operational Requirements 11-6
            Premission Planning 11-7
            Helocasting 11-9
            ERDS or K-Duck 11-11
            ERDS Operation 11-20
            Rolled or Tethered Duck Operations 11-21
            Recovery Operations 11-23

Appendix A	WEIGHTS, MEASURES, AND CONVERSION TABLES	A-1
Appendix B	MOON PHASES	B-1
Appendix C	REPORTS AND REQUESTS	C-1
Appendix D	PIBAL SYSTEM	D-1
Appendix E	MALFUNCTION REPORT	E-1
Appendix F	JUMP PROCEDURES AND JUMPMASTER CHECKLISTS	F-1
Appendix G	FAST-ROPE TROOP BRIEFING AND OPERATIONAL CHECKLIST	G-1
Appendix H	CASTMASTER BRIEFING	H-1
	GLOSSARY	Glossary-1
	REFERENCES	References-1
	INDEX	Index-1

Figures

Figure 1-1. Resupply mission 1-9
Figure 1-2. Typical air resupply mission 1-12
Figure 3-1. Computation of open quadrant and aircraft track (desired heading) 3-8
Figure 3-2. Level turning radius required for one-approach DZs and LZs (medium aircraft) 3-9
Figure 3-3. Level turning radius for STOL or light aircraft 3-9
Figure 3-4. Area DZs 3-11
Figure 3-5. Obstacles and reference points (area DZ) 3-12
Figure 3-6. Example of D = RT computation (parachutists only) 3-16
Figure 3-7. Example of D = RT computation (parachutists and cargo) 3-16
Figure 3-8. Example of T = D/R computation 3-17
Figure 3-9. Determination of the RP by WSVC 3-18
Figure 3-10. D = KAV 3-19
Figure 3-11. Surface wind measurement and MEW 3-20
Figure 3-12. Marking the RP 3-21
Figure 3-13. Mask clearance ratio 15:1 3-23
Figure 3-14. L and T marker patterns 3-25
Figure 3-15. GMRS day and night markings 3-26
Figure 3-16. Day CARP DZ markings 3-27

Contents

Figure 3-17. Night CARP DZ markings 3-27
Figure 3-18. Code letter with masked GMRS letter 3-28
Figure 3-19. Army VIRS offset 3-28
Figure 3-20. Example communication sequence between GTA controller and pilot 3-30
Figure 3-21. MFF DZ markings 3-31
Figure 3-22. DZ mission briefing checklist 3-33
Figure 4-1. Helicopter landing point dimensions 4-4
Figure 4-2. Sample landing points for multiple size 3 aircraft 4-4
Figure 4-3. Slope landing rules 4-6
Figure 4-4. Determining ground slope 4-7
Figure 4-5. Day and night LZ obstruction angle 4-8
Figure 4-6. Y-marked LZ pattern 4-10
Figure 4-7. Platform LZs for rotary-wing aircraft 4-11
Figure 4-8. Preparing mountainous terrain landing pads for rotary-wing aircraft 4-12
Figure 4-9. Basic airfield layout 4-14
Figure 4-10. Training LZ size and night marking requirements for light and medium STOL aircraft 4-17
Figure 4-11. Marking land LZ for light and medium STOL aircraft 4-18
Figure 4-12. Marking land LZ for medium aircraft 4-19
Figure 4-13. Level turning radius required for light and medium STOL aircraft 4-20
Figure 4-14. Level turning radius required for heavy aircraft 4-20
Figure 4-15. Obstacle clearance 4-22
Figure 4-16. Light and medium STOL aircraft wing-tip clearance 4-23
Figure 4-17. Extended STOL LZ for one-direction landing and takeoff of light and STOL aircraft 4-24
Figure 4-18. Land LZ marked with beacons 4-25
Figure 4-19. Snow LZ (STOL aircraft) 4-32
Figure 5-1. MH-6M helicopter 5-2
Figure 5-2. MH-6M and AH-6M aircraft dimensions 5-4
Figure 5-3. MH-6M and AH-6M aircraft dimensions and turning radius 5-4
Figure 5-4. AH-6M helicopter 5-5
Figure 5-5. AH-6M plank system for aircraft weapons configurations 5-6
Figure 5-6. AH-6M weapons variation 5-6
Figure 5-7. AH-6M safety approach areas 5-9
Figure 5-8. MH-60L helicopter 5-10
Figure 5-9. MH-60L FRIES bar 5-12
Figure 5-10. MH-60L defensive armed penetrator 5-12
Figure 5-11. Armament options for the MH-60L DAP 5-13
Figure 5-12. MH-60K helicopter 5-16
Figure 5-13. MH-60K with M134 minigun window-mounted field of fire 5-18
Figure 5-14. MH-60K dimensions and turning radius 5-20
Figure 5-15. MH-60K dimensions for strategic airlift preparation 5-21

Figure 5-16. MH-60K aircraft capabilities 5-22
Figure 5-17. MH-47G helicopter 5-23
Figure 5-18. MH-47E helicopter 5-28
Figure 5-19. MH-47E aircraft capabilities and dimensions 5-32
Figure 5-20. MH-47E capabilities of cargo areas 5-33
Figure 5-21. MH-47E capabilities of turning radius 5-34
Figure 5-22. MH-47E maximum package size 5-35
Figure 5-23. MH-47E ramp door maximum package size 5-36
Figure 5-24. MH-47E compartment data 5-37
Figure 5-25. MH-47E fitting capabilities 5-38
Figure 5-26. MH-47E passenger seating and litter placement 5-39
Figure 6-1. MC-130E 6-12
Figure 6-2. AC-130H 6-14
Figure 6-3. AC-130U 6-14
Figure 6-4. MC-130P refueling two MH-53Js 6-19
Figure 6-5. EC-130J 6-21
Figure 6-6. CV-22 Osprey 6-24
Figure 6-7. HH-60G Pave Hawk 6-25
Figure 7-1. C-23B/B+ Sherpa 7-1
Figure 7-2. C-23B/B+ static-line retrieval system 7-2
Figure 7-3. C-23B/B+ static-line exit procedures 7-4
Figure 7-4. C-27A Spartan 7-5
Figure 7-5. CASA-212 7-7
Figure 7-6. CASA-212 seating configuration for combat-equipped parachutists ... 7-8
Figure 7-7. Equipment for static-line airdrop 7-11
Figure 7-8. UV-18B Twin Otter 7-13
Figure 7-9. Seating configuration with seats and without combat equipment 7-14
Figure 7-10. Seating configuration without seats or combat equipment 7-14
Figure 7-11. Seating configuration without seats and with combat equipment ... 7-15
Figure 7-12. Twin Otter rigged for static-line jump 7-16
Figure 7-13. U-21A Ute 7-18
Figure 7-14. In-flight seating configuration 7-19
Figure 7-15. Example of anchor line installation 7-20
Figure 7-16. C-208B Caravan 7-21
Figure 8-1. Rigging of the A-7A cargo slings 8-2
Figure 8-2. Rigging of the A-21 cargo bag 8-5
Figure 8-3. Rigging of the A-22 cargo bag 8-5
Figure 8-4. Poncho-expedient parachute 8-7
Figure 8-5. Rigging knots 8-8
Figure 9-1. SPIES equipment 9-5
Figure 9-2. D rings attached to the SPIES rope 9-6

Figure 9-3. Suspension slings connected by Type IV link 9-9
Figure 9-4. SPIES rope connected to cargo hook and sling 9-9
Figure 9-5. Placement of wood clock 9-10
Figure 9-6. Placement of Type IV link 9-10
Figure 9-7. Placement of four pairs of snap links 9-11
Figure 9-8. Snap links connected to sling and ring, gates reversed 9-11
Figure 9-9. Snap links alternating on strap 9-12
Figure 9-10. Recovery rope tied to SPIES rope 9-12
Figure 9-11. Recovery rope connected to SPIES rope and helicopter 9-12
Figure 9-12. Rigging complete 9-13
Figure 9-13. SPIES rigged on CH-46/47 and CH-53 helicopters 9-14
Figure 10-1. Fast-rope extraction loops 10-5
Figure 10-2. SPIES harness 10-6
Figure 10-3. Belay devices used to lower equipment during FRIES operations 10-6
Figure 10-4. UH/MH-60 mounting bracket for FRIES (inside aircraft) 10-9
Figure 10-5. Fast-rope attachment point extended (outside aircraft) 10-9
Figure 10-6. MH-47 aft-mounted FRIES bars (inside aircraft) 10-9
Figure 10-7. FRIES mount over the forward door (with and without hoist) 10-10
Figure 10-8. Fast rope attached to aft FRIES bar (bar retracted) 10-10
Figure 10-9. MH-47 with two aft fast ropes 10-10
Figure 10-10. Fast rope attached to forward mount 10-11
Figure 10-11. MH-60 rigged for fast roping 10-21
Figure 10-12. Fast-rope attachment point opened 10-21
Figure 10-13. Fast rope attached and release bar seated 10-22
Figure 10-14. Quick-release pin inserted 10-22
Figure 10-15. Fast rope rigged on an MH-6 10-23
Figure 10-16. Rucksacks upright on cargo deck 10-28
Figure 10-17. Belay device SMC ladder 10-28
Figure 10-18. Belay device figure eight 10-28
Figure 10-19. Double lowering line routing using figure eight 10-29
Figure 10-20. Lowering line routing using the Sky Genie 10-29
Figure 11-1. Helocasting 11-9
Figure 11-2. K-duck fixture 11-12
Figure 11-3. Quick-pin assemblies 11-13
Figure 11-4. CRRC harness 11-14
Figure 11-5. CRRC harness nose strap 11-14
Figure 11-6. CRRC harness crow's feet 11-14
Figure 11-7. Crow's foot with bow strap ratchet attached 11-15
Figure 11-8. Nose strap and crow's feet connection 11-17
Figure 11-9. Belly support strap connected to sling band 11-18
Figure 11-10. Fuel tank and engine in place 11-19

Figure 11-11. Sling band tie 11-19
Figure B-1. Moon phases B-2
Figure B-2. Tides as related to moon phases B-3
Figure E-1. USASOC airborne operations flash report E-2
Figure E-2. USASOC malfunction officer/NCO checklist E-4
Figure F-1. JM checklist for C-27A F-9
Figure F-2. JM checklist for CASA-212 .... F-11
Figure G-1. Format for fast-rope troop briefing G-1
Figure G-2. Fast-rope operations checklist G-3
Figure H-1. Recommended CM briefing format H-1

Tables

Table 1-1. MFF parachuting altitudes 1-2
Table 2-1. Rates of descent 2-8
Table 3-1. Standard DZ size criteria 3-4
Table 3-2. Minimum DZ criteria for single-ship SO missions on marked DZ 3-6
Table 3-3. Minimum DZ criteria for single-ship SO missions on unmarked DZ 3-7
Table 3-4. Point of impact location using CARP 3-14
Table 3-5. Aircraft drop speeds of T-10B parachute 3-16
Table 3-6. Aircraft drop altitudes in feet 3-19
Table 3-7. USAF forward throw data 3-20
Table 3-8. Wind limitations during peacetime and wartime operations 3-36
Table 4-1. Minimum airfield criteria (standard) for USAF aircraft 4-14
Table 4-2. Minimum airfield criteria (special use) for USAF aircraft 4-14
Table 4-3. Standard ATC light signals 4-26
Table 4-4. Water and air temperature 4-28
Table 4-5. State of sea (Code Table 3700) extract 4-29
Table 5-1. MH-6M aircraft capabilities 5-3
Table 5-2. AH-6M aircraft capabilities 5-9
Table 5-3. MH-60L aircraft capabilities 5-15
Table 5-4. MH-60K aircraft capabilities 5-19
Table 5-5. MH-47G external cargo hooks 5-25
Table 5-6. MH-47G aircraft capabilities 5-26
Table 5-7. MH-47E external cargo hooks 5-29
Table 5-8. MH-47E aircraft capabilities 5-31
Table 5-9. SOAR aircraft capabilities matrix 5-40
Table 6-1. Aircraft type classification 6-2
Table 6-2. Aircraft capabilities 6-4
Table 6-3. Aircraft communications capabilities 6-8
Table 6-4. Aircraft navigation capabilities 6-8

Table 6-5. Additional capabilities	6-9
Table 6-6. AC-130 defensive equipment	6-17
Table 6-7. AC-130 target categories	6-17
Table 6-8. AC-130 ammunition loads	6-17
Table 6-9. Osprey characteristics	6-24
Table 6-10. USAF rotary-wing aircraft capabilities	6-28
Table 6-11. USAF rotary-wing aircraft communication capabilities	6-29
Table 6-12. USAF rotary-wing aircraft navigation capabilities	6-29
Table 6-13. USAF rotary-wing aircraft air transportability	6-29
Table A-1. Linear measure	A-1
Table A-2. Liquid measure	A-1
Table A-3. Weight	A-1
Table A-4. Square measure	A-2
Table A-5. Cubic measure	A-2
Table A-6. Temperature	A-2
Table A-7. Approximate conversion factors	A-2
Table A-8. Area	A-3
Table A-9. Volume	A-3
Table A-10. Capacity	A-3
Table A-11. Statute miles to kilometers and nautical miles	A-3
Table A-12. Nautical miles to kilometers and statute miles	A-4
Table A-13. Kilometers to statute and nautical miles	A-4
Table A-14. Yards to meters	A-5
Table A-15. Meters to yards	A-5
Table D-1. Wind speed in knots for a 10-gram helium balloon	D-1
Table D-2. Wind speed in knots for a 30-gram helium balloon	D-2
Table F-1. Standard time warnings	F-1
Table F-2. Modified time warnings	F-2
Table F-3. Standard jump commands	F-3
Table F-4. Rotary-wing standard jump commands	F-5
Table F-5. CASA-212 jump procedures	F-6
Table F-6. C-208B jump procedures	F-7

Preface

Field manual (FM) 3-05.210, *Special Forces Air Operations*, defines the current United States (U.S.) Army Special Forces (SF) concept of planning, coordinating, and executing air operations. This FM provides techniques and procedures for air operations supporting all SF missions and collateral activities.

PURPOSE

As with all doctrinal manuals, FM 3-05.210 is authoritative but not directive. It serves as a guide and does not preclude SF units from developing their own standing operating procedures (SOPs) to meet their needs. The techniques and procedures discussed in this FM provide a base from which to develop unit procedures to cope with a special mission or area requirement. This publication focuses on—

- Premission preparation, including rehearsals and briefbacks.
- Significant actions, considerations, and decisions that can determine the success of a mission.
- Use of equipment and indigenous assets in various air operations.
- Types of airborne missions, airdrops, and training requirements.
- Drop zones (DZs), landing zones (LZs), pickup zones (PZs), and markings.
- Standard, special operations (SO), and nonstandard aircraft.
- Rigging and inspection of aircraft for fast roping, rappelling, bundle drops, special patrol infiltration and extraction system (SPIES), ladder operations, helocasting, hard ducks, rolled ducks, K-ducks, and personnel drops.

SCOPE

The primary users of this manual are commanders and operational personnel at the team (Special Forces operational detachment A [SFODA]), company (Special Forces operational detachment B [SFODB]), and battalion (Special Forces operational detachment C [SFODC]) levels. This FM is specifically for SF; however, it is also intended for use Armywide to improve the integration of SF into the plans and operations of other forces.

APPLICABILITY

Commanders and trainers should use this and other related manuals in conjunction with command guidance to plan and conduct successful air operations. This manual is not intended as a stand-alone reference and is meant to supplement FM 3-21.220, *Static Line Parachuting Techniques and Tactics*, FM 90-26, *Airborne Operations*, FM 3-05.60, *Army Special Operations Forces Aviation Operations*, and United States Special Operations Command (USSOCOM) Manual (M) 350-6, *Special Operations Forces Infiltration/Exfiltration Operations*. This publication applies to the Regular Army, the Army National Guard (ARNG)/Army National Guard of the United States, and the United States Army Reserve (USAR) unless otherwise stated.

ADMINISTRATIVE INFORMATION

The proponent for this publication is the United States Army John F. Kennedy Special Warfare Center and School (USAJFKSWCS). Submit comments and recommended changes to Commander, USAJFKSWCS, ATTN: AOJK-DTD-SF, Team IV, Fort Bragg, NC 28310-9610.

Unless this publication states otherwise, masculine nouns and pronouns do not refer exclusively to men.

Note. Numbers in this publication (meters to feet and kilometers to miles) have been rounded off to make computations easier and act as an aid to memorizing these numbers. Care has been taken to make sure no critical dimensions are less than the critical dimensions listed in applicable manuals. In addition, many of the measurements used by Soldiers involved in air operations are U.S. standard terms rather than metric. Appendix A consists of conversion tables that may be used when mission requirements or environments change.

This page intentionally left blank.

Chapter 1

Special Forces Air Operations

Special Forces air operations are characterized by penetration flights into hostile or politically sensitive areas to infiltrate, resupply, and exfiltrate SF operational elements. SF air operations include airborne, airland, rappelling, fast-rope insertion and extraction system (FRIES), SPIES, ladder, helocast, external raft delivery system (ERDS), rolled and tethered duck, soft duck, hard duck, and various other operations. Missions are normally flown during hours of darkness or periods of limited visibility by a variety of special operations, conventional, or nonstandard aircraft. Air support provided by U.S., allied, or host nations may depend on the mission, situation, availability, and capability of the aircraft and aircrew.

MISSION

1-1. The types of air missions used to support special operations forces (SOF) are—
- Infiltrate, resupply, and exfiltrate.
- Combat search and rescue (CSAR).
- Personnel recovery.
- Materiel pickup and delivery.
- Surveillance and reconnaissance (visual, photographic, and electronic).
- Airborne radio retransmission.
- Close air support (CAS) or interdiction within assigned capabilities.
- Diversionary tactics.
- Psychological Operations (PSYOP) loudspeaker or leaflet sorties.
- Military air deception tactics.

CHARACTERISTICS OF SPECIAL FORCES AIR MISSIONS

1-2. SF air missions are normally unescorted, single aircraft missions flying at minimum clearance altitude. In daytime, this altitude is below 500 feet (ft); at night, at or below 1,000 feet. SF air operations, especially in an unconventional warfare (UW) environment, employ as many of the following characteristics as possible:
- Frequent course changes (doglegs) en route to and departing from the LZ, PZ, or DZ.
- Predetermined flight track from the initial point (IP) to the LZ, PZ, or DZ.
- Arrival at the LZ, PZ, or DZ within a designated time limit, track, and drop altitude.
- Deliveries are, as determined by the capability of the delivery system, technique used, parachute performance characteristics, or terrain limitations. Military free fall (MFF) parachuting missions range from 3,500 feet above ground level (AGL) to 25,000 feet mean sea level (MSL). Personnel may use parachutes with more positive opening at lower altitudes (Table 1-1, page 1-2).
- Airdrops on a release point (RP) that a reception committee has computed and marked or on an unmarked DZ when the navigator has determined the RP. The drop occurs during a single pass over the DZ.
- Maintaining track, altitude, and airspeed (power settings) for a designated distance and time to avoid compromising the LZ, PZ, or DZ after making the drop.

Chapter 1

- Sorties to overfly the primary and alternate LZs, PZs, or DZs. When conditions prevent the aircraft from using the primary LZ, PZs, or DZ, it proceeds to the alternate LZ, PZ, or DZ to try to complete the infiltration.
- Airland delivery missions when joint special operations areas (JSOAs) expand and come under some degree of friendly control. Aircraft normally follow a straight-in approach to the LZ from the IP.
- Fixed-wing gunship CAS or interdiction operations.

Table 1-1. MFF parachuting altitudes

	Minimum	Maximum
Exit Altitude (in ft)	5,000 AGL	35,000 MSL
Opening Altitude (in ft)	3,500 AGL	25,000 MSL

Note. Opening above 25,000 ft MSL exceeds the MC-4 and MC-5 parachute design specifications. The U.S. Navy MTI-XS/Sea Level (SL) maximum deployment altitude is 18,000 ft MSL.

TYPES OF SPECIAL FORCES AIR OPERATIONS

1-3. SFODs use a variety of infiltration and exfiltration techniques from rotary-wing, tiltrotor, or fixed-wing aircraft. Each aircraft has its own advantages, disadvantages, requirements, and criteria. The deploying detachment should determine the best method to infiltrate and exfiltrate its area of operations (AO) according to its capabilities, needs, and mission.

ROTARY-WING AIR OPERATIONS

1-4. SFODs use rotary-wing aircraft to conduct rappelling, fast rope, SPIES, ladder, helocast, K-duck, or rolled and tethered duck operations. Rotary-wing aircraft have limited use for resupply missions. Compared with fixed-wing aircraft, rotary-wing aircraft are usually slower, have less range, have less cargo capacity, and are more vulnerable to antiair defenses.

Tiltrotor-Aircraft Operations

1-5. The CV-22 Osprey is a tiltrotor aircraft that combines the vertical takeoff, hover, and vertical landing qualities of a helicopter with the long-range, fuel efficiency, and speed characteristics of a turboprop aircraft. Its mission is to conduct long-range infiltration, exfiltration, and resupply missions for special operation forces. Disadvantages include noise, landing area and availability of aircraft.

Airborne Operations

1-6. Airborne operations from rotary-wing aircraft include static-line or MFF jumps over both land or water. SFODs can use rotary-wing aircraft when conducting water jumps that involve a combat rubber raiding craft (CRRC).

Airland Operations

1-7. Airland operations require no additional equipment or support for the infilling team (SFODA) and is the preferred method.

Rappelling, FRIES, SPIES, and Ladder Operations

1-8. SFODs use rappelling, FRIES, SPIES, and ladder operations where a suitable LZ or PZ is not available and when an airborne operation is not feasible. These operations—
- Require additional equipment to off-load or onload the aircraft.
- Require additional training and rehearsals.

- Limit the number of troops infiltrated and exfiltrated.
- Require SO aircraft with correct anchoring systems.
- Refer to USSOCOM M 350-6.

Helocasting

1-9. SFODs use helocasting to infiltrate personnel from a rotary-wing aircraft into water. Rolled and tethered duck and ERDS operations involve the infiltration of troops and a CRRC into water DZs. (USSOCOM M 350-6 provides more details.)

FIXED-WING AIR OPERATIONS

1-10. SFODs use fixed-wing aircraft for infiltration, resupply, or exfiltration missions. When SFODs use fixed-wing aircraft, the primary infiltration technique is by airborne operation. Airborne operations can be either static line or MFF parachuting. Although fixed-wing aircraft can infiltrate and exfiltrate personnel by airland operations, the increased requirements of an LZ versus a DZ usually make this option impractical. From fixed-wing aircraft, SFODs can drop supplies with or without a parachute. Fixed-wing aircraft can fly at longer ranges, have more cargo capacity, are faster, and can fly at higher altitudes.

Static-Line Operations

1-11. Static-line operations allow for the infiltration of a large number of personnel and large quantities of supplies into small areas in a very short time. Ground personnel may assist these operations; however, their assistance is not a requirement. Some aircraft have the capability to drop personnel and small boats on water DZs.

MFF Parachuting Operations

1-12. MFF parachuting operations allow the clandestine insertion of small groups of Soldiers into denied territory using aircraft flying at high altitude to avoid detection and enemy air defenses. These operations can be conducted on small DZs.

1-13. If the situation permits, the aircraft can land to off-load or onload personnel and supplies. This type of airland operation requires a secure area that fits the LZ specifications found in Chapter 4. Airland operations require rapid and extensive support and logistics.

AIRBORNE INFILTRATION TECHNIQUES

1-14. SFODs infiltrate personnel and supplies into an AO by using static-line or MFF parachutes. There are several different techniques employed during airborne operations. Each technique uses a variety of equipment and applications. Each of the different techniques has its own advantages, disadvantages, and criteria for use. Not all aircraft are capable of conducting all of the various techniques.

STATIC-LINE OPERATIONS

1-15. There are several variations of static-line airborne operations. FM 3-21.220 and United States Army Special Operations Command (USASOC) Regulation (Reg) 350-2, *Training Airborne Operations*, prescribe the tactics, techniques, procedures, and requirements for standard static-line parachute operations.

GROUND-MARKED RELEASE SYSTEM

1-16. SFODs use ground-marked release systems (GMRSs) for covertly infiltrating a small unit into a denied area without using radio communications. A trained reception committee on the DZ marks the RP with specified ground markings. The pilot and jumpmaster (JM) use the ground markings as a reference. The JM directs the GMRS, spots the ground marker from the aircraft and, if all conditions are correct and safe for the drop, commands GO. Normally, SFODs use GMRS airborne operations to infiltrate into a

Chapter 1

denied area where partisans or a resistance force is available to provide a reception committee on the DZ. The basic steps of the operation are as follows:

- The reception committee arrives at the DZ and establishes security in accordance with (IAW) the unit SOP.
- The reception committee computes the location of and sets up panels or lights on the DZ without displaying them. The committee uses a prearranged ground marking.
- The reception committee displays the panels or lights for exactly 4 minutes—2 minutes before and 2 minutes after the time on target (TOT).
- If the reception committee does not display the proper predesignated signal, the JM aborts the jump.
- If the jump is aborted, the Special Forces operational detachment (SFOD) and the reception committee will designate an alternate time or DZ.
- Actions of the infiltrating unit and the reception party are IAW unit SOP.

1-17. All JM duties are the same as outlined in FM 3-21.220, except—

- When using GMRS, the JM should spot out of the left side of the United States Air Force (USAF) aircraft.
- The method for spotting out of USAF aircraft is for the primary JM to spot out of the left side and, if desired or required, for the assistant jumpmaster (AJM) to spot out of the right side.
- The JM will estimate when the aircraft is 10 seconds out from the RP (ground marking) and will give the command STANDBY to the first parachutist.
- When conducting GMRS airborne operations, the JM maintains control of the static lines as the parachutists exit the aircraft. As the JM controls the static lines, the safety, who is wearing a USAF immediate deployment bail out parachute, observes the stick and is able to move freely, should a problem arise with one of the parachutists. This operation differs from standard airborne operations where the safety controls the parachutists' static lines.
- The aircrew will illuminate the green light at a predetermined time. If conditions are safe, the JM will command GO.

1-18. Advantages of the GMRS include the following:

- The navigator and JM have visual markings from which to work.
- Parachutists have a ground reference.
- The reception committee establishes security on the DZ.
- Personnel maintain radio silence throughout the complete air operation.
- Personnel are on the ground to assist parachutists.

1-19. Disadvantages of the GMRS include the following:

- A trained reception committee must mark the DZ.
- The reception committee must correctly compute the RP and IP.
- The size of the reception committee and the distance between markings increase security risks.
- Sufficient visual conditions are required for the JM to see the markings on the DZ.

COMPUTED AIR RELEASE POINT

1-20. When using the computed air release point (CARP), the navigator is responsible for getting the plane to the DZ at the right heading and altitude. The navigator activates the green light. The green light is the signal for the JM to begin the drop. The JM should ensure the aircraft is at the correct DZ, heading, and altitude before releasing the parachutists or bundles. This type of operation may or may not require a reception committee to mark the DZ. An unmarked DZ with no reception committee is a blind drop.

Note. All JM duties are the same as outlined in FM 3-21.220.

1-21. The advantages of CARP are as follows:
- CARP can be used during limited visibility when the JM is unable to see the DZ or other identifying terrain features.
- A reception committee is optional. If one is used, it does not require training and is smaller than that required for GMRS.
- If the DZ is marked, the markings are smaller and require less area than GMRS. CARP decreases the chance the enemy or others will see markings or lights.

1-22. Disadvantages of CARP include the following:
- The accuracy of the drop is dependent on the USAF navigational equipment.
- If there is no reception committee, the infilling detachment will not have any security or assistance on the ground.

BLIND DROP

1-23. A blind drop is a type of CARP jump with an unmarked DZ and no reception committee. Blind drops provide infiltration of a small reconnaissance element or SF team into a denied area to conduct SO missions. These operations are single-ship airborne operations conducted in a single pass without communications or assistance.

1-24. Blind drops are used when—
- SFODs are operating in a unilateral role; for example, when they conduct operations against selected targets without the support of a resistance force.
- The enemy situation prevents normal marking and recognition signals.

1-25. Blind drops require the following:
- Under instrument meteorological conditions (IMC), an MC-130 or other aircraft will be equipped with the adverse weather aerial delivery system (AWADS). During training, air-to-ground communications are required when conducting blind drops with AWADS-equipped C-130s or MC-130s.
- Under visual conditions, blind drops can be conducted from non-AWADS-equipped aircraft, to include short takeoff and landing (STOL) aircraft. Air Mobility Command (AMC) special operations low-level (SOLL) aircrews will use a night vision goggle (NVG) at night.
- For U.S. Army training drops, a minimum ceiling of 500 feet above drop altitude is required for fixed- and rotary-wing aircraft. The ground force commander or user may waive this requirement with proper authorization from the Army command. However, the airborne commander must identify a minimum ceiling before the mission is flown. During operational missions, the airborne commander in coordination with the air mission commander, or pilot in command, determines ceiling and visibility minimums. For joint exercises, USAF personnel are authorized to use Army minimums. When dropping equipment and the ceiling is less than 600 feet AGL, all personnel must be clear from the DZ not later than (NLT) 5 minutes before the scheduled equipment airdrop TOT and must remain clear until the airdrop is completed.

1-26. The JM follows the same blind drop procedures and requirements as for CARP. He also makes sure the parachutists know that the DZ might be obscured and that it will be unmarked.

1-27. Advantages of blind drops include the following:
- Blind drops can be conducted during limited visibility when the JM is unable to see the DZ or other identifying terrain features.
- The enemy or others have no chance of seeing any markings or lights.
- A reception committee is not required.
- USAF combat control team (CCT) or special tactics team (STT) support on the DZ is not needed.

1-28. Disadvantages of blind drops include the following:
- The accuracy of the drop depends completely on aircraft navigational equipment.
- Depending on the type of aircraft used, blind drops may be limited to favorable astronomical and weather conditions in the objective area.
- There is no security on the ground.
- There is no assistance available for the detachment.

MARKED POINT OF IMPACT

1-29. A marked point of impact (PI) airborne operation has a reception committee that uses a predesignated block letter or light signal to mark the desired PI on the DZ. This marking is the only one required. The navigator computes the RP according to the marked PI and flies as if the operation were a CARP jump. Following are the procedures for conducting a marked PI operation:
- The parachutists and JM procedures are the same with two exceptions. First, the JM briefs the parachutists on what will be the marking signal. Second, the unit SOP dictates the actions of the parachutists.
- The reception committee marks or illuminates the DZ for 2 minutes before and 2 minutes after the designated TOT.
- If the proper predesignated signal is not displayed, the jump is aborted.
- Alternate time and DZ are as per premission coordination between the detachment and the reception committee.
- Actions of the infiltrating unit and the reception party are per the operations plan (OPLAN).

1-30. Advantages of a marked PI operation include the following:
- A reception committee is not required to figure out the RP.
- Parachutists have a target for which to aim.
- A reception committee establishes security on the DZ.
- Personnel can maintain radio silence throughout the complete air operation.

1-31. Disadvantages of a marked PI operation include the following:
- Visual conditions must allow the navigator to see the signal to compute the RP.
- There is an increased security risk because of the signal on the ground.
- The operation may require multiple passes.
- The accuracy of the drop depends wholly on USAF navigational equipment.

VERBALLY INITIATED RELEASE SYSTEM

1-32. Verbally initiated release system (VIRS) is a procedure to provide verbal steering guidance to an aircraft and call the release when it reaches a predetermined point on the ground. The reception committee calculates the RP on the ground and places whoever is giving directions to the aircraft at the RP. This person then guides the aircraft to a spot directly overhead and radios the aircraft to release the parachutists or cargo. Commands used when conducting VIRS should be as short and concise as possible so as not to interfere with the approach of the aircraft to the DZ. For example, the person on the ground commands—
- LEFT TURN and RIGHT TURN to align the aircraft on the desired inbound heading. Direction changes are given in relation to the direction of flight.
- STOP TURN when aircraft is on course.
- STAND BY to the aircraft about 5 seconds before release or as prebriefed.
- EXECUTE when the aircraft reaches the predetermined point on the ground. This command is transmitted three times.

1-33. The advantage of VIRS is that the DZ requires no marking and rapid adjustment can be made as the wind changes.

1-34. The disadvantages of VIRS include the following:
- The procedure requires radio communications.
- Personnel require extensive experience with performing the procedure.
- A trained reception committee must correctly compute the RP and PI.
- The reception committee must be able to see the aircraft at all times during the inbound approach.

Rough-Terrain Airborne Infiltration

1-35. The purpose of a rough terrain airborne operation is to parachute an element into an area that has no suitable DZ. It is a jump into an unprepared, mountainous, rocky, or wooded DZ. This technique minimizes terrain considerations and gives the commander maximum latitude in DZ selection. Rough-terrain airborne infiltration is not normally limited to favorable astronomical and weather conditions. A reception committee is not necessary for rough-terrain airborne infiltration. When conducting rough-terrain jumps, all parachutists will wear the parachutist rough-terrain system (PRTS) to provide protection. The PRTS consists of the following:
- A camouflaged rough-terrain suit constructed of material resistant to penetration or tearing.
- Pliable attachable or detachable pads to protect vital body areas.
- A smoke jumper parachutist helmet with face, eye, and maximum impact protection.
- A lowering device 100 to 170 feet long capable of lowering up to 360 pounds.
- Gloves that will protect the hands during the landing, allow the parachutists to maintain a firm grip on their equipment, and protect the parachutist's hands when lowering to the ground.

Note. Parachutists must receive training on the use of PRTS. This training will address tactical employment of the PRTS, procedures for adjusting the parachute harness, equipment inspection, and rigging and derigging of equipment. JM procedures remain unchanged except to make sure the parachutists have received training on the PRTS during sustained airborne training.

1-36. Advantages of this technique are as follows:
- A trained reception committee is not needed.
- Selection of DZ is less restrictive.
- The RP does not have to be very accurate.

1-37. Disadvantages of this technique are as follows:
- Extensive training of all parachutists is required.
- Injuries are more common.
- A longer time is required on DZ to recover personnel and equipment.
- There is no security or assistance without a reception committee.
- There is more equipment to cache.
- Parachute may not be able to be recovered leaving signature of team's presence.
- Additional time required for assembly.

Military Free Fall

1-38. Units use MFF parachuting infiltrations when enemy air defense and detection systems prevent a low-altitude penetration or when mission needs demand a clandestine insertion. FM 3-05.211, *Special Forces Military Free-Fall Operations*, prescribes the technical and procedural guidance for MFF operations.

Chapter 1

1-39. The characteristics of MFF parachuting operations are as follows:
- Flights at altitudes above normal sight and sound or offset from the DZ.
- Ram-air parachute system, which is a high-performance gliding system, is used. The ram-air parachute system has an air speed of 20 to 30 miles per hour (mph) and is highly maneuverable.
- Exit altitude between 5,000 feet AGL and 30,000 feet MSL during training.
- Opening altitude between 4,000 feet AGL and 25,000 feet MSL during training.
- Parachutist drops on a visually marked RP, an unmarked DZ by using the high altitude release point (HARP), or a visible preselected RP to reach the desired ground impact point.
- Can be conducted on DZs too small for static-line infiltration.

1-40. The type of MFF parachuting operation used will depend on the enemy defenses and political considerations. The two types of MFF parachuting operations are as follows:
- *High Altitude Low Opening (HALO).* The parachutists exit from the aircraft at altitudes up to 35,000 feet MSL, free-fall to 4,000 feet AGL, and deploy their parachutes. This technique requires the aircraft to fly within several kilometers of the DZ.
- *High Altitude High Opening (HAHO).* The parachutists exit from the aircraft at altitudes up to 35,000 feet MSL, free-fall to 25,000 feet MSL or lower, and deploy their parachutes. Depending on the altitude and winds, the parachutists can cover up to 30 kilometers ground distance under canopy.

1-41. The two types of release for MFF parachuting are as follows:
- *JM-Directed.* The JM calculates the RP by using wind and altitude information. He then directs the aircraft to that spot and releases the parachutists under safe conditions.
- *HARP.* This release is similar to a CARP jump in that the navigator computes the RP by using the HARP formula. Once the aircraft arrives at the HARP, the aircrew activates the green light and the JM gives the go signal.

ADVERSE WEATHER AERIAL DELIVERY SYSTEM

1-42. AWADS is a multipurpose, self-contained tactical navigation system. It greatly improves mission aircraft capability to infiltrate and resupply personnel and equipment into blind, unmarked DZs in adverse weather or darkness. During SF training missions, SF and USAF personnel can conduct AWADS operations safely in IMC with a minimum 300 feet AGL ceiling and minimum visibility of 1 kilometer. The following procedures apply when using AWADS for infiltration:
- If the mission requires an AWADS IMC airdrop option, the SF and USAF personnel plan the drop as a CARP release. In this situation, marking the PI is optional. However, a marked PI provides the crew with the capability to make a visual CARP release in case of AWADS failure in visual conditions.
- If DZ RP markings are used, the ceiling and weather conditions must permit visual sighting. If DZ RP markings are used and a drop using a ground-marked RP is desired, an AWADS IMC airdrop option cannot be made because the GMRS and AWADS CARP procedures are incompatible.
- The JM briefs the parachutists on the psychological effect of exiting the aircraft in or just above the clouds. The parachutist perceives a false sense of excessive aircraft speed that may cause him to hesitate in the door.
- The parachutist has limited time for DZ orientation, which may lead to dispersion problems.
- The minimum drop altitude during instrument flight rules will be 500 feet above the highest obstruction (AHO), 5.5 kilometers on either side of the DZ centerline from the DZ entry point to the DZ exit point. DZ entry or exit points are defined as follows:
    - **Entry point.** A geographical point on the DZ run-in course for which mission planners establish drop altitude, drop airspeed, and stable flight conditions. This point will normally be at least 11 kilometers before PI.

Special Forces Air Operations

- **Exit point.** A geographical point on the DZ departure course (extended DZ centerline) at or before which the departure maneuver is to be performed. This point is normally no closer than 3.7 kilometers past the trailing edge of the DZ. The mission planners select the DZ exit point and the departure profile, ensuring at least a 1,000-foot altitude separation from all obstructions within 9.25 kilometers of the DZ departure flight path.

Note. Minimum ceiling or visibility restrictions do not apply to AWADS or MC-130 Combat Talon aircraft for actual contingency or combat operations. For combat, ceiling and visibility minimums can be as low as zero ceiling and zero visibility for both aircraft. Minimum IMC resupply drop altitude for the Combat Talon is 250 feet AGL.

AERIAL RESUPPLY

1-43. An airdrop involves all types and methods of air-to-ground delivery of equipment and supplies from an aircraft in flight (Figure 1-1). The airdrop is one of the best and fastest means of resupply. In some cases, it may be the only means of resupply available to the commander. Because of the uniqueness of the UW environment, SFODs have different methods of ordering, preparing, and obtaining supplies.

Figure 1-1. Resupply mission

TYPES OF RESUPPLY

1-44. The deploying detachment plans resupply missions in conjunction with the special operations task force (SOTF) during isolation phase. After the infiltration, the SOTF or advanced operations base (AOB) coordinates the aerial delivery of automatic, emergency, or on-call resupply missions to deployed operational elements IAW the preplanned schedule or as per any changes sent by the deployed detachment. Preplanned automatic resupply and emergency resupply provide operational elements with immediate equipment and supplies until routine on-call supply procedures can be established by the AOB, depending on the element size.

Automatic Resupply

1-45. Planning for an automatic resupply mission occurs before infiltration. Planning includes delivery time, location, contents, and the DZ marking and authentication to be used. Automatic resupply is delivered after successful infiltration and radio contact is established unless canceled, modified, or rescheduled by the deployed operational element. Automatic resupply replaces expended, lost, or damaged equipment. Automatic resupply also augments equipment that could not be carried during the initial infiltration, and it serves to reinforce U.S. support of the resistance movement.

On-Call Resupply

1-46. The on-call resupply mission is based on operational needs after the SOTF and the SF element establish communications with each other. On-call resupply requests consist of expendable supplies and major equipment items that are not consumed at a predictable rate. The special operations theater support element, SOTF, or AOB holds these supplies in readiness for immediate delivery following specific mission requests.

Emergency Resupply

1-47. Personnel plan the emergency resupply mission before infiltration. It includes a specified delivery time, a provisional location (to be confirmed), a rigged emergency resupply bundle, and the DZ marking and authentication to be used.

1-48. The emergency resupply bundle is flown—
- When radio contact has not been made between the operational element and the SOTF or AOB within a specified time after infiltration IAW unit SOP.
- When communications is lost between a deployed SF element and the SOTF or AOB for a predetermined consecutive number of scheduled radio contacts IAW unit SOP.
- Upon receiving a call from the deployed detachment.

1-49. If forced into continuous movement and unable to reach the emergency resupply DZ, the SF element selects and reports emergency DZs at the first opportunity. If, during this time, an element misses a certain number of radio contacts, the supporting air unit delivers the resupply on the last reported DZ. The emergency resupply bundle contains mission-essential equipment and supplies to restore the operational capability and survivability of the SF element and indigenous assets.

1-50. Normally, the emergency resupply bundle contains the following:
- Communications equipment with extra batteries.
- Radar transponders (RTs) or other marking devices.
- Survival and medical supplies.
- Selected weapons, ammunition, and demolition items.
- Food and water.

TYPES OF AIRDROP

1-51. The types of airdrops are as follows:
- Free drop.
- High-velocity drop.
- Low-velocity drop.
- High-altitude airdrop resupply system (HAARS) container delivery system (CDS).
- High-speed, low-level aerial delivery system (HSLLADS) drop.
- Delayed-opening airdrop.

Free Drop

1-52. A free drop is the delivery of certain nonfragile items of equipment or supply from a slow-flying aircraft at low altitude without the use of parachutes or other retarding devices. Normally, the special packaging required for fragile items greatly limits this technique. Free drops are most effective when the aircrew drops the supplies into a river, stream, or other body of water and ground forces can immediately recover the supplies.

High-Velocity Drop

1-53. High-velocity drop is a procedure in which the drop velocity is greater than 30 feet per second but lower than free drop velocity. It is used to deliver certain items of supply that are specially packed and rigged in containers having layers of energy-dissipating material attached to the underside and a stabilizing device. The stabilizing device, such as a ring-slot parachute, minimizes oscillation of the load and creates just enough drag to keep the load upright during descent so that it will land on the energy dissipater.

Low-Velocity Drop

1-54. Low-velocity drop is a procedure in which the drop velocity is less than 30 feet per second. Low-velocity drop is the delivery of supplies from an aircraft using cargo parachutes. Such loads are specially prepared for airdrop either by packing the items in air-droppable containers or by lashing them to air-droppable platforms. The parachute riggers attach cargo parachutes to the load or the platform to slow the descent of the load and to ensure minimum landing shock.

HAARS Container Delivery System

1-55. HAARS CDS allows for the airdrop of containers at altitudes between 2,000 feet AGL and 25,000 feet MSL. Using this system, the aircraft drops the containers from the drop altitude and they free-fall, with a drogue chute to maintain upright position, to a preset altitude. Time-delay devices such as power-actuated reefing line cutters, barometric opening devices, or timers deploy the parachutes. These devices delay the opening of the cargo parachute at low altitudes to permit a good ground dispersion pattern. The aircrew employs high-altitude bombing techniques combined with HARP computations when using this airdrop system.

HSLLADS Drop

1-56. The HSLLADS drop was developed for airdrop resupply from the MC 130 Combat Talon flying at 250 knots and as low as 250 feet AGL. This system employs a modified container using A-21 covers and a modified 22- or 28-foot extraction parachute. This system can deliver up to four cargo containers weighing a minimum of 250 pounds each (but not exceeding a total of 2,200 pounds) at delivery altitudes ranging from 250 to 750 feet AGL. A "slingshot" ejection system ejects the cargo load over the RP.

Delayed-Opening Airdrop

1-57. The JM attaches an automatic-opening device (AOD) or timer to the parachute and bundle. He free-drops the bundle from the aircraft, and the AOD or timer deploys the parachute at the preset altitude or time.

METHODS OF AIRDROP

1-58. The five methods of airdrop currently employed are as follows:
- Door loads.
- Wing loads.
- Gravity.
- Extraction.
- External transport.

Chapter 1

Door Loads

1-59. Personnel slide or push bundles out of the aircraft door or tail ramp opening. This method is suitable for free, low-velocity, and high-velocity drops. The size of the opening in the aircraft and the capability of personnel to eject the bundle limit the load in size and weight.

Wing Loads

1-60. Personnel rig loads in containers attached to shackles on the underside of the aircraft wings. The load-carrying capacity of the aircraft and the type container and its asymmetrical flight characteristics limit the size, weight, and shape of the load.

Gravity

1-61. Personnel release load-restraining ties to allow the load to slide out of the cargo compartment of the aircraft that is flying in drop altitude. The nose of the aircraft is slightly elevated.

Extraction

1-62. Personnel use a drogue parachute on platform loads. Use of the drogue parachute extracts platform loads from the aircraft cargo compartment.

External Transport

1-63. Personnel hang loads from a hook clevis on a helicopter flown to the delivery site. Personnel use the free, low-velocity, or high-velocity method to drop the load.

Sequence of Resupply

1-64. A typical air resupply mission involves a particular sequence of actions by the detachment, joint special operations task force (JSOTF), and the air support unit. Figure 1-2 shows this sequence.

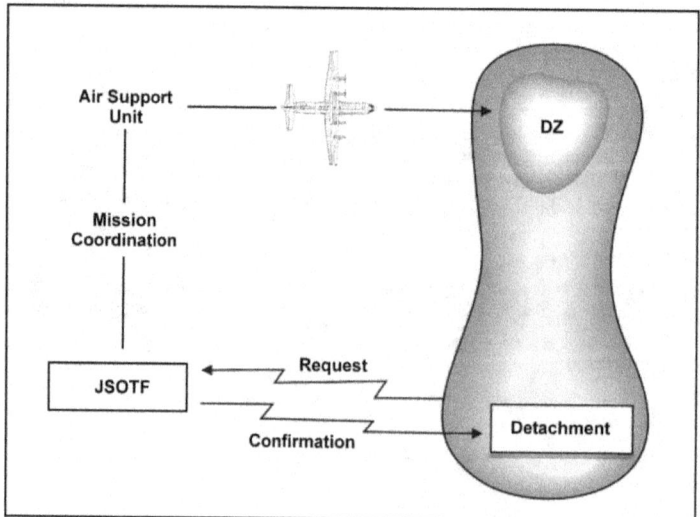

Figure 1-2. Typical air resupply mission

1-65. The detachment—
- Identifies and reports DZ, PZ, or LZ sites.
- Transmits DZ, PZ, or LZ data and resupply requests to the JSOTF.

1-66. The JSOTF—
- Processes DZ, PZ, or LZ data and resupply requests.
- Coordinates the mission with the air support unit.
- Transmits the mission confirmation message to the operational element.
- Prepares and delivers supplies and personnel to the departure site.

1-67. The air support unit—
- Prepares the mission confirmation data for the JSOTF.
- Receives and loads the supplies and personnel.
- Executes the air delivery mission.

1-68. Upon receipt of the mission confirmation message from the JSOTF, the detachment—
- Organizes the reception committee.
- Secures and marks the DZ, PZ, or LZ.
- Receives the personnel or supplies.
- Removes and distributes the incoming supplies.
- Sterilizes the DZ.

COMBAT CONSIDERATIONS

1-69. Various techniques and concepts that are used in combat are not used in training because of safety reasons. USASOC Reg 350-2 contains detailed information.

This page intentionally left blank.

Chapter 2
Premission Preparation

Successful air operations depend on detailed air mission planning, preparation, coordination, and rehearsals. Infiltration and exfiltration of SFODs are normally joint operations that require coordination with the supporting unit that will execute the infiltration and exfiltration. Pre-mission planning should include joint briefings and rehearsals between the aircrew and the deploying operational element. Each group must know the sequence of events and responsibilities under normal and emergency conditions to ensure efficient mission completion and to increase the chance of survival.

PLANNING CONSIDERATIONS

2-1. Regardless of the type of mission planned, mission planners must consider:
- Mission, enemy, terrain and weather, troops and support available–time available and civil considerations (METT-TC) (Army).
- Operations security (OPSEC).
- Security in the AO.
- Transportation available.
- Time for preparation, training, and rehearsals.
- Special equipment and delivery systems.
- Door bundles and CDSs.
- Reception committees.
- Safety.

METT-TC

2-2. METT-TC is considered in all aspects of mission planning, to include infiltration and exfiltration.

Mission

2-3. The mission may require rapid deployment into the AO, thereby dictating the most expeditious method for infiltration. In other cases, however, mission success may depend mainly on maintaining secrecy with rapid execution of secondary importance. The mission statement may or may not specify means of infiltration and exfiltration. The element being infiltrated decides the means when not specified in the mission statement.

Enemy

2-4. The enemy threat, capabilities, disposition, security measures, and air detection or defense systems affect the means selected for infiltration or exfiltration. Order of battle affects the routes, communications procedures and capabilities, external exfiltration capabilities, and sources of resupply. The SOTF must look at their teams' capabilities and air assets available to better assist the team with this phase of the mission.

Terrain and Weather

2-5. Mission planners consider terrain and bodies of water when selecting the method of infiltration or exfiltration. Terrain affects personnel, the need for special equipment, altitude selection, approach and egress routes, landing areas for mission aircraft, DZs, and beach landing sites (BLSs). Air infiltration routes that provide terrain masking are preferred in static-line parachute operations and airland operations. Seasonal weather conditions effect infiltrations and exfiltrations. Mission planners also consider meteorological conditions. If parachuting with self-contained underwater breathing apparatus (parascuba) techniques are used, certain situations impact entry or recovery techniques. For example, periods of reduced visibility or high surface winds with their effect upon surf conditions may prohibit the use of parachutes, inflatable boats, or surface or subsurface swimming. These same conditions generally favor land infiltration or exfiltration. The AWADS reduces the impact of weather as a limiting factor for air infiltrations. When considering the enemy situation, light conditions will determine time available for infiltration or exfiltration. The infilling element performs air operations, regrouping, and movement on the ground after infiltration during favorable periods of sunrise and sunset, moonrise and moonset, moon phase (Appendix B), and twilight.

Troops and Support Available

2-6. The number of personnel to be infiltrated, their training, and the amount of equipment carried may be limiting factors that affect how the operational element will infiltrate. SFODs should be proficient in a variety of infiltration and exfiltration techniques. SFODs may require special or refresher training for some infiltration means such as MFF, water operations, mounted operations, or others. A need for special skills may call for the use of attachments whose physical stamina and capabilities may be limiting factors. Mission planners must also consider availability of supporting forces and special equipment required.

Time Available

2-7. Mission planners consider the distance to and from the objective area. For infiltration, they consider the distance from the departure area to the objective area; for exfiltration, the distance from the objective area to the recovery area. The considerations for infiltration and exfiltration are the range, speed, capabilities, and capacity of the aircraft used.

Civil Considerations

2-8. Mission planners must consider the local civilians in the AO to avoid flying over populated areas or having to move near populated areas during movement from the infiltration area to the objective area.

OPERATIONS SECURITY

2-9. OPSEC is a command responsibility. Commanders must consider OPSEC in all staff efforts—intelligence, communications-electronics, logistics, administration, and maintenance—to provide maximum protection for an operation. They must integrate OPSEC throughout every SF mission from initial planning through postexecution stages to keep the enemy from learning the following:

- Plan—how, when, where, and why we will do something.
- Execution—how, when, where, and why we are doing it.
- After action—how, when, where, and why we did it, and how we can do it better in the future.

2-10. OPSEC consists of four main categories of security measures. They are—

- *Signal security.* This category includes communications security and electronics security.
- *Physical security.*
- *Information security.*
- *Deception.*

2-11. All categories are interrelated; mission planners need to analyze each of them as a group for use in an operation. Army Regulation (AR) 530-1, *Operations and Signal Security*, and AR 25-2, *Information Systems Security*, provide OPSEC guidance.

SECURITY IN THE AREA OF OPERATIONS

2-12. Security is of prime importance because of the visibility of reception operations and the vulnerability of SF and indigenous assets engaged in these operations. Observable operational patterns and activities may give reception site information to the enemy. Avoid contact with the enemy if possible. Use proper counterintelligence measures. Use of area DZs and RTs greatly enhances the safety of reception operations. However, mission planners should consider the electronic warfare (EW) threat.

TRANSPORTATION AVAILABLE

2-13. The transportation means selected for the delivery or recovery depends on the specific needs of the mission. When selecting the transportation means, mission planners should consider the following:

- What type of aircraft meets mission needs?
- What type of navigational system is in the aircraft?
- Are electronic countermeasures in place?
- What is the skill level of the aircrew?
- To which unit is the aircraft assigned?
- Is aircraft dedicated?
- Is a backup aircraft designated and, if so, is it the same type?
- What is the range and carrying capacity of the aircraft?
- What are the weather limitations?
- Does the aircraft possess aerial refueling capability?

Base the selection of the aircraft on the capabilities, limitations, and availability of the mission support platform. Chapters 3 and 4 provide information about the various platforms.

TIME FOR PREPARATION, TRAINING, AND REHEARSALS

2-14. The amount of time available for the team to prepare and train for an infiltration affects the method that they may use. A team should use an infiltration technique that has the best chance of allowing undetected infiltration if they have the amount of time needed to prepare for it.

2-15. Rehearsals are the best means for determining flaws in procedures or errors in planning. Mission planners must thoroughly coordinate all procedures to be used. When training, the SFOD should use the exact type of aircraft they will use for infiltration, if possible. The detachment's rehearsals should occur under terrain, astronomical, hydrographical, and meteorological conditions close to those seen in the AO. The more complex the procedures, the greater the need for rehearsals.

SPECIAL EQUIPMENT AND DELIVERY SYSTEMS

2-16. The type and method of air-to-ground delivery depends on the specific needs of the mission. Any infiltration or exfiltration that requires special equipment for the SFOD or specially equipped aircraft will usually require additional time for acquiring the equipment, inspections, coordination, and rehearsals. Special considerations include the following:

- Specialized teams such as HALO and self-contained underwater breathing apparatus may require additional preparation time to rehearse their infiltration technique.
- The capabilities of personnel augmenting a team may reduce the team's infiltration options.
- High-altitude and rough-terrain drops need special equipment and may limit cargo loads.

Chapter 2

- Mission planners using DZ surveys and map reconnaissance acquire DZ intelligence for the supporting air unit to accurately figure offset aiming points when using AWADS aircraft.
- Programming and navigational planning for an IMC mission are more time-consuming and demanding than for a visual drop.

Equipment Selection

2-17. The selection of equipment and supplies to be carried on the initial infiltration will be based on the—
- Type of infiltration.
- Need for security.
- Enemy threat.
- Size of the resistance force and situation in the JSOA.
- Means of transportation selected and method of insertion.
- Distance, terrain, and signal propagation conditions.
- Weight and bulk of equipment.
- Communications equipment compatibility.
- Equipment availability.
- Potential for external resupply.
- Time between resupplies.

Equipment Preparation

2-18. Equipment reliability in the field depends primarily upon the care the unit takes in preparing the equipment for transportation and for its use. The using element must inspect and test all equipment before infiltration.

Packing and Rigging

2-19. The type of resupply aircraft available will dictate the type of resupply bundle packed by the detachment. The SFOD inventories, packs, and rigs all planned resupply bundles before infiltration. The infilling element and the next higher element coordinate with the SOTF or AOB to ensure proper marking and identification of bundles. See FM 3-21.220 for instructions on padding and packaging equipment.

DOOR BUNDLES AND CONTAINER DELIVERY SYSTEM

2-20. The infilling element uses airdrop containers and door bundles during infiltration or resupply operations when more equipment than can safely be carried by the detachment members is required. The infilling element rigs equipment and supplies as door bundles, packed in airdrop containers, or properly prepared for other approved types or methods of airdrop. This procedure permits the parachutist to jump unencumbered by excess equipment. However, equipment loss will occur if the infilling element does not recover containers. The type of platform being used may limit the use of door bundles or CDSs. The element should use airdrop containers only when there is an adequate reception party or in low-level drops of 500 to 700 feet where dispersion is not a problem.

RECEPTION COMMITTEE

2-21. A reception committee is any organized group that meets the infilling detachment at the infiltration site to assist them. The presence or absence of a reception committee may influence the infiltration method in several ways. Their presence may allow for the use of an area, marked area, or rough terrain DZs as opposed to an unmarked DZ. The presence or absence of a reception committee may help determine the amount of accompanying equipment and follow-on resupply missions. Sterilization of the infiltration site and disposal of air items are less of a problem when a reception committee is present than when conducting a blind or rough terrain drop. Chapter 3 contains more information about reception committees.

Safety

2-22. The safety considerations for training operations and combat operations are different. During training, safety is the primary consideration for all operations. During combat operations, the commander may waive or change many of the safety regulations and requirements to fit the tactical situation. However, safety should still be an important consideration when planning an air operation. The operation is futile if a high percentage of the unit is incapacitated and unable to complete the mission.

EMERGENCY PROCEDURES

2-23. During premission planning, the SFOD establishes the emergency recognition signals and procedures for use with the supporting air unit and the SOTF or AOB. The SFOD considers such emergency procedures as mission aborts, emergency landings, and evasions. Although most signals and procedures are unit SOP, mission planners must take care that the adjacent and ground forces and reception committee have a complete understanding of the use of recognition signals and procedures to be followed. The using unit carefully checks the electronic equipment used in assembly and recognition procedures before departure to ensure proper functioning and adequate power sources. The deployed detachment communicates with the SOTF or AOB by using message formats IAW the current Signal Audio Visual Service Supplement (SAVSERSUP). Appendix C consists of formats that apply to aerial missions.

Note. These message formats may be tailored by the SOTF/JSOTF/AOB to better fit mission requirements.

Mission Aborts

2-24. The operation order states that the commanders involved in the mission are jointly responsible for deciding whether to proceed with or to abort the mission. When deciding whether to abort or proceed, commanders should consider weather, lack of or improperly displayed identification markings, or incorrect authentication signals. There are two courses of action (COAs) available:
- Abort the primary reception, proceed to a preselected alternate DZ or LZ, and conduct infiltration there as planned.
- Abort the entire mission and return to the SOTF or AOB.

Emergency Landings

2-25. Before the SFOD boards the aircraft, the loadmaster briefs emergency landing procedures. However, the infilling unit and aircrew coordinate emergency landing procedures before then so everyone has a chance to become familiar with the procedures and to eliminate conflicts with detachment SOP. Although the exact actions for egressing the aircraft and assembling are standard, the actions immediately following will depend on where the aircraft has landed or crashed. The infilling unit and aircrew develop contingencies for landing or crashing in—
- Friendly-controlled areas.
- Neutral-controlled areas.
- Enemy-controlled areas.

Note. After an emergency landing and depending on the situation, the SFOD and aircrew may have to initiate the evasion plan of action.

EN ROUTE EVASION PLAN OF ACTION

2-26. A vital part of premission planning is the development of a viable en route evasion plan of action. Such a plan enhances the survivability of the aircrew and the SF element in case of emergency evacuation of the plane over or in hostile areas.

2-27. The mission commander is responsible for—
- Checking all factors bearing on survivability.
- Devising an evasion plan that is feasible and executable.
- Briefing all mission members before departure.

PHASES

2-28. Each mission will present unique problems when planning for evasion. The evasion plan consists of two phases.

Phase I

2-29. Phase I is the portion of the mission from the initial penetration of enemy-controlled territory to the LZ or DZ. The infilling unit and aircrew jointly plan for this phase and consider the presence of the aircrew. The SFOD commander establishes criteria during isolation to determine the minimum requirements for continuing with the mission. The senior SF survivor takes charge. When deciding whether to continue, the senior SF survivor considers the capabilities, experience, and expertise of the surviving personnel and presence or absence of mission-essential equipment. In the event that the SFOD continues with the mission, the senior aircrew survivor will execute a COA for aircrew survivors that will not interfere with the assigned mission. If the SFOD decides to abort and initiate evasion, the aircrew moves with them. The detachment and aircrew should avoid the aircraft crash site and the aircraft flight path before ground contact. Therefore, they will avoid discovery by hostile forces responding to the crash or detection reports.

Phase II

2-30. Phase II commences with the infiltration of the SFOD and the departure of the aircraft. The evasion plan only considers the deploying detachment. The senior SF survivor determines to proceed with the assigned mission if enough of the team members have survived.

OTHER CONTINGENCIES

2-31. The deploying detachment and supporting aviation unit should also plan for other contingencies. They should plan for ground delays, weather conditions, aborts or cancellations, early or late arrival at the LZ or DZ, rescheduled air missions, and mission compromise.

Ground Delays

2-32. Planned mission routes will determine the length of delay that can be incurred in meeting the established TOT. If departure is delayed and the routes can be safely altered to arrive on time, participants should execute the mission.

Weather Decision

2-33. The commanders of the SOTF or AOB and the airlift control element (ALCE) jointly make the final decision on operational delays or weather cancellations. Their decision is based on existing weather minimums.

Aborts or Cancellations

2-34. When a mission is aborted or canceled while en route, the supporting aircraft returns to the launch base or a designated alternate site. The SF commander attempts immediate contact with the SOTF or AOB for further orders.

Early or Late Arrival at LZ or DZ

2-35. Missions not accomplished because of early or late arrival at or over the primary objective area will proceed to the alternate site. The mission confirmation message provides the alternate site.

Rescheduled Air Missions

2-36. Under the delay provisions, the SOTF commander reschedules missions not accomplished for any reason. If delay provisions are not prescribed, the SOTF commander must submit a new mission request.

Mission Compromise

2-37. If infiltrating during an operation related to direct action (DA) or special reconnaissance (SR), mission planners establish plans for the direct fire support and immediate exfiltration of the detachment under distress.

JOINT MISSION BRIEFING

2-38. Face-to-face coordination between the SFOD commander, JM, and aircrew takes place during isolation. This coordination is critical to ensure everyone knows exactly what is planned and expected. The joint mission briefing takes place before the briefback. At this briefing, the infilling unit briefs the final infiltration plan. The aircrew must attend the briefing. At a minimum, the aircrew members attending should include the aircraft commander and primary loadmaster. The navigator's attendance is at the discretion of the aircraft commander. A briefing for the pilot and JM normally takes place at planeside. At that time, they receive any changes or updates to the mission.

RATES OF DESCENT

2-39. SF Soldiers are carrying increasingly heavier combat equipment loads on airborne operations. The combination of heavier loads and less dense atmosphere at higher elevations causes a sharp increase in the jumper's rate of descent with the MC1 series or T-10B parachute. The sharp rate of descent may cause unnecessary injuries. USASOC direct reporting unit (DRU) commanders will not allow training with combat loads that cause the rate of descent of the jumper to exceed 22 feet per second (ft/sec). Table 2-1, page 2-8, provides the rates of descent to assist the DRU commander.

2-40. Table 2-1 is based on the following:
- Ft/sec increase per 1,000 feet gained.
- Ft/sec increase per 25 pounds added suspended weight.

2-41. In Table 2-1—
- Elevation is expressed in feet above MSL.
- Total suspended weight (pounds [lb]) is the total weight of the jumper, combat equipment, and air items.
- Rate of descent is expressed in ft/sec. Depending upon the total weight and relative air density, the average rates of descent for different canopies are as follows:
  - MC-6, 10 ft/sec.
  - MC1-1B, 18 to 22 ft/sec.
  - MC1-1C, 14 to 18 ft/sec.
  - T-10C, 19 to 23 ft/sec.
- Rates of descent in ft/sec for the T-10B and MC1-1B main parachutes at various elevations are provided.

Table 2-1. Rates of descent

Elevation (ft above MSL)	Total Suspended Weight (lb)								
	200	225	250	275	300	325	350	375	400
10,000	17.3	18.3	19.3	20.3	21.3	22.3	23.3	24.3	25.3
9,000	17.1	18.1	19.1	20.1	21.1	22.1	23.1	24.1	25.1
8,000	16.9	17.9	18.9	19.9	20.9	21.9	22.9	23.9	24.9
7,000	16.7	17.7	18.7	19.7	20.7	21.7	22.7	23.7	24.7
6,000	16.5	17.5	18.5	19.5	20.5	21.5	22.5	23.5	24.5
5,000	16.3	17.3	18.3	19.3	20.3	21.3	22.3	23.3	24.3
4,000	16.1	17.1	18.1	19.1	20.1	21.2	22.1	23.1	24.1
3,000	15.9	16.9	17.9	18.9	19.9	20.9	21.9	22.9	23.9
2,000	15.7	16.7	17.7	18.7	19.7	20.7	21.7	22.7	23.7
1,000	15.5	16.5	17.5	18.5	19.5	20.5	21.5	22.5	23.5
MSL	15.3	16.3	17.3	18.3	19.3	20.3	21.3	22.3	23.3
NOTE: Rate of descent represented in ft/sec.					Above average rate of descent.		Exceeds acceptable rate of descent.		

Note. The rate of descent for the MC1-1C/SF10A is slightly lower than Table 2-1 shows for the T-10B and MC1-1B main parachutes. For planning purposes, the above rates of descent will also apply to the MC1-1C/SF10A main parachute.

2-42. Other factors that unit commanders should consider when deciding the amount of equipment their troops will carry on airborne operations are—
- Forecasted winds affecting lateral speed (winds at altitude).
- Proficiency of jumpers.
- Air temperature (colder temperatures cause a faster rate of descent).
- Night operations (limited visibility conditions).
- DZ conditions.

Chapter 3

Drop Zones

A DZ is any designated area where personnel and equipment may be delivered by means of parachute or free drop. During premission planning, the SFODA selects DZs for infiltration, resupply, and other uses. When planning, the SFODA uses all available intelligence resources, satellite imagery, and maps. The commanders of the SOTF and the supporting air unit jointly approve the selected DZs. After the SF operational element infiltrates into the AO, the element must confirm and report additional DZ data for use by the SOTF or AOB and the supporting air unit. SF personnel select and mark DZs to be used for future reception operations that are generated by the mission. The ground unit commander (GUC) selects the DZ with a location that best supports all aspects of the tactical operational plan. For tactical training, mission planners check the USAF assault zone availability report (AZAR) for an approved DZ within the tactical area. If the selected DZ is not on the AZAR, mission planners conduct a tactical assessment.

SELECTION OF DROP ZONE

3-1. Mission planners select the DZ that is best suited for the mission, using a variety of criteria. The selection criteria may vary for each mission. Mission planners must weigh all the factors and consider the advantages and disadvantages of each factor. Often the DZ that is selected is not perfect but simply the best available. The SFODA will designate primary and alternate DZs during mission planning. The SOTF and supporting air unit approve the DZs designated by the SFODA. When selecting a DZ, the SFODA should consider as many of the below-listed factors as possible:

- Support of the mission.
- Type of airborne operation.
- Type of supporting aircraft.
- Infiltration route.
- Security.
- Safety.
- Weather and astronomical conditions.
- Size of the DZ.
- Other characteristics of the DZ.

SUPPORTING THE MISSION

3-2. Some of the main considerations when selecting the DZ that best supports the mission are as follows:
- Security.
- Distance from target or area of interest.
- Terrain between DZ and target.
- Locations of enemy units and built-up areas.
- Amount of time that the detachment has for movement.
- Amount of equipment that the team members are carrying.

Normally, the detachment will identify all possible DZs that support the mission and then determine which ones are the best.

3-3. The supporting air unit provides for airdrop accuracy and air safety. The ground unit oversees establishment operation, safety off the DZ, and for elimination or marking of ground hazards associated with the DZ. The JM provides for airdrop accuracy when JM-directed release procedures are used. The using forces will take responsibility for injured personnel and damaged equipment that could result from using a DZ that does not meet the minimum-sized criteria.

TYPES OF AIRBORNE OPERATIONS

3-4. Airborne operations are classified as static line, MFF, or airdrop. The types of static-line airborne operations include personnel, CDS, and heavy equipment drops. A DZ that may be suitable for one type of airborne operation may not be suitable for another type because of restrictions on the size, support requirements, number of personnel to be dropped, and other factors. The more personnel or bundles being dropped by static-line parachute, the greater the DZ size.

SUPPORTING AIRCRAFT

3-5. Mission planners should consider the following factors when determining the capabilities of the supporting aircraft:

- Carrying capacity of the aircraft.
- Type of aircraft.
- Aerial refueling capability.
- Weather limitations.
- Navigational systems.
- Electronic countermeasures.
- Level of skill of the aircrew.
- Unit from which the aircraft comes.

Mission planners must also determine whether or not the aircraft is dedicated to the mission and if there is a backup aircraft. If there is a backup aircraft, is it the same type? If the capabilities of the backup aircraft are different from those of the primary aircraft, will the plan have to be changed to accommodate the backup aircraft?

Note. Before setting up a DZ, the reception committee needs to know the type of aircraft, drop speed, and altitude.

INFILTRATION ROUTE

3-6. Mission planners select the primary and alternate DZs so that the aircraft can overfly them in order—primary then alternate—without making major course corrections. Air routes to and from the DZ should deconflict with other air operations, restrictive terrain, or man-made objects (television or radio towers). The route to the DZs should not overfly built-up areas, enemy defenses, or the objective.

SECURITY

3-7. The DZ must provide maximum security from the enemy threat. It should be located as far away from enemy positions and civilian areas as possible. The approach and exit routes must be concealed from observation or secured against interdiction. Additionally, the DZ should be near areas suitable for caching supplies and disposing of air delivery equipment.

SAFETY

3-8. A DZ that does not meet all criteria for the safety of the infiltrating personnel but meets all air safety criteria may be used for cargo drops. To ensure the airdrop is safe, the equipment and personnel can be recovered, and personnel and equipment can be employed to accomplish the mission, the DZ should be as free of obstacles as possible. Cargo airdrop altitude should not exceed 1,000 feet. Examples of obstacles are—

- Trees 35 feet or higher impeding recovery of personnel or equipment.
- Water 4 feet deep within 1,000 meters (m) from any edge of the DZ.
- High-tension wires that are carrying active current of 50 volts or greater. (Can the power be turned off before the drop?)
- Any other conditions that may injure parachutists or damage equipment (inactive electric wire, barbed wire fences, swamps, ditches, gullies, and so forth).

METEOROLOGICAL CONDITIONS

3-9. Meteorological conditions impact airborne operations. Mission planners must consider seasonal weather and other meteorological conditions in the operations area.

SIZE OF THE DZ

3-10. There are different DZ dimension (in meters) criteria for different types of airborne operations and different types of USAF aircraft. Table 3-1, pages 3-4 through 3-6, provides the minimum sizes in peacetime for standard USAF aircraft. The DZ sizes in Table 3-1 must be followed unless a waiver is issued. Tables 3-2 and 3-3, pages 3-6 and 3-7, respectively, provide the minimum DZ sizes for USAF SO aircraft. During contingency or wartime missions, the SF element commander may waive DZ sizes. However, size requirements remain a joint responsibility of the air component commander (ACC) or Commander, Air Force special operations forces, and the supported force commander. Commanders will use minimum size criteria when selecting DZs to support the ground tactical plan during training:

- *Width.* The DZ width should allow for minor computation errors in wind drift.
- *Length.* The absolute minimum DZ length depends on the ground dispersion pattern formed by the number of jumpers or cargo containers to be dropped. This pattern generally parallels the line of flight of the aircraft along the long axis of the DZ.

Note. For USAF unilateral operations, the 70-meter increase in length required for each additional parachutist may be computed from the point of impact rather than added to the total length of the minimum size DZ for one parachutist. Alternating door exit procedures for training (ADEPT) do not apply to unilateral USAF operations. Air Force Instruction (AFI) 13-217, *Drop and Landing Zone Operations,* details all USAF requirements.

3-11. The SO DZ size criteria are dependent on the type of aircraft making the drop and whether or not the DZ is marked (Table 3-2). A marked DZ is defined as a DZ that has a PI or RP marked with a precoordinated visual or electronic signal. The PI is marked for CARP drops, and the RP is marked for GMRS drops. Standard DZ markings are raised angle markers (RAMs), VS 17 marker panels, visible lighting systems, and light beacons. Virtually any type of lighting or visual marking system is acceptable if all participating units are briefed and concur. Day marking or visual acquisition devices include, but are not limited to, colored smoke, mirror, railroad flares, and any reflective or contrasting marker panel (for example, space blanket). In some cases, geographical points may be used. Night markings or acquisition aids may include a light gun, flares, fire pots, railroad flares, flashlights, chemical lights, and infrared (IR) lighting systems. Electronic navigational aid (NAVAID) markings (zone marker [ZM], SST-181, tactical air navigation [TACAN], and so on) may be used for either day or night operations and placed as directed by mission requirements.

Chapter 3

Table 3-1. Standard DZ size criteria

CDS Using C-130				
Note: For visual formations (day and night), increase width by 92 m (46 m on each side).				
	Width[1, 2]	Number of Containers		Length[2]
		Single Row	Double Row	
Altitude (AGL) to 600 ft	400 yds/366 m	1	1 to 2	400 yds/366 m
		2	3 to 4	450 yds/412 m
		3	5 to 6	500 yds/457 m
		4	7 to 8	550 yds/503 m
		5 to 8	9 to 16	700 yds/640 m
		9 to 12	10 to 24	850 yds/777 m
Altitude (AGL) Above 600 ft	Add 40 yds/36 m to width and length for each 100 ft above 600 ft (add 20 yds/18 m to each side of DZ, 20 yds/18 m to each end).			

CDS/LCADS-LV Using C-17				
	Width[1, 2]	Number of Containers		Length[2]
		Single Row	Double Row	
Altitude (AGL) to 600 ft	450 yds/412 m	1	1 to 2	590 yds/540 m
		2	3 to 4	615 yds/562 m
		3	5 to 6	665 yds/608 m
		4 to 8	7 to 16	765 yds/700 m
		9 to 14	17 to 28	915 yds/837 m
		15 to 20	29 to 40	1065 yds/974 m
Altitude (AGL) Above 600 ft	Add 40 yds/36 m to width and length for each 100 ft above 600 ft (add 20 yds/18 m to each side of DZ, 20 yds/18 m to each end).			

High-Velocity (HV) CDS/HV-LCADS (Using 12-, 22-, or 26-ft Ring-Slot Parachutes)		
	Width[1, 2]	Length[2]
Altitude (AGL) to 3,000 ft	580 yds/530 m	One row of containers 660 yds/604 m.
		Add 50 yds/46 m to the trailing edge for each additional row of containers.
Altitude (AGL) Above 3,000 ft	Add 25 yds/23 m to each side and 100 yds/91m to each end for every 1,000-ft increase in drop altitude.	

Drop Zones

Table 3-1. Standard DZ size criteria (continued)

HAARS CDS		
	Width[1, 2]	Length[2]
Altitude (AGL) to 3,000 FT	500 yds/457 m	1 to 8 containers: 1200 yds/1,098 m
		9 or more containers: 1900 yds/1,739 m
Altitude (AGL) Above 3,000 ft	Add 25 yds/23 m to each side and 50 yds/46 m to each end for every 1,000-ft increase in drop altitude.	

HSLLADS/High-Speed Kit		
	Width[1, 2]	Length[2]
Altitude (AGL) to 3,000 ft	300 yds/274 m	600 yds/549 m

Recovery Kit		
	Width[1, 2]	Length[2]
MC-130	200 yds/183 m	200 yds/183 m
AWADS	400 yds/366 m	400 yds/366 m
C-130	400 yds/366 m	400 yds/366 m

Personnel		
	Width[1, 2]	Length[2]
Altitude (AGL) to 1,000 FT	600 yds/549 m	1 Parachutist 600 yds/549 m
		Additional parachutist: Add 75 yds/69 m to the trailing edge for each additional parachutist.
Altitude (AGL) Above 1,000 ft	Add 30 yds/28 m to width and length for each 100 ft above 1,000 ft (Add 15 yds/14 m to each side of the DZ; 15 yds/14 m to each end).	

Heavy Equipment (HE)		
	Width[1, 2]	Length[2]
Altitude (AGL) to 1,100 ft	600 yds/549 m	1 platform: 1000 yds/915 m
	Additional platforms: For the C-130, add 400 yds/366 m to the trailing edge for each additional platform. For the C-17, C-5, add 500 yds/457 m to the trailing edge for each additional platform. Add 30 yds/23 m to each side and 50 yds/46 m to each end for every 1,000-ft increase in altitude.	
Altitude (AGL) Above 1,100 ft	Add 30 yds/28 m to width and length for each 100 ft above 1,100 ft (Add 51 yds/14 m to each side of the DZ; 15 yds/14 m to each end).	

[1] Adjustments on width are as follows:
  a. For day visual formations, increase width by 100 yds/92 m (50 yds/46 m on each side).
  b. For C-130 station-keeping equipment (SKE) AWADS formation, increase width by 400 yds/366 m (200 yds/183 m on each side).
  c. From official sunset to sunrise, increase width by 1000 yds/912 m for single-ship visual drops (50 yds/46 m on each side) or 200 yds/184 m for visual formations (100 yds/92 m on each side).

Table 3-1. Standard DZ size criteria (continued)

[2] For C-17 DZ size adjustment (more than one may be required):
  a. For visual formations (day and night), increase width by 100 yds/92 m (50 yds/46 m on each side).
  b. For night visual airdrop, increase width an additional 100 yds/92 m (50 yds/46 m on each side). DOES NOT APPLY TO AIRCRAFT PERFORMING GLOBAL POSITIONING SYSTEM (GPS) DROPS.
  c. For SKE HE/CDS formation, increase width by 400 yds/366 m (200 yds/183 m on each side).
  d. C-17s require SKE to perform personnel formation airdrop. For personnel formations performing GPS drops below 1,000 ft AGL, the DZ width using center PI is 1,128 m for two-ship elements and 1,638 m for three-ship elements. When using offset PI, minimum width is 1,000 m for two-ship elements and 1,183 m for three-ship elements. Above 1,000 ft, add 30 yds/28 m to width and length for each 100 ft above 1,000 ft.
  e. Single-ship IMC drops have no adjustment below 1,000 ft. Above 1,000 ft, add 30 yds/28 m to width and length for each 100 ft above 1,000 ft.

Table 3-2. Minimum DZ criteria for single-ship SO missions on marked DZ

Personnel			
Note: For each additional parachutist, add 70 m to the DZ length.			
Type Drop	MC-130	AWADS	C-130[1]
CARP	300 yds/275 m x 300 yds/275 m	600 yds/550 m x 600 yds/550 m	600 yds/500 m x 600 yds/550 m
GMRS	300 yds/275 m x 300 yds/275 m	300 yds/275 m x 300 yds/275 m	300 yds/275 m x 300 yds/275 m
HSLLADS/High-Speed Kit			
CARP and GMRS	300 yds/275 m x 600 yds/550 m	NA	NA
Recovery Kit			
CARP and GMRS	200 yds/180 m x 200 yds/180 m	400 yds/365 m x 400 yds/365 m	400 yds/365 m x 400 yds/365 m
CDS	Use the same dimensions as for a conventional platform, but add 60 yds/50 m for each additional container.		
HE	Use the same dimensions as for a conventional platform.		

[1] For all C-130s, add 400 yds/365 m to DZ length for each additional platform.

NOTE: The above minimum DZ sizes are inclusive of the 100-m safety zones that are required by the Army at each end of the DZ.

Table 3-3. Minimum DZ criteria for single-ship SO missions on unmarked DZ

Type Drop	MC-130	AWADS	C-130
Personnel	400 yds/365 m x 600 yds/550 m	600 yds/550 m x 600 yds/550 m	600 yds/550 m x 600 yds/550 m
Note: For each additional parachutist, add 70 m to the DZ length.			
HSLLADS	300 yds/274 m x 600 yds/550 m	NA	NA
Recovery Kit	200 yds/183 m x 200 yds/183 m	200 yds/183 m x 200 yds/183 m	200 yds/183 m x 200 yds/183 m
CDS/CRS	Use the same dimensions as for a conventional platform, but add 62 yds/50 m for each additional container.		
HE	Use the same dimensions as for a conventional platform.		

NOTES:
1. For C-130, add 400 yds/365 m to DZ length for each additional platform.
2. The minimum DZ sizes are inclusive of the 100-m safety zones that are required by the Army at each end of the DZ.
3. For all blind drops, add 40 yds/37 m to each end of the DZ for each 100-ft increase in altitude above the minimum DZ altitude for the load being dropped.
4. During SO airdrops, the minimum DZ sizes shown above will normally apply unless precluded by mission requirements.

Note. DZs that are unmarked or obscured by weather are considered unmarked. The navigators confirm the DZ location and determine the RP by radar and onboard navigational equipment. The navigator determines drop accuracy by considering such factors as terrain, usable radar targets, and chart and equipment accuracy. The SO element for which the drop is being made will take responsibility for the accuracy of the airdrop when the DZ size does not meet the minimum criteria.

3-12. There is no minimum size for MFF DZs. However, an area 50 meters by 100 meters is recommended. The JM or commander must consider the experience level of the parachutists.

OTHER CHARACTERISTICS OF THE DZ

3-13. Depending on the type of airborne operation, a DZ should have the following characteristics. Some DZs may be limited or have restrictions because of their characteristics.

- *Shape.* Square or circular DZs are preferable. These DZ shapes permit a wider choice in selecting the aircraft approach track.
- *Ground Surface.* Ground surfaces should be reasonably level and relatively free of obstructions, such as rocks, trees, fences, and power lines. Swamps, paddies, and marshy ground may be used. However, they can hinder recovery operations. DZs located at elevations greater than 4,000 feet above sea level require special attention because of the increased rate of parachute descent at these altitudes (Appendix D).
- *Terrain.* Flat or rolling terrain is desirable. Mission planners select sites in mountainous or hilly country containing large valleys or level plateaus for mission security. If a DZ is located on a relatively steep slope, mission planners must plan to have the aircraft flown parallel to the ridgeline to make the drop. Avoid the use of cultivated fields. Small valleys or pockets surrounded by hills are difficult to locate from the air. Do not select them except in unusual

Chapter 3

circumstances. The surrounding area must be relatively free of obstacles that could interfere with safe flight.

- *Night Operations.* Rising ground or hills more than 500 feet higher than the surface of the site should be no closer than 6 kilometers and must be reported. Regardless of good illumination, high terrain still constitutes a hazard to aircraft since darkness greatly reduces height perception. Report navigational obstacles more than 98 feet high for operational drops at altitudes less than 400 feet AGL.
- *Approach Quadrants.* It is important that the site should have one or more open approach quadrants free of terrain or vegetation masks that could block the aircrew's vision of the DZ marking during the final approach of the aircraft. There should be an open approach quadrant of at least 45 degrees that allows the air support unit a choice when determining its approach track from the IP (Figure 3-1).
- *Approach Path.* A single, clear line of approach is acceptable for medium aircraft under two conditions:
  - There is a clear, level turning radius of at least 4 kilometers on each side of the DZ.
  - There is a clear, level turning radius as prescribed by current regulations of the supporting air unit for the type aircraft to be used.

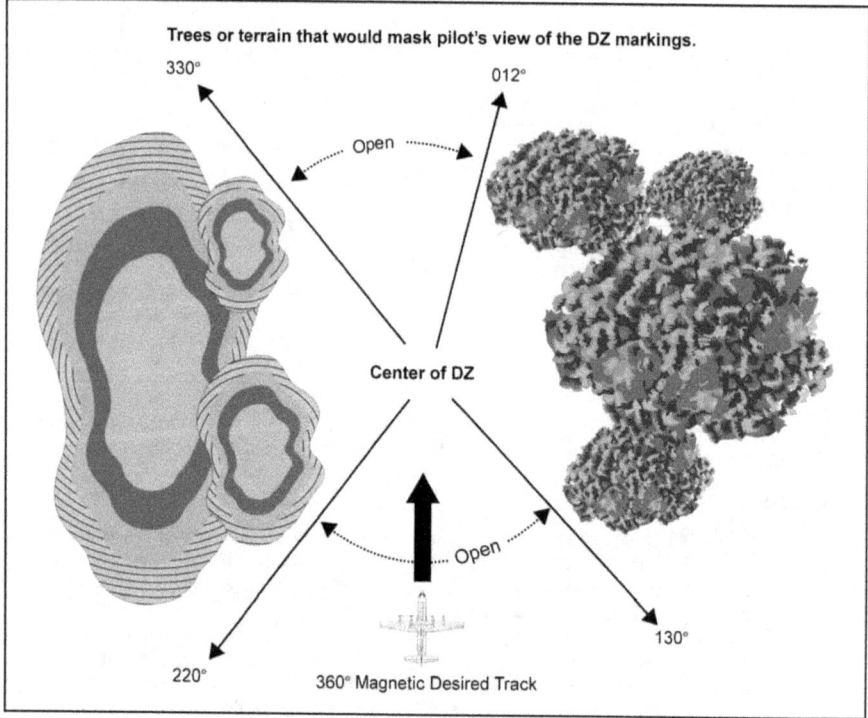

Figure 3-1. Computation of open quadrant and aircraft track (desired heading)

Note. For STOL and light aircraft, the distance must be 2 kilometers (Figures 3-2 and 3-3, page 3-9).

- *Access to Area.* The unit must have access to and from the DZ to recover equipment or conduct troop movement. DZs with no roads leading to them or next to a river with no bridges are examples of impeded access to areas.

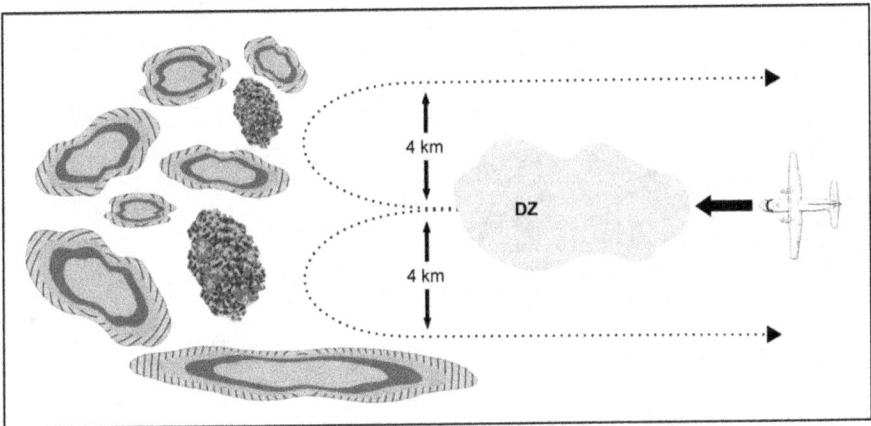

Figure 3-2. Level turning radius required for one-approach DZs and LZs (medium aircraft)

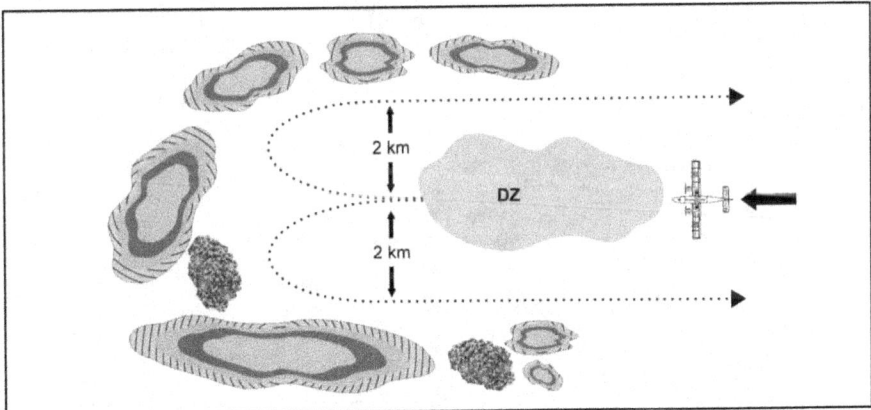

Figure 3-3. Level turning radius for STOL or light aircraft

TYPES OF DROP ZONES

3-14. SF units use many types of DZs. A DZ can be classified according to the type of mission to be flown, intended purpose of the DZ, or DZ markings. The different classifications are frequently combined (for example, primary infiltration DZ, alternate resupply DZ, and so on). Mission planners will select some DZs based solely on map reconnaissance, satellite imagery, and available intelligence before the detachment infiltrates the area. Mission planners will select other DZs once the detachment has infiltrated the area and conducts reconnaissance operations.

Primary Drop Zone

3-15. A primary DZ is the intended DZ for any type of airborne operation. It is usually the DZ that best satisfies the selection requirements and supports the mission. However, because of mission requirements and security, the primary DZ may not be the best DZ to use. One reason a primary DZ may not be the best is that the enemy may also be aware that it is the best DZ. The primary DZ may be manned by a reception committee and marked or unmarked. Primary DZs are further classified by their intended purpose; for example, primary infiltration, primary resupply, primary emergency resupply, and so on.

Alternate Drop Zone

3-16. An alternate DZ is selected for every mission. Usually it is the next-best DZ for that mission and must meet the same criteria. The alternate DZ is used whenever something prevents the primary DZ from being used. Things that prevent the use of the primary DZ include enemy situation, unsafe weather conditions, lack of recognition signal from DZ, inability to locate DZ, and other factors that may be determined by the SOTF or SFODA. Alternate DZs are located as close as possible to the primary track of the aircraft to preclude excessive aircraft maneuver and to minimize the possibility of enemy detection. The infiltration aircraft should normally plan to overfly the alternate DZ as a matter of course. The unit SOP / mission planning and situation will determine how and why the detachment will use their alternate DZ. If possible, a skeleton reception committee will man the alternate DZ. The alternate DZ may be marked or unmarked. Alternate DZs are further classified the same as primary DZs.

Infiltration Drop Zone

3-17. Mission planners select primary, alternate, and contingency infiltration DZs during the initial planning of a SF mission. The selection of an infiltration DZ must satisfy the requirements of the infiltrating element and the supporting air unit. Unless other units have been to the area and surveyed the DZs, mission planners will select DZs based on map reconnaissance, satellite imagery, and all available intelligence resources.

Resupply Drop Zone

3-18. The number and type of resupply DZs selected during pre-mission preparation will depend on the mission type and duration. There are three types of resupply operations for SF operations: automatic, on-call, and emergency. While mission planners may not plan for automatic and on-call resupply operations, they will always plan for an emergency resupply operation during mission preparation. The detachment will conduct a reconnaissance of these DZs to make sure they support the mission as planned. It may also be necessary for the detachment to change or add resupply DZs after infiltration, using message format GRAZE. The conduct of an automatic or on-call resupply operations is usually IAW unit SOP. The resupply can be scheduled either before infiltration or on call. Once the detachment is ready for their resupply, they send a coded message to the SOTF or AOB requesting a resupply bundle. If everything is a go for the resupply mission, the SOTF or AOB sends a mission confirmation message to the operational element. The detachment will provide surveillance on the DZ and mark it IAW unit SOP and the precoordinated plan at the appropriate times. Unless specified otherwise, resupply bundles will not be dropped unless the proper authentication signal is displayed on the DZ. Bundle or equipment recovery operations will be conducted IAW unit SOP. During an emergency resupply operation, unit SOP will dictate whether the resupply bundle will be dropped with or without a recognition signal displayed.

Unmarked Drop Zone

3-19. Commanders use unmarked DZs for a pre-planned blind drop by parachute of personnel, emergency resupply, and other operations, as needed. The ground or auxiliary forces commander assigns observers to keep the DZ under constant surveillance before and during the scheduled drop time. After the drop, the observers alert the operational element. Then the DZ is rapidly cleared and sterilized. Unmarked DZs are—
- Sometimes limited to specific astronomical conditions, because of visibility, depending on aircraft type.
- Sometimes odd shapes and sizes.

Drop Zones

- Located with identifiable terrain features.
- Located in isolated or remote areas away from the enemy threat.
- Reasonably close to planned evacuation routes.

3-20. The pilot or navigator computes the RP after visual or radar sighting of the DZ. If bad weather or limited visibility in the DZ area prevents the drop but the terrain allows a safe drop close to the objective, the drop may be conducted on the nearest field along the line of flight of the aircraft. However, the field cannot be more than 3 kilometers from the original DZ. The pilot will advise the personnel to be dropped in such instances.

AREA DROP ZONE

3-21. The area DZ is well adapted for use with pre-planned automatic resupply drops where mission planners frequently select DZs by map reconnaissance. The area DZ consists of a prearranged flight track over a series of acceptable drop sites located not more than 1 kilometer on either side of the track. Points A and B establish the line-of-flight path (Figure 3-4). The distance between these two points will not exceed 28 kilometers and will have no changes in ground elevation over 295 feet. The mission request identifies Points A and B by coordinates. The drop procedures are as follows:

- The reception committee establishes a marked DZ at any location along the line of flight between Points A and B.
- The reception committee displays the DZ markings no longer than 10 minutes, beginning 2 minutes before the scheduled arrival time of the aircraft over Point A until 8 minutes past or until all jumpers and cargo have landed.
- The aircraft arrives at Point A at the scheduled time and proceeds toward Point B at drop airspeed and altitude.
- Once the JM or pilot locates and identifies markings, the drop is made.

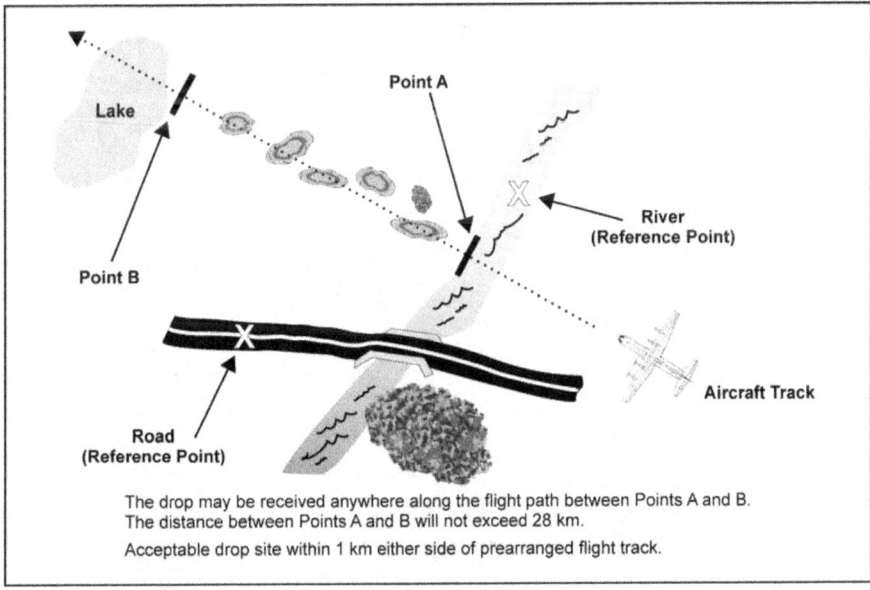

Figure 3-4. Area DZs

Chapter 3

- If an RT is used for area DZ identification, it should be positioned to mark Point A and turned on before the TOT of the aircraft, as jointly agreed upon by the commanders concerned. The transponder should be on for 15 minutes or until the reception committee observes the first deployed parachute. A RT may also be used on an area DZ to mark the desired PI.
- The normal DZ report format is used to report area DZs except the reception committee leader (RCL) or his representative gives the locations of Points A and B (including reference points). The RCL or his representative does not report open quadrants.
- Obstacles not shown on the issued map in reference to Points A or B are reported when they are over 90 meters above the level of the terrain and within 4 kilometers on either side of the line of flight (Figure 3-5).

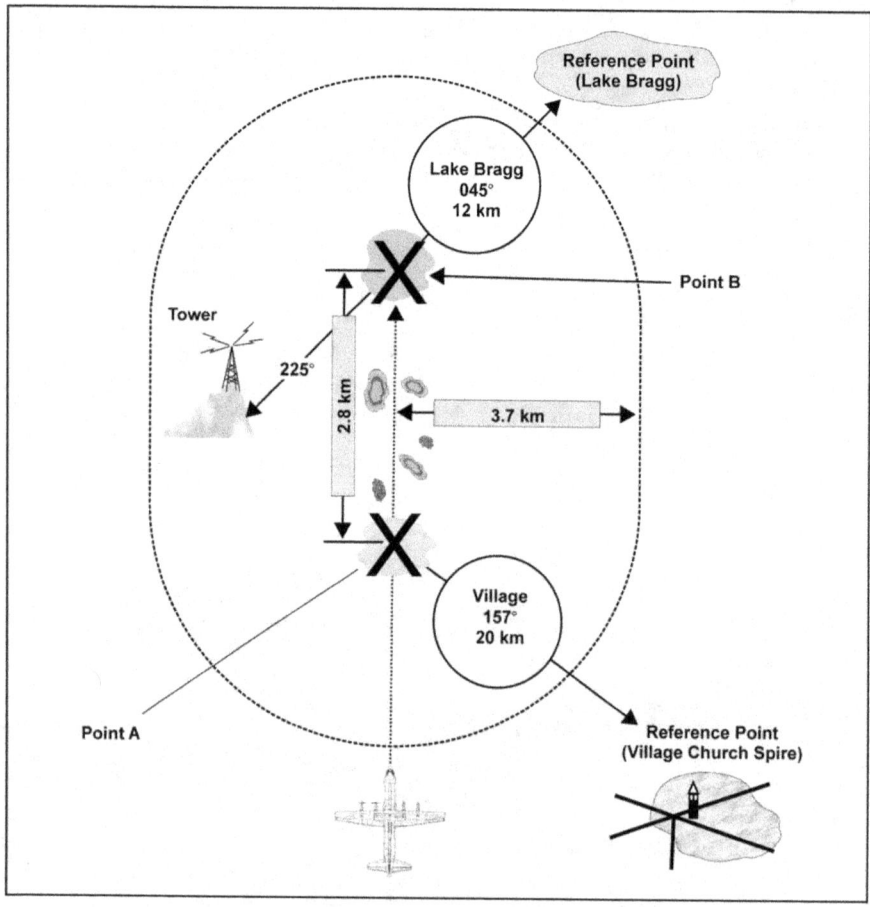

Figure 3-5. Obstacles and reference points (area DZ)

Note. For C-17 operations, ground personnel relay PI coordinates to the aircrew no later than 15 minutes before the TOT.

3-12   FM 3-05.210   27 February 2009

Military Free-Fall Drop Zone

3-22. The maneuverability of the ram-air parachute system allows for greater flexibility in the selection of MFF DZs. For planning, the following should be considered:
- Minimum DZ dimensions should be about 100 meters by 50 meters (about the size of a football field), but there is no minimum required size. The experience and capabilities of the jumpers determine the minimum required size.
- The DZ may be physically located on the objective.
- Parachutists can maneuver away from obstacles at low altitude.
- The DZ should be relatively level.
- The number of parachutists is not a selection factor.

Rough Terrain Drop Zone

3-23. The rough terrain DZ is the easiest to select. It can be any piece of isolated or rugged terrain safe from enemy observation and enemy use of antiair defense units. Some factors to consider include:
- There is no preferred shape. Any piece of terrain may be used.
- A DZ that is in isolated or rugged terrain is preferable.
- The DZ can be near base areas, thereby easing movement and increasing initial supply capability.

Water Drop Zone

3-24. Suitable bodies of water may be used, but water DZs require rapid recovery procedures. Select water that is at least 10 feet deep and clear of underwater obstructions to that depth. The surface must be clear of all floating debris, moored craft, and protruding obstacles. For personnel drops, current speed should not exceed 1 meter per second. For a safe water jump, the minimum temperature is 50 degrees Fahrenheit.

POINT OF IMPACT

3-25. The PI on the DZ is the location where the first jumper or bundle to exit the aircraft should land. On most USAF-surveyed DZs, the PI for a particular type load is predetermined. Its surveyed location can be found on Air Force (AF) Form 3823 (Drop Zone Survey). The location of the PI is based on the designated track of the aircraft, prevailing winds, and other factors. During training, the PI should be located along the DZ centerline. However, because of the tactical situation, the PI may be located near a wood line and the commander may waive the 100-meter buffer zone required during training operation. The PI location for GMRS or VIRS can be 100 meters in from the leading edge of the DZ and centerline. Table 3-4, page 3-14, lists the distances that the PI must be from the leading edge of the DZ when using CARP.

3-26. The location of the PI may be adjusted for SO or to meet specific mission requirements. The JM must brief all participants before the operation. The PI location may be adjusted for aircrew acquisition training. The PI may be located anywhere within the DZ boundaries as long as the minimum required DZ requirements for the aircraft are met and all participants have been briefed.

Random Points of Impact

3-27. When mission requirements dictate, the random points of impact placement option may be used. This option may be exercised in two ways:
- *Option One.* The mission commander will notify the reception committee at least 24 hours in advance that the random points of impact placement will be used. When the DZ is established, the reception committee will randomly select a point on the DZ and establish that point as the PI for the airdrop. In this case, the reception committee will ensure that the DZ minimum size requirements for the load being dropped are met and that the entire DZ falls within the surveyed boundaries.

Chapter 3

- *Option Two.* The mission commander or supported force commander may request the DZ established with the PI at a specific point on the DZ. The commander makes this request at least 24 hours in advance. The requester will ensure the minimum DZ size criteria are met for the type load being dropped and that the entire DZ falls within the surveyed boundaries. These procedures will only be used during visual meteorological conditions (VMC) operations.

Table 3-4. Point of impact location using CARP

Type of Drop	Type of Aircraft	Day	Night
For Single Aircraft [1, 2]			
CDS [3]	C-130	200 yds/183 m	250 yds/229 m
CDS [3]	C-17	225 yds/206 m	275 yds/251 m
Personnel	All	300 yds/275 m	350 yds/320 m
Equipment	All	500 yds/457 m	550 yds/503 m
For Multiple Aircraft [1, 2]			
Personnel	All	300 yds/275 m	350 yds/320 m
Equipment	All	500 yds/457 m	550 yds/503 m

NOTES:
[1] PI may be adjusted for SO or to meet specific mission requirements. JM must brief participants.
[2] Location of PI may be adjusted for aircrew acquisition training. The PI may be located anywhere within the surveyed DZ boundaries as long as the minimum required DZ size for that type of aircraft fits within the boundaries. All participants must be briefed when using this option.
[3] For HV CDS and HAARS, position the PI in the center of the DZ.

MULTIPLE POINTS OF IMPACT

3-28. Multiple points of impact (MPI) airdrops are authorized if all personnel involved have been properly briefed. An MPI airdrop is an aerial delivery method that allows for the calculated dispersal, both laterally and longitudinally, of airdropped loads to predetermined locations on a DZ. The DZ must meet the minimum size requirements for each PI, and mission planners must provide the precise location of each PI to aircrews. Ground personnel can mark MPIs using standard markings or nonstandard markings dictated by the tactical situation. If the points are placed laterally, the DZ width must be increased accordingly to meet the distance criteria from the DZ edge to the PI. When using a water DZ, this manner of placement reduces the effects of wake turbulence across the DZ.

> *Note.* A formation of C-17s conducting a personnel airdrop may require offset (laterally displaced) PIs. When required, offset PIs will be 250 yards left and right of the centerline personnel PI. Offset PIs for night personnel drops will be marked with flanker lights.

DROP ZONE REPORTS

3-29. Each SF operational element reconnoiters its AO as soon as possible (ASAP) after infiltration. This reconnaissance is to select sites for DZs and to confirm, reassess, or refute the sites selected during premission planning. The importance of DZ reporting by the operational element is to identify and send data to the SOTF on DZ locations for current or future use. DZ data can be sent separately or as a part of a specific mission request. The messages from the operational elements to the SOTF are concise and use precise message formats. Each SF operational element, using its signal operating

instructions (SOI), reports data on each DZ site in a mission request or in an information report. The DZ data each operational element sends must include the following:
- Code name (from SOI, or as appropriate).
- Location (universal transverse mercator [UTM] grid designation and coordinates of the center of the DZ).
- Track of recommended aircraft approach.
- Obstacles.
- Reference point(s).

3-30. Other data, such as the time of drop, services or items desired, and alternate DZs can be included in a mission request. Prevailing requirements, as stated in the SOI, determine message format and content.

DROP ZONE FORMULAS FOR GMRS AND VIRS

3-31. In planning a DZ operation, computing the RP is extremely important. The RP is the exact point on the DZ over which jumpers exit the aircraft. The JM locates the RP marker in relation to the desired PI by using a backward planning sequence. The PI must be determined or known before locating the RP. To locate the RP, the JM must know or compute three factors—dispersion, wind drift, and forward throw.

DISPERSION

3-32. Dispersion is the length of the pattern formed by the impact of the parachutists or containers. The PI for the first parachutist or container depends on how the calculated dispersal pattern fits into the available DZ space. When considering the DZ selection, it may be necessary to compute the dispersion of the planned drop to ensure the DZ can support the operation in one pass. First, compute the ground dispersion pattern to determine the absolute minimum length of the DZ. It is the computed horizontal distance depending on and formed by the PI of the first jumper or container to the impact point of the last jumper or container as determined by the known number of incoming jumpers or containers. It generally parallels the line of flight of the aircraft. This computation provides the reception committee with accurate data that will ensure the aerial delivery lands within the usable limits of the DZ. The formula for computing the dispersion is $D = RT$ (dirt) formula.

3-33. The $D = RT$ formula is used to compute ground dispersion pattern for helicopters, STOL, and nonstandard aircraft. To use this formula, some con-versions and mathematics are required. To compute this formula, at least two of the values must be known or determined. The values for the formula are:
- D = unknown dispersion pattern in meters.
- R = ground speed (GS) of aircraft in meters per second (MPS). To find the aircraft GS, convert aircraft airspeed (expressed in knots) to GS (MPS). Do this by multiplying the aircraft airspeed by 0.51 (1 knot equals 0.51 MPS). Refer to Table 3-5, page 3-16, for assistance in estimating aircraft airspeeds.
- T = time, in seconds, required for the aircraft to release its cargo. To determine the time over the DZ that is needed to release a parachutist or equipment, use the following factors:
    - Allow 1 second for each parachutist to exit the aircraft. However, do not include the time for the first parachutist out the door (10 parachutists require 9 seconds). Mathematically, this is represented as $10 \times 1 - 1$. When parachutists and bundles are dropped, a bundle is first out the door and 1 second is not subtracted for the first parachutist that follows the bundle.
    - Allow 3 seconds per bundle to exit the aircraft. Do not include the time for the first bundle out the door (3 bundles would require 6 seconds). Mathematically represented: $3 \times 3 - 3$.
    - Personnel jumping with T-10 parachutes may exit both doors at the same time. The door with the most parachutists is used to calculate the time required.

Chapter 3

Table 3-5. Aircraft drop speeds of T-10B parachute

Types of Aircraft	Drop Speeds
UH-60 Blackhawk	64 to 75 Knots (Optimum 70 Knots)
CV-22 Osprey (USAF)	80 to 110 Knots (Optimum 90 Knots)
CH-46/53 (U.S. Marine Corp [MC]) Sea Stallion	80 to 110 Knots (Optimum 90 Knots)
CH-47 Chinook	80 to 110 Knots (Optimum 90 Knots)
MC-130 Combat Talon I and II	70 to 90 Knots
C-5 Galaxy	130 Knots
C-17 Globemaster III	130 Knots
C-27A (Aeritalia G-222)	125 Knots
CASA-212	90 to 110 Knots

For personnel drops, add a 100-meter (328 feet) safety factor to each end of the computed ground dispersion pattern. This safety factor is not necessary in combat situations. Figures 3-6 and 3-7 below, and Figure 3-8, page 3-17, provide examples of D = R/T and T = D/R computations.

What dispersion would be required for 8 parachutists when jumping from an aircraft flying at a drop speed of 90 knots?	STEP 1:	Solve for R (ground speed of aircraft). R = knots x 0.51 90 knots x 0.51 = 45.90 meters per second (MPS).
	STEP 2:	Solve for T (time in seconds needed for aircraft to drop personnel and/or equipment). T = number of parachutists x 1 - the first parachutist. (8 x 1 - 1) = 7 seconds.
	STEP 3:	Solve for D (dispersion). D = 45.90 MPS x 7 seconds = 321.30 m = 322 m. (Always round up to the nearest whole number.)
	STEP 4:	Compute Safety Factor. Always add 100 m to each end of computed ground dispersion pattern. 322 + (100 m x 2) = 522 m.

Figure 3-6. Example of D = RT computation (parachutists only)

What dispersion would be required for 12 parachutists and 2 cargo containers when jumping from both doors of an aircraft flying at a drop speed of 110 knots?	STEP 1:	Solve for R. R = 110 knots x 0.51 = 56.1 MPS.
	STEP 2:	Solve for T. Divide the load in half, and figure the most used door: six parachutists and one cargo container. T = Six parachutists x 1 second = 6 seconds. One bundle x 3 seconds = 3 seconds. 6 + 3 = 9 seconds. Subtract 3 seconds for the bundle (the bundle will be in the door). T = 6 seconds.
	STEP 3:	Solve for D. D = 56.1 MPS x 6 seconds = 336.6 m = 337 m (round up).
	STEP 4:	Compute Safety Factor. Always add 100 m to each end of the computed ground dispersion pattern. 337 + 200 = 537 m.

Figure 3-7. Example of D = RT computation (parachutists and cargo)

Drop Zones

The dispersion formula D = RT can be reversed to form the time formula, T = D/R. T is the time the aircraft is over the DZ in seconds, D is the length of the DZ in meters, and R is the ground speed of the aircraft in MPS. When solved, this formula provides the time that the aircraft will spend on its pass and, therefore, the maximum number of parachutists that can exit over the DZ. If a DZ less than the required length must be used, compute the flight time over the DZ to determine how much of the load can be released in one pass. Use the T = D/R formula.	T = Time, in seconds, available for the aircraft to release its cargo. D = Maximum length of the dispersion pattern. DZ length -100 m from each end. In combat, use the entire DZ. R = Convert the aircraft airspeed (expressed in knots) to its ground speed (expressed in MPS) as in the D = RT formula (knots x 0.51). Round up the answer to the next whole number. T = D/R. Divide D by R. Any fractional answer is rounded down to the next whole number.
In a COMBAT situation, how many parachutists from a CH-47 (drop speed of 90 knots) can land on a 750-meter DZ during each pass?	T = time over DZ in seconds. D = DZ length = 750 m. R = 90 knots x 0.51 = 45.9 (round up) = 46 MPS. **Solution:** T = D/R D/R = 750 m divided by 46 MPS = 16.3 seconds (round down) = 16. T = 16 seconds over DZ x 1 parachutist per second + 1 parachutist (the first parachutist exiting the aircraft does not affect the number of seconds spent over the DZ) = 17 parachutists. Thus, 17 parachutists can land on the 750-meter-long DZ per pass.

Figure 3-8. Example of T = D/R computation

Note. Mass exit procedures, or exiting at the same time out of two doors, will be employed only when using T-10B parachutes.

WIND DRIFT

3-34. Wind drift is the horizontal distance a jumper or container will travel with the wind from the point of parachute opening through descent to the PI on the DZ. The RP is located a calculated distance upwind from the PI. Computing the wind drift distance is the second step in determining the RP. There are two choices for determining wind drift—the wind streamer vector count (WSVC) method or the D = KAV formula.

WSVC Method

3-35. The JM uses the WSVC (Wind Streamer) method (Figure 3-9, page 3-18) to determine the RP from the air. Normally, the JM executes this method, which does not require markings on the DZ. The WSVC method should not be used for tactical employment, since the aircraft is required to make multiple passes over the DZ. The steps for the WSVC method are as follows:

- *Streamer Drop.* On the first aircraft pass over the desired PI, the JM drops a streamer from the aircraft. The aircraft then turns to allow the JM to keep the streamer in sight. The pilot adjusts his route so that the flight path is over the streamer on the ground and the desired impact point (DIP) (in a straight line).
- *Count.* As the aircraft passes over the streamer, the JM begins a count, stopping the count directly over the impact point. He immediately begins a new count. When that count equals the first count, the aircraft is over the RP for the first parachutist.
- *Aircraft Flight Adjustment.* The pilot then maneuvers the aircraft to fly along the axis of the DZ and over the RP. The pilot may make slight adjustments based on how the parachutists land on the DZ.

Chapter 3

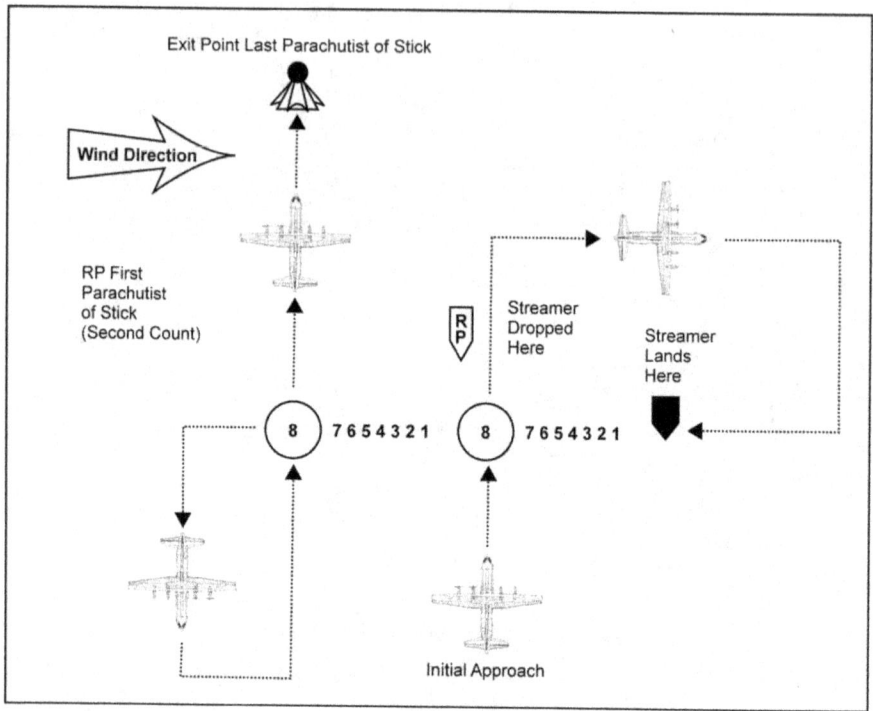

Figure 3-9. Determination of the RP by WSVC

Note. If aircraft must be shut down for a long period, the JM throws another wind drift indicator at the last RP to make sure the RP is still valid.

D = KAV Formula

3-36. When using the D = KAV formula to compute wind drift, base the wind drift on three factors: the wind velocity, the aircraft drop altitude, and the constant factor for the type of parachute used. Figure 3-10, page 3-19, shows an example computation using the D = KAV formula.
- D = unknown drift of a parachute. The drift is measured in meters.
- K = constant factor that represents the typical drift characteristic for a type of parachute. It represents the lateral wind drift, in yards, for each 100 feet of altitude loss in a 1-knot wind. Those constants are—
  - 1.5 for CDS, door bundles, and heavy equipment (heavy drop or HE).
  - 3.0 for personnel parachutes.
  - 2.4 for tactical training bundles.

Drop Zones

Note. The K-factor of a parachute is based on the flight characteristics of the parachute, not on its mode of use. Chapter 2 of FM 10-500-3, *Airdrop of Supplies and Equipment Rigging Containers*, states the K-factor for the T-10 parachute used in the cargo mode is the same (3.0) as for personnel drops using the T-10 parachute. When receiving parachutists and containers in the same drop, use the K-factor for personnel parachutes.

- A = drop altitude expressed in hundreds of feet AGL. For example, 800 feet is converted to 8. The mission and situation determine the drop altitude, or the drop altitude is known (airborne commander dictates). Refer to Table 3-6 for aircraft drop altitudes.
- V = wind velocity in knots. Obtain the mean effective wind (MEW) measurement using the pilot balloon (PIBAL) or surface wind velocity measured with an authorized anemometer. Two options are available for determining wind velocity: the surface wind measurement and MEW (Figure 3-11, page 3-20).

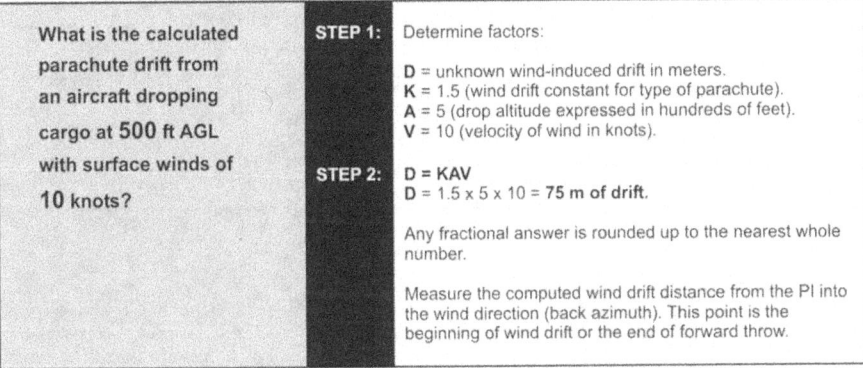

What is the calculated parachute drift from an aircraft dropping cargo at 500 ft AGL with surface winds of 10 knots?	STEP 1:	Determine factors: D = unknown wind-induced drift in meters. K = 1.5 (wind drift constant for type of parachute). A = 5 (drop altitude expressed in hundreds of feet). V = 10 (velocity of wind in knots).
	STEP 2:	D = KAV D = 1.5 x 5 x 10 = 75 m of drift. Any fractional answer is rounded up to the nearest whole number. Measure the computed wind drift distance from the PI into the wind direction (back azimuth). This point is the beginning of wind drift or the end of forward throw.

Figure 3-10. D = KAV

Table 3-6. Aircraft drop altitudes in feet

	DAY (AGL)	NIGHT (AGL)
All Services, Rotary Wing		
Personnel	1,500	1,500
Bundles	300	500
USAF Aircraft (Troop Carrier) [1]		
Personnel (Combat)	600 to 800[2]	600 to 800[2]
Personnel (Training)	1,250	1,200
Door Bundles	300	500
HE	1,100	1,100

[1] AWADS or SKE drop altitude is 500 ft above highest obstacle that falls within 3 miles either side of DZ flight path. For CDS, minimum drop altitude is 300 ft AGL for G-14 parachutes and 400 ft AGL for all other parachutes.

[2] During contingency and wartime operations, the airborne commander, with the commander of the AF units, will determine the drop altitude for personnel and equipment drops. The joint task force commander will make the final decision. For planning purposes, the combat jump altitude is 650 ft AGL. **Minimum combat jump altitude is 475 ft AGL for the MC1-1B/C and 435 ft AGL for the T-10B/C.**

Surface Wind Measurement	Use an anemometer to measure wind velocity. Report wind direction in magnetic degrees and wind speed in knots. Report the direction from which the wind is coming. Readings may be in knots or mph, depending on the type used. Multiply mph by 0.87 to convert to knots. Divide knots by 0.87 to convert to mph. When wind velocities are below 10 knots, the direct substitution of mph in the wind drift formula gives sufficient and accurate results.
Mean Effective Wind	A theoretical wind of constant speed and direction that extends from the ground to a designated altitude. When required, the DZ safety officer (DZSO) determines the MEW by timing the ascension of a helium-filled balloon to a predetermined altitude and measuring the angle of drift. Will not necessarily correspond to the speed and direction of the wind at a particular altitude or level. An indicator of the drift line and distance an airdropped object can travel. The PIBAL system determines the MEW. This system should be used when possible, as it is more reliable than surface wind measurement, which measures surface wind velocity only. (Refer to Appendix D for determining the MEW by using the PIBAL system.)

Figure 3-11. Surface wind measurement and MEW

Note. Either the AN/PMQ-3A or commercial anemometers authorized by the United States Army Infantry School messages are recommended for use. The school messages authorizing commercial anemometers are date-time group (DTG) 101000Z MAR 94, subject: *Use of Anemometers During Airdrop Operations*, and DTG 212000Z OCT 94, subject: *Use of Turbometer During Static Line Airdrop Operations*. A command-initiated risk assessment will determine the safety of other anemometers.

FORWARD THROW

3-37. Forward throw is the effect that inertia has on a falling object. When an object leaves an aircraft, it is traveling at a speed equal to the speed of the aircraft. The parachutist (or bundle) continues to move in the direction of flight until the dynamics of parachuting takes effect. He makes adjustments for this factor by moving the RP the appropriate distance in the direction of the approach of the aircraft. The forward throw is usually a constant distance that is based on the type of aircraft and drop. The forward throw for—

- Personnel using USAF high-performance aircraft is a constant of 230 meters.
- Personnel and equipment using STOL or rotary-wing aircraft is to divide the drop speed of the aircraft in half. This yields the forward throw in meters; for example, an aircraft flying at 70 knots would have a forward throw of 35 meters.
- Equipment within a USAF aircraft can be determined by the distances shown in Table 3-7.

Table 3-7. USAF forward throw data

Aircraft Elements	C-130	C-17/C-5	C-27
Personnel	250yds/230 m	250 yds/230 m	250 yds/230 m
HE	500yds/458 m	720 yds/658 m	NA
CDS	550 yds/503 m	750 yds/686 m	550 yds/503 m
Door Bundles	250 yds/230 m	250 yds/230 m	250 yds/230 m
Tactical Training Equipment	160 yds/147 m	NA	160 yds/147 m

DETERMINING THE RELEASE POINT

3-38. Once the dispersion, wind drift, and forward throw are calculated or known, mission planners can determine the RP with the addition of the wind direction and drop heading. The drop heading on all DZs depends on two factors: the long axis and prevailing winds. Both are considered when the situation

Drop Zones

permits. However, the long axis is the primary concern. With a GMRS or CARP DZ, mission planners can obtain drop heading from the AF Form 3823.

3-39. A circular or random approach DZ does not need a set drop heading. The mission commander gives the aircrew and the DZ commander the drop heading no later than 24 hours in advance of the airdrop operation.

3-40. Mission planners calculate the RP in the following manner:
- Determine the location of the surveyed PI on the DZ to ensure there is enough space for the dispersion of the personnel or cargo. point
- Determine the direction from which the wind is blowing, and move in that direction (upwind). Move the distance determined by the wind drift calculations. This position should be directly below the spot where the first parachutist's parachute should open.
- Compute the back azimuth of the direction of flight of the aircraft, and move the distance of the established forward throw. This position is the RP (Figure 3-12).

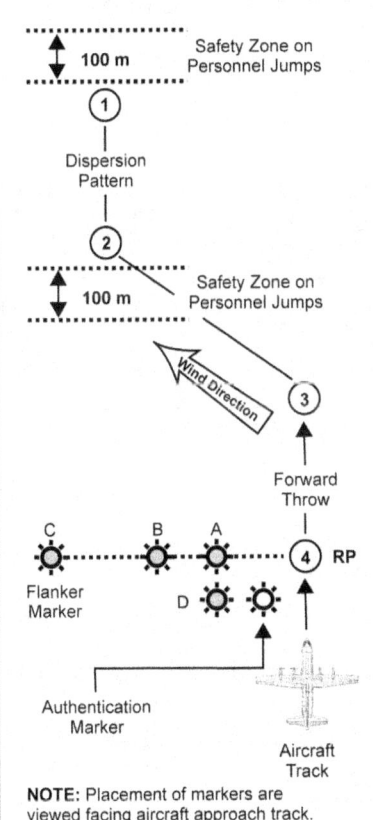

LEGEND
1. Impact point of last parachutist or container. Distance 1 to 2: Computed ground dispersion pattern (D = RT).

2. Desired impact point of first parachutist or container. Distance 2 to 3: Computed wind drift (D = KAV).

3. Distance 3 to 4: Compensation for forward throw. (Medium aircraft: Constant 220 m and STOL or helicopter: 0.5 x aircraft speed.)

4. Location of RP.

TO MARK THE RP
1. After selecting the desired impact point of the last parachutist (Point 1), compute the ground dispersion pattern. Starting at Point 1, face into the direction of aircraft approach and pace off the computed distance (Point 1 to 2).

2. At Point 2, face into the wind direction and pace off the computed wind drift distance (Point 2 to 3).

3. At Point 3, face into the direction of the aircraft approach and pace off the constant forward throw distance (Point 3 to 4).

4. You are now standing at or under the RP (Point 4).

5. To position DZ markers: At Point 4, face into the direction of aircraft approach, position Marker A 100 m to your right and Marker B 50 m to the right. Position Marker C (flanker marker) 150 m to the right of Marker D 50 m forward of Marker A into the direction of aircraft approach. Position authentication marker 15 m to the left of Marker D.

NOTE: Placement of markers are viewed facing aircraft approach track.

Figure 3-12. Marking the RP

3-41. When using high-velocity aircraft and free drops for aerial resupply, the method for determining the RP is different as follows:

- Compute the ground dispersion pattern in the same manner as above.
- Disregard wind drift. Wind conditions do not affect these type drops.
- Without the restraint of a parachute, compensate for forward throw by moving the RP in the direction of aircraft approach a distance in meters equal to the drop altitude of the aircraft. For example, if drop altitude is 600 feet, measure off 183 meters.

DROP ZONE MARKINGS

3-42. The types of marking systems used to help in the marking of DZs are identified below. The two types of markers used by SF are visual ground markers and NAVAIDS.

3-43. SF personnel can use visual ground markers to identify the PI, RP, or both for the airdrop. When using ground markings, SF personnel employ lights or panels in a distinctive configuration IAW unit SOI. They use NAVAID markers, radar beacons, ZMs, and TACAN to identify DZs for specially equipped delivery aircraft that have onboard equipment to read the signals. These systems provide the greatest amount of security to a DZ marking party. They may be used for either day or night operations and placed as directed by mission requirements. SF personnel may use just one system or a combination of systems. For example, they may use a NAVAID to ensure the aircraft arrives in the vicinity and then use visual markers on the DZ.

VISUAL GROUND MARKERS

3-44. There are two categories of visual ground markers—night or day. At night or during periods of limited visibility, SF personnel use visible light sources as markers. When mission requirements dictate and aircrews are qualified and equipped, SF personnel may substitute infrared lights for overt lights by using the DZ marking patterns. When selecting a light source, SF personnel should consider the atmospheric and terrain conditions and security in the area. Strong light emissions may be necessary to penetrate haze or ground fog. Whatever light source SF personnel select, all lights must be of the same type and have equal light emission to form a distinctive pattern. In daytime, SF personnel should use standard or improvised panels. The color they select must contrast sharply with the ground and vegetation background colors in the area. Whenever security permits, SF personnel may use smoke grenades or smoke pots to augment or replace panels. During day operations when SF personnel are using only smoke, they place smoke grenades or pots at the RP. During night operations, a white air traffic control light may be used to mark the RP. When setting up the markers, SF personnel should use the proper code letter for the jump and set up the markers at the proper distance.

3-45. SF personnel should place the markers so that they are visible only from the direction of the approaching aircraft. They should appropriately hood selected light sources, screen the light sources on three sides, or place light sources in pits to reduce side glow. SF personnel aim the lights at the flight path of the aircraft. They position the panels at an angle of about 45 degrees from the horizontal to present the maximum surface toward the approaching aircraft. They also place the markers where obstacles will not mask the pilot's line of sight.

3-46. As a guide, SF personnel should use a mask clearance ratio of 15 to 1 (1 unit of vertical clearance to 15 units of horizontal clearance). For example, if a DZ marker must be placed near a terrain mask, such as the edge of a forest on the DZ approach track with 10-meter high trees, the marker would require 150 meters of horizontal clearance from the trees. This applies to static-line jumps only (Figure 3-13, page 3-23). Red smoke, red flares, scrambled panels, or the absence of a planned signal indicates no drop to the aircraft. Since the aircraft is required to fly along the markings on the DZ, these markings must be visible to the aircrew.

Drop Zones

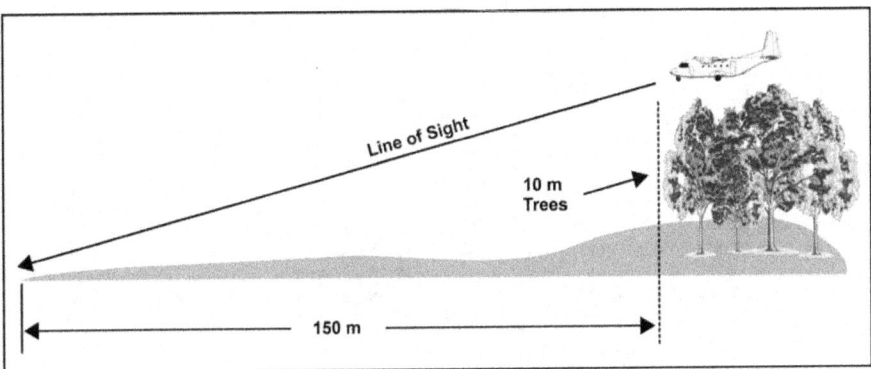

Figure 3-13. Mask clearance ratio 15:1

RADAR BEACONS

3-47. Radar beacons, such as the AN/PPN-18 or -19 or the SST-181 series, can be used in a direct mode (placed on the DZ) or in an offset mode (off the DZ). Radar beacons work on the line-of-sight principle. Heavy precipitation, dense foliage, and terrain masking will weaken the signal from a radar beacon. Also, other beacons, radar, electronic countermeasures (ECM), and electronic counter-countermeasures (ECCM) equipment working within the signal and frequency ranges of the beacon can disrupt the beacon. When used in an offset mode in a low-threat environment (when the aircraft can ingress at 1,000 feet AGL or above), the beacon must be between 1 and 25 kilometers from the DZ. When used in a high-threat environment (when the aircraft must ingress at 1,000 feet AGL or below), the beacon must be within 10 kilometers of the DZ. C-130 aircraft require a collocated pair of tuned I-band (SST-181) beacons for radar beacon airdrops. For CARP airdrops, the beacons should be at the computed RP. For MFF airdrops, the beacons should be on the PI.

3-48. Personnel should place the beacon as high as possible off the ground, but never less than 1 meter above the ground, and at least 2 meters from any metallic, electrical, or magnetic-radiating object. During offset use, the angular difference between the aircraft approach heading and the beacon-to-target azimuth should be less than 45 degrees for an omnidirectional beacon and 25 degrees for directional beacon. (These figures are total left or right of the aircraft approach heading.) The beacon should be beyond the long axis of the DZ in respect to the aircraft approach heading. This beacon positioning prevents the aircraft from turning off before completely flying the DZ. Joint Publication (JP) 3-09.1, *Joint Laser Designation Procedures*, provides further information.

3-49. Radar beacons require varied turn-on times for varied aircraft types. These times are 10 minutes before and 2 minutes after TOT for AMC aircraft and 15 minutes before and 5 minutes after TOT for Air Combat Command aircraft. In extremely cold weather, the operator should add an additional 10 minutes to these warm-up times.

Note. For C-17 operations, ground personnel relay PI coordinates to the aircrew no later than 15 minutes before the TOT.

3-50. To conduct beacon operations, the operator and the mission aircrew need to know the following:
- Beacon type and code.
- Beacon location (8-digit UTM grid coordinates).
- Beacon elevation in feet MSL.
- DZ location (8-digit UTM grid coordinates).
- DZ elevation in feet MSL.
- Beacon to DZ offset magnetic azimuth.

Chapter 3

- Beacon to DZ offset range in meters.
- TOT.
- Requested aircraft heading.
- Desired results.
- Target description.

ZONE MARKER

3-51. Ground personnel should place the ZM TPN-27 within 1,500 yards of the PI. For maximum accuracy, the ZM should be as close to the PI as possible. If line-of-sight considerations preclude placement of the ZM at the briefed location, relocate it and advise the aircraft on initial contact of the new location relative to the PI. During night airdrop operations, the ZM should be visually marked with a light to identify it as a hazard to parachutists and to prevent accidental destruction of the ZM by vehicular traffic.

TACTICAL AIR NAVIGATION

3-52. The aircrew uses TACAN to guide the aircraft to the DZ. However, ground personnel should not place TACAN on a DZ as an airdrop aid. Use of visual ground markers are recommended with TACAN.

STATIC-LINE DROP ZONE MARKING PATTERNS

3-53. Personnel should use primary and alternate marking patterns to identify the RP for night and day drops. The types of markings are based on whether or not the drop is GMRS or CARP.

GROUND-MARKED RELEASE SYSTEM

3-54. The GMRS uses markings known as the four-panel inverted L, six-panel T, or seven-panel H. The GMRS also uses an alternate marking for the inverted L (Figure 3-14, page 3-25).

Marking Placement for Inverted L

3-55. Figure 3-14 shows the placement of four panels. How to place each panel follows:
- From the RP, move 100 meters to the left (perpendicularly) of drop heading, and use a VS-17 panel for Panel A (corner panel). Emplace the VS-17 panel with the long axis of the panel parallel with the drop heading. Elevate the panel at a 45-degree angle toward the approaching aircraft. With the panel elevated, the aircrew and JM can visually identify the DZ.
- From Panel A, continue in the same direction for 50 meters and place Panel B (alignment panel), as described above.
- From Panel B (alignment panel), continue in the same direction for 150 meters and place Panel C (flanker panel) as described above.
- Return to Panel A, move 50 meters on a back azimuth of the drop heading for the location of Panel D (approach panel). Emplace Panel D as described above.
- Position an authentication marker 15 meters to the right of Panel D (as viewed from the approaching aircraft).
- At night, replace all panels with a white light. Lights may be shielded on three sides or placed in pits.

Marking Placement for T or H Pattern

3-56. The T or H pattern is recommended for C-17 and C-5 airdrops because of aircraft side-angle vision limitation. The H pattern is also recommended for circular DZs. The T or H pattern is the only marking authorized for use with USAF aircraft. The T or H pattern requires the following:
- The pattern is displaced for 4 minutes, beginning 2 minutes before until 2 minutes past scheduled drop time or upon observing the first deployed parachute.
- During static-line drops, the pilot aligns the aircraft as accurately as possible 100 meters to the right of the right-hand row of markers.
- The drop occurs when the aircraft is adjacent to the last marker in the right-hand row.

Drop Zones

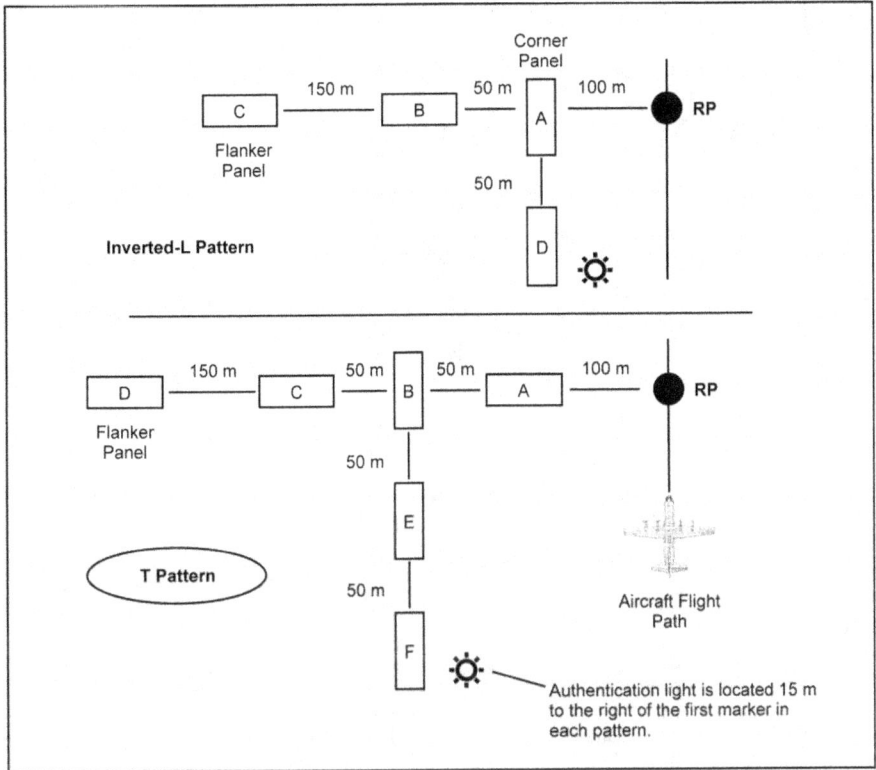

Figure 3-14. L and T marker patterns

GMRS Alternate Marking

3-57. The SOI specifies a daily pattern that uses the required number of markers to form the prescribed pattern. The distance between markers will be 50 meters (164 feet). The flanker panel is always placed 150 meters (495 feet) to the left of the upper left panel of the pattern as viewed from the approaching aircraft. (Use the flanker panel regardless of drop altitude.) An authentication marker is positioned 15 meters (49 feet) to the right of the lower right marker of the pattern (day and night) as viewed from the approaching aircraft (Figure 3-15, page 3-26).

> *Note.* If any portion of the inverted L falls within the 15:1 mask clearance ratio of obstacles on the approach end of the DZ, a code letter (H, E, A, or T) or far panel may be placed on the departure end of the DZ for CDS or bundle drop, if coordinated during drop zone support team (DZST) and aircrew mission briefing. This far marking is on line with Panel A to allow the aircrew to begin alignment on the RP until the inverted L comes into view. If a code letter is used, it can be used to distinguish the DZ from other DZs in the area.

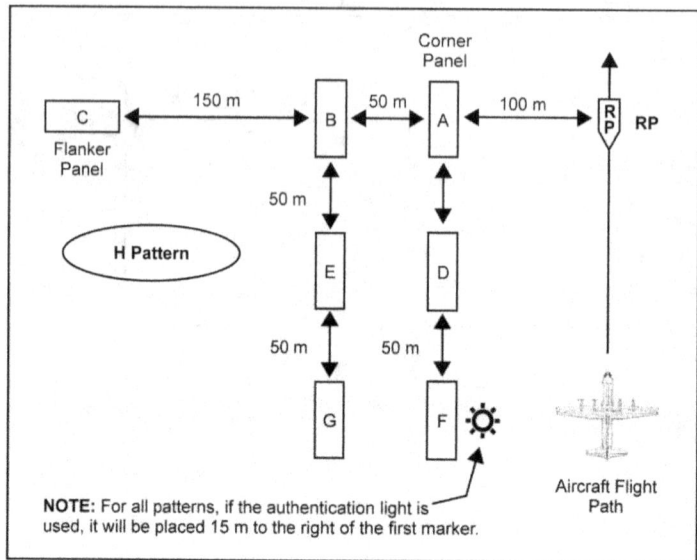

Figure 3-15. GMRS day and night markings

CARP DROP ZONE MARKINGS

3-58. During day operations, the DZSO or personnel on the ground will mark the PI with a RAM or block letter. The standard RAM is a triangular-shaped marker constructed of bright orange material, 6 feet wide (minimum) at the base and 6 feet high (minimum), displayed at a 60-degree angle into the direction of flight. If authentication is required, a block letter will be used instead of the RAM. The block letter will be placed at the base of the RAM. Authorized letters for PI markings are, J C, A , R, and S. The block letters H and O are authorized for circular DZs. The block letters should be aligned into the surveyed DZ axis or into the aircraft line-of-flight, if different from the survey. The minimum size for block letters is 35 feet by 35 feet (11 x 11 meters) and will consist of at least nine marker panels (Figures 3-16 and 3-17, page 3-27, and Figure 3-18, page 3-28).

3-59. During night operations, the PI will be marked with a block letter and flanker lights. The block letter will be set up in the same manner as for day operations except that instead of the panels, at least nine white lights, with a recommended minimum output rating of 15 candela, are used. Flanker lights will be white and placed 250 meters to the left and right of the PI. For actual personnel airdrops, an amber rotating beacon will be placed 1,000 meters from the PI on the trailing edge of the DZ. The DZ identification must be coordinated and briefed to the ground party and the aircrew. When mission requirements dictate and aircrews are qualified and equipped, IR lights may be substituted for overt lights.

VERBALLY INITIATED RELEASE SYSTEM

3-60. Units use a VIRS when normal drop procedures are not tactically feasible. The reception committee—
- Determines where the RP is located.
- Places the designated traffic controller to guide the aircraft at the RP.
- Gives verbal steering guidance to the pilot to align the aircraft over that point.
- Tells the pilot to release the jumpers.

Drop Zones

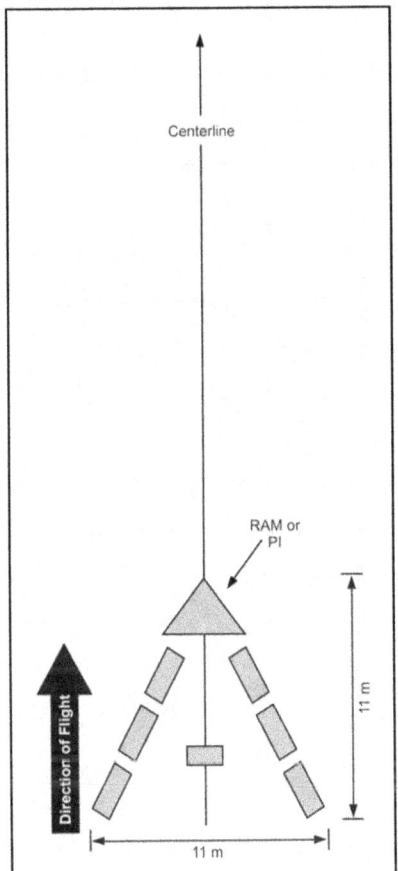

Figure 3-16. Day CARP DZ markings

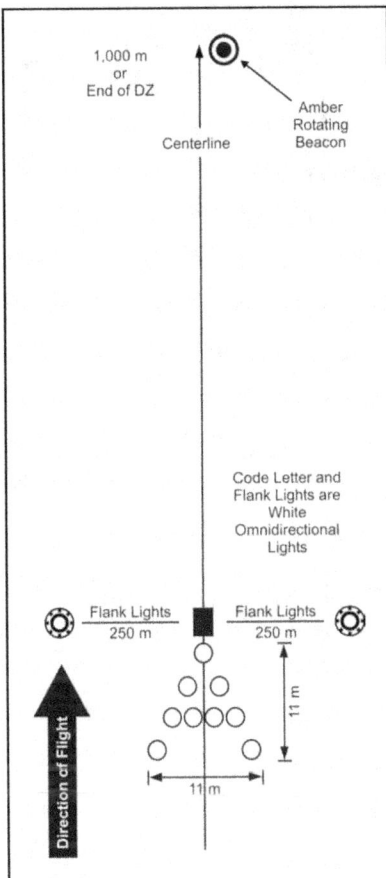

Figure 3-17. Night CARP DZ markings

This method allows the conduct of the operation with a minimum amount of prior DZ information and coordination. The RP may be unmarked or marked with a code letter, smoke, single panel, or a light (Figure 3-19, page 3-28).

DAY DROP ZONE MARKING

3-61. During daylight, personnel use the precoordinated visual signal or mark the RP IAW with the unit SOI. They emplace the code letter (H, E, A, or T) with the base panel of the letter at the RP and oriented to the aircraft track. VS-17 panels placed together form the code letter. Each letter is two panels high and one panel wide. A flank panel is placed 200 meters to the left of the code letter or at the edge of the DZ (whichever is less) and aligned with the base panel. A far panel is placed 500 meters from the base panel or at the edge of the DZ (whichever is closer) and on line with the drop heading. Both the far and flank panels consist of a single VS-17 panel. These panels may also be elevated at a 45-degree angle to improve visibility.

Chapter 3

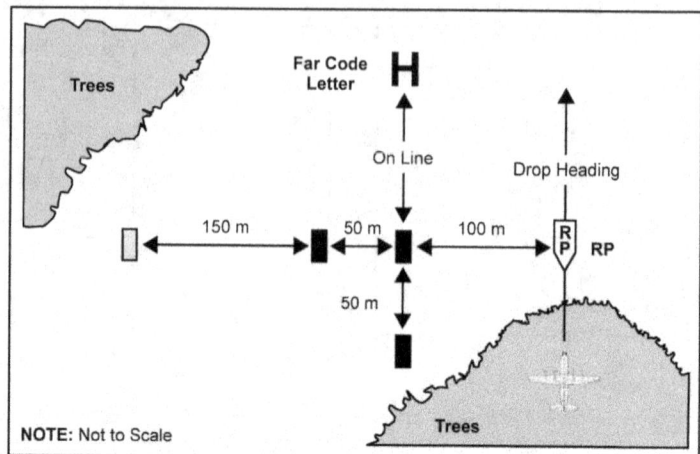

Figure 3-18. Code letter with masked GMRS letter

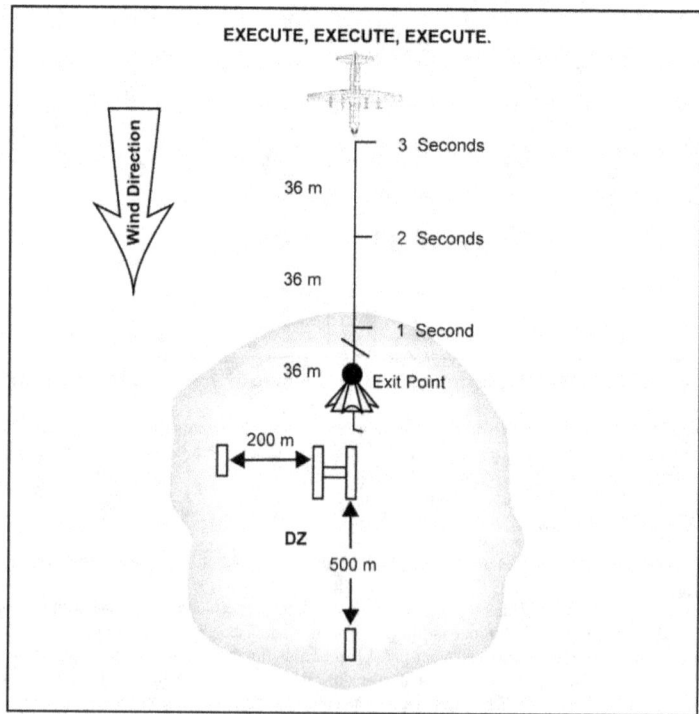

Figure 3-19. Army VIRS offset

Drop Zones

Day Drop Zone Marking

3-62. During daylight, personnel use the precoordinated visual signal or mark the RP IAW with the unit SOI. They emplace the code letter (H, E, A, or T) with the base panel of the letter at the RP and oriented to the aircraft track. VS-17 panels placed together form the code letter. Each letter is two panels high and one panel wide. A flank panel is placed 200 meters to the left of the code letter or at the edge of the DZ (whichever is less) and aligned with the base panel. A far panel is placed 500 meters from the base panel or at the edge of the DZ (whichever is closer) and on line with the drop heading. Both the far and flank panels consist of a single VS-17 panel. These panels may also be elevated at a 45-degree angle to improve visibility.

Night Drop Zone Markings

3-63. The procedures for establishing the DZ are the same for night operations except that white light is used for the code letter and far or flank markings. Each code letter is four lights high and three lights wide. There is a distance of 5 meters between each light in the code letter. The far and flank lights are signal lights. Also, a white-and-red lens air traffic control (ATC) (SE-11) light should be located at the RP. Lights may be shielded on three sides or placed in pits to prevent enemy ground observation.

Guidance Procedures

3-64. The reception committee leader or a designated person acts as the ground-to-air (GTA) controller. The GTA controller guides the jump aircraft over the DZ on the proper drop heading, and at the proper altitude and drop speed. He makes sure the parachutists exit the aircraft at the proper RP. Once the parachutists have exited the aircraft, the GTA controller must then clear the aircraft from the control zone. Figure 3-20, page 3-30, shows an example conversation between the GTA controller and the pilot.

MFF DROP ZONE MARKING PATTERN

3-65. For MFF, the DZ selection criteria are the same as static line except there is no minimum size. The JM must consider the level of training for the jumpers when figuring the size of the DZ. A football-size (50 meters by 100 meters) DZ is recommended for training. FM 3-05.211 and USASOC Reg 350-2 provide additional MFF information. Smoke, panels, wind socks, lights, and other marking techniques are used to mark MFF DZs. Visual markings that identify the DZ are generally ineffective for MFF operations, because MFF parachutists jump from a high exit altitude during HALO jumps and high exit altitude and distance during HAHO jumps. DZs are normally identified by location in reference to major terrain features. MFF parachutists assemble under canopy on a designated group leader. The group leader's PI is the PI for the remainder of the group.

3-66. The following DZ markings can be used, when the tactical situation permits, to indicate wind direction to the parachutists under canopy (Figure 3-21, page 3-31):

- Smoke, panels, wind socks, and so forth, are effective during daylight. Panels should be arranged in the form of an arrow pointing into the wind.
- At night, two lights (one red and one green) can be used. When using two lights, the red light should be placed 15 meters upwind of the green light. Parachutists land over the green light facing the red light. White lights should not be used at night to avoid confusion with buildings, streetlights, and vehicles.

3-67. The RP for MFF operations is normally identified by location in relation to major terrain features. Computations for determining the RP are dependent upon the availability of altitude wind data. (FM 3-05.211 provides the method of computation.) At night, a colored light may be used to mark a known RP. This light should be bright enough to be easily identified from a high altitude; for example railroad fuzes.

Chapter 3

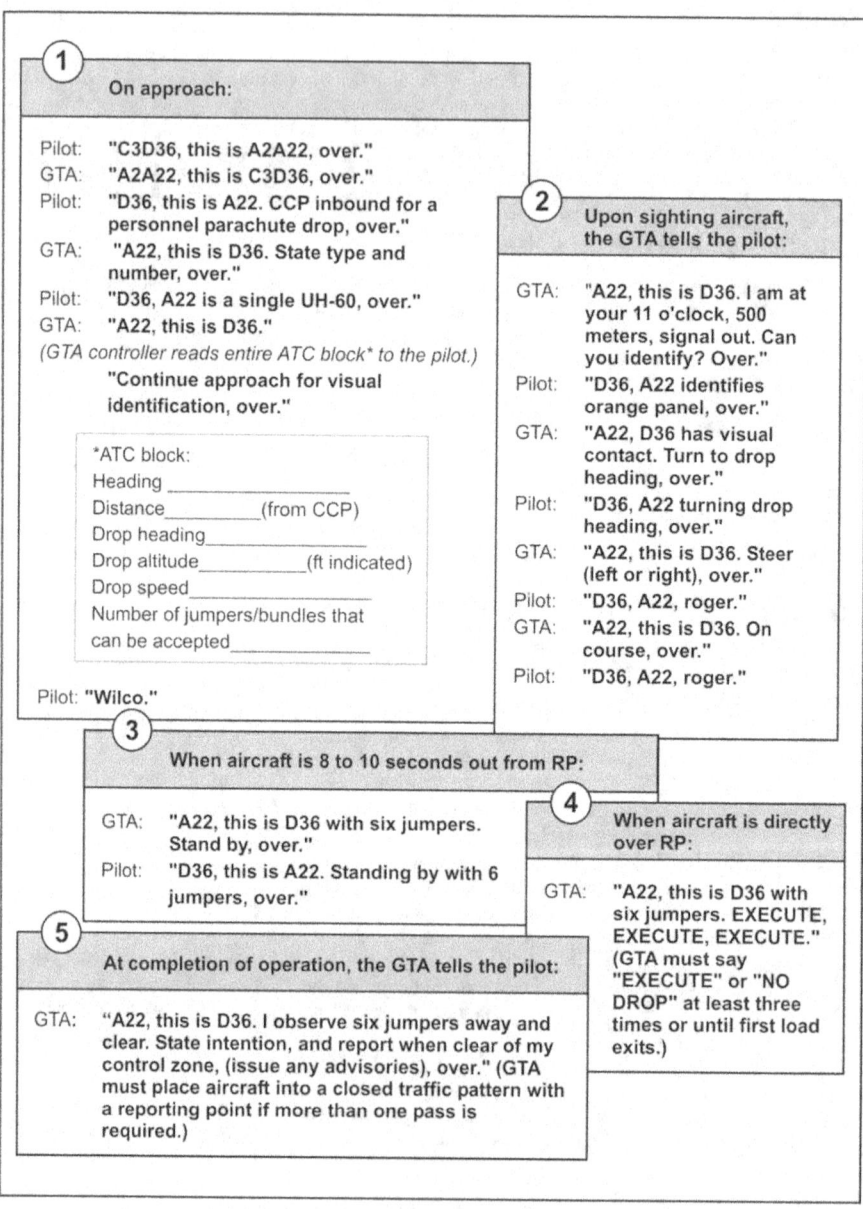

Figure 3-20. Example communication sequence between GTA controller and pilot

Drop Zones

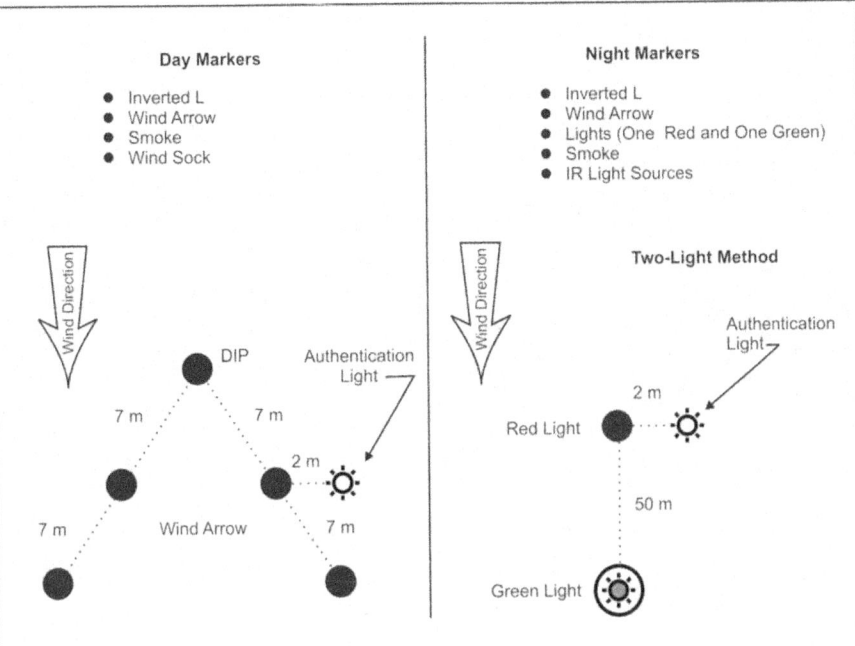

Figure 3-21. MFF DZ markings

3-68. Other techniques for marking MFF DZs are variations of lighting sources and other devices. These include, but are not limited to—

- Radar beacons placed at the PI with individual parachutists using compatible tracking devices.
- IR chemical lights with parachutists using AN/PVS-7orAN/PVS-14 NVGs.
- Navigation Satellite Timing and Ranging GPS portable terminal worn by parachutists.

ORGANIZATION AND OPERATIONS OF A DROP ZONE

3-69. The DZST operates the DZ. In a UW environment, the DZST is referred to as a reception committee. There are differences between the organization and operation of a DZ for training and for real-world tactical operations. In peacetime, airborne operations are conducted IAW all applicable operational and safety regulations. Wartime airborne operations may require that these regulations be modified or waived to fit the tactical situation. The Army may conduct DZ operations jointly with the AF STT or solely. Even a strictly Army operation, if it uses general purpose forces, may involve differences between how general purpose and SOF operate and conduct the operations due to training, mission requirements, and other

Chapter 3

factors. USASOC Reg 350-2 lists the SOF requirements for operating a DZ during training. AFI 13-217 governs the Air Forces procedures, techniques, and requirements for operating DZs.

3-70. Army DZSTs have the primary mission of supporting wartime airdrops for battalion-size units and below, and peacetime airdrops of personnel, CDS, and heavy equipment for operations involving four or fewer aircraft. With some exceptions, these primary mission airdrops are limited to day or night visual conditions. The DZST also maintains the secondary mission of supporting other types of airdrops. The secondary missions may include:

- Wartime force projection.
- Sustainment of personnel, equipment, and CDS peacetime airdrops under AWADS and IMC conditions.
- VMC formation drops with four or more aircraft.

DZST AND DZST TEAM LEADER

3-71. The size of the DZST varies, depending on the type and complexity of the mission. The senior member of the DZST functions as the drop zone support team leader (DZSTL). In operations in which the STT is not present, the DZSTL has overall responsibility for the conduct of operations on the DZ. He represents the airborne and airlift commanders. The DZSTL assumes all the responsibilities normally associated with the USAF STT and Army DZSO. USASOC Reg 350-2 covers all of the qualifications and responsibilities of the DZST. The DZST primarily—

- Conducts premission coordination.
- Evaluates the DZ for suitability and safe operating conditions.
- Makes sure all DZ markings are properly displayed.
- Operates all visual acquisition aids.
- Makes sure no-drop signals are relayed to the aircraft.

3-72. Once the DZSTL has been assigned a mission, he must conduct accurate premission coordination. The recommended DZST and aircrew mission-briefing checklist, Figure 3-22, page 3-33, reflects the minimum essential information that must be addressed and confirmed by the DZSTL. Normally, peacetime drops should employ every acquisition aid and safety device available, including air-to-ground radio communications, PIBAL, MEW measurement, ATC light gun, and smoke or flares. During contingency or wartime operations, limited airdrop support equipment is available. Therefore, it is important for premission coordination and briefings to be comprehensive with respect to visual signals (drop cancellation, postponement, and authentication procedures). The coordination must be timely to ensure the DZST has enough time for planning and for moving to and establishing the DZ.

ESTABLISHMENT AND OPERATIONS OF A DROP ZONE

3-73. The way the DZ is to be set up is determined by the type of airdrop that is to take place and the capabilities and equipment of the reception committee and the aircraft. Some marking methods, for example, CARP and GMRS, require markings to be placed on the DZ. However, the WSVC method and blind drops require no markings on the DZ. Other marking methods, such as VIRS and NAVAID may or may not require markers or equipment on the DZ. To become operational, essential personnel are to be located on or at the DZ for controlling, marking, medical evacuation (MEDEVAC), wind readings, and malfunctions. Examples include the following:

- CARP is usually used for a joint airborne operation of more than three troop carrier aircraft. USAF aircraft only in conjunction with STT or a qualified DZST use CARP. The DZST may consist of single-Service personnel (Army) or AF (special tactics squadron [STS]) and Army personnel. If STS personnel are manning the DZ, they are responsible for the airdrop.
- For small joint airborne operations involving four or fewer troop carrier aircraft, CARP or GMRS can be used. If GMRS is being used, only an Army DZST is required. If rotary-wing and small fixed-wing aircraft are employed, either VIRS or WSVC methods can also be used.

Drop Zones

> - DZ name, location, and joint air attack team (JAAT) mission sequence number verified.
> - TOT(s) block time (no-drop procedures, for example, racetrack).
> - Current DZ survey AF Form 3823 (date) verified (most current and recertified within 5 years).
> - Type drop: heavy equipment, personnel, CDS, or combination.
> - Type release: VIRS, CARP, GMRS, AWADS, or visual.
> - Type parachutes.
> - Ground quick disconnects.
> - Number of parachutists and bundles.
> - Number and type of aircraft.
> - DZ information.
> - Markings and signals:
>   - Panels and lights.
>   - Block letter identification.
>   - Smoke and flares.
> - Emergency no-drop procedures.
> - Mission cancellation indication.
> - DZ support capabilities:
>   - Radios available and radio frequencies.
>   - Visual acquisition aids available.
>   - NAVAIDs available.
>   - MEW equipment.
>   - Airspace coordination verified.
>   - Aircraft (mission) commander's name, unit of assignment, and telephone number.
>   - DZSTL name, rank, unit of assignment, and telephone number.
>   - Drop score, incident, and accident reporting procedures.

Figure 3-22. DZ mission briefing checklist

DZSTL LOCATION

3-74. The location of the DZSTL varies depending on the type of operation. For personnel drops, the DZSTL is located at the PI. For CDS drops, the DZSTL is located 150 yards to the 6 o'clock position of the PI. For HE, free drops, high-velocity, and AWADS drops with a ceiling of less than 600 feet, the DZSTL is located off of the DZ.

NO-DROP COMMUNICATION TO AIRCRAFT

3-75. No-drop conditions are relayed to the aircraft using pre-coordinated smoke, flares, forming the code letter into two parallel bars perpendicular to flight, or the absence of a planned signal. Forming the code letter into an "X" indicates mission cancellation or scrambling the panels.

Note. The type of marking used is coordinated in the premission briefing.

Chapter 3

EQUIPMENT

3-76. The DZSTL should maintain an inventory of basic equipment to support an airdrop mission. This inventory is as follows:

- VS-17 panels.
- Smoke (red, yellow, and green).
- White, steady lights, preferably Whalen.
- ATC gun, SE-11 light gun, or four-cell magnesium light flashlight.

Note. ATC gun requires a special power source and plug to function properly.

- Signal mirror.
- Binoculars.
- NVG for night operations.
- Anemometers—AN/PMQ-3A or commercial anemometers.
- Compass.
- Signal flares.
- PIBAL system with helium source.

Note. Premission coordination and mission complexity may require other items of equipment and signals.

RECEPTION COMMITTEE

3-77. Any sort of organized force that is supporting an unconventional warfare airborne operation at the DZ is considered a reception committee. Once established in the JSOA, SF team members organize, train, and supervise indigenous assets to assist and conduct future air reception operations IAW SF training and doctrine and unit SOP. The reception committee has the same purpose as the DZST. However, because of the tactical situation, it has additional functions and is not required to meet the same peacetime requirements of the DZST. The reception committee should take as many safety precautions as possible without jeopardizing the mission or security. It also—

- Provides operational security.
- Emplaces or operates the marking system.
- Maintains surveillance of the reception site before and after each operation. (Although the tactical situation and available assets will determine actual surveillance times, reception sites should be observed for a minimum of 24 hours before and after the operation.)
- Recovers incoming personnel and supplies.
- Moves supplies to designated distribution points or cache sites.
- Sterilizes the reception site to maintain secrecy, to preclude compromise of the mission, and to ensure the success of future operations.

3-78. The reception committee normally consists of five parties for air reception operations. Small reception committees may combine the functions of two or more parties. For example, the command and marking parties may be combined.

Command Party

3-79. The command party consists of an RCL, an SF advisor, radio operators, and messengers. These personnel—

- Control and coordinate all committee actions.
- Provide medical support, when necessary.

Marking Party

3-80. The marking party includes the personnel required for the type of marking system to be used. Marking party personnel—
- Emplace or operate the marking system.
- Assist in recovering personnel and supplies.
- Assist in sterilizing the site.

Security Party

3-81. Security party personnel provide surveillance and security of the DZ. They—
- Provide surveillance of the DZ before and after the operation.
- Provide early warning. Prevent or delay enemy interference.
- Normally include an inner and outer security element. The inner element is placed around the perimeter of the reception site to conduct delaying or holding actions. The outer element sets up outposts, roadblocks, and ambush sites along approach routes to stop or delay enemy movement.
- May be increased by members of the auxiliary who provide surveillance and information on enemy activities or movements and who conduct limited diversionary attacks or ambushes.
- Provide security during transfer of personnel and supplies from the reception sites.

Recovery Party

3-82. The recovery party includes personnel required for the scheduled number of incoming personnel or supply bundles. As a minimum, two persons are assigned for each parachutist or bundle. Party members—
- Recover, guide, and deliver incoming personnel or bundles to the collection point.
- Position personnel from the DIP along the length of the dispersion pattern. Assign specific members or groups to track the descent of each parachutist or supply bundle to make sure they are immediately recovered and to preclude loss during darkness.
- Position a separate recovery detail at the exit end of the DZ to track and locate parachutists or supply bundles. This detail also serves to determine the exact line of flight of the aircraft, thus making a sweep of the DZ easier should the delivery be disrupted or lost.
- Employ a signal system that precludes undue noise or movement.
- Sterilize the reception site.

Transport Party

3-83. The transport party consists of members of all, or part of, the command, marking, and recovery parties. Members of the transport party move supplies received to designated distribution points or cache sites.

WIND LIMITATIONS

3-84. The DZSO and assistant DZSO measure surface winds. Table 3-8, page 3-36, depicts wind limitations during peacetime and wartime operations. There are no altitude wind limitations for airdrops. If the JM does not receive surface wind data, he will not have enough information for a drop. For example, the JM's decision to drop personnel depends on altitude winds. JMs should be cautious when altitude winds are greater than 30 knots. Winds on the DZ are measured using the AN/PMQ-3A anemometer or two commercial anemometers (one each for the DZSO and the assistant DZSO). Non-recommended anemometers should be used only after a command-initiated risk assessment is completed. Regardless of the method or device used to measure DZ winds, the airborne commander makes sure the winds on the DZ do not exceed 13 knots during static-line personnel airdrops.

Chapter 3

Table 3-8. Wind limitations during peacetime and wartime operations

Peacetime Operations	
Personnel	
Static Line (Land)	13 Knots
Static Line (Water)	18 Knots
MFF	18 Knots
CDS	
If Using G-13 or G-14 Parachutes	20 Knots
HE Drops	
With Ground Quick Disconnects	17 Knots
Without Ground Quick Disconnects	13 Knots
Wartime Operations	
The joint task force commander will set limits.	

THE 10-MINUTE WINDOW

3-85. On multiple-aircraft operations or single-aircraft operations using more than 2,100 meters of DZ, the surface wind is measured from two points on the DZ. For single operations using less than 2,100 meters of DZ, the wind is measured from only one location, normally the PI or RP. Beginning 12 minutes before TOT, the DZSO begins a constant monitoring of the surface wind by using an anemometer.

3-86. If the surface wind exceeds allowable wind limits, the aircraft is notified of a no-drop, and a new 10-minute (min) window is established. If the wind remains within limits during this new window, the drop takes place as planned. If the wind exceeds allowable limits during the new window, a no-drop signal is relayed to the pilot and the entire procedure starts again. A no-drop signal may be relayed to the aircraft by radio, smoke, flares, scrambled panels, or another planned signal.

IDENTIFICATION

3-87. Identification is the method in which the aircrew and the reception committee identify themselves to one another. Means of identification may include time, approach, and marking pattern.

Air-to-Ground

3-88. The aircraft identifies itself to the reception committee by—
- Arriving in the objective area within the specified time limit, usually 2 minutes before to 2 minutes after scheduled drop time.
- Approaching at designated drop altitude and track.

Ground-to-Air

3-89. The reception committee identifies itself to the aircraft by—
- Displaying the correct marking pattern within the specified time limit.
- Using the proper authentication code signal.
- Setting the proper code on the RT or laser target designator.

AUTHENTICATION

3-90. Authentication is the procedure by which the aircrew and the RCL identify themselves to each other. There is no standard authentication system for UW reception operations. During mission planning, the commanders concerned agree upon the authentication system to be used. The SOI prescribes authentication procedures. For security purposes, authentication procedures are changed on a predetermined schedule.

3-91. Authentication between the aircraft pilot or navigator and the RCL may be accomplished by using a coded light source, panel signal, radio contact, RT, or combinations thereof. These may be used individually or with the marking pattern.

3-92. When using an RT or LTD for authentication, the ground and air commanders concerned jointly agree upon positioning, codes, and turn-on or turn-off times during mission planning.

DROP ZONE STERILIZATION

3-93. The reception committee must take specific actions to ensure DZ sterilization. The parachutists must also perform certain actions to help sterilize the DZ and allow for a quick departure from the DZ.

3-94. To ensure sterilization, the reception committee, when one is used, must—
- Police the area thoroughly.
- Remove all evidence or signs of occupancy, such as crushed undergrowth, heel scuffs, trails, and human waste.
- Recover all rigging straps and other air delivery equipment.
- Assign an individual at the recovery collection point to account for the air items and packages as recovery teams bring them off the DZ or LZ.
- Provide a two- or three-man surveillance team, preferably from the supporting auxiliary element, to maintain a close watch on the DZ or LZ area for enemy activity during the 48 hours following the drop.

3-95. To assist in sterilization, the individual parachutist must—
- Recover all parachute items, straps, bundles, and equipment worn on the drop.
- Bury unwanted air items separately, preferably at the base of thick bushes.
- Erase drag marks, footprints, and impact marks, if possible.
- Avoid trampling or crushing vegetation. When moving off the DZ, bypass plowed areas and grassy fields.
- Prevent accidental compromise of the operation by avoiding paths and roads and by moving cross-country to the assembly point.

POSTMISSION REQUIREMENTS

3-96. In a training operation, immediately following the operation, several reports must be forwarded to a higher headquarters. In a UW situation, the detachment will notify the SOTF or AOB of the success or failure of the air operation IAW unit SOP.

3-97. Most of the training reports are self-explanatory and require little time to complete. Appendix E includes information about the following:
- USASOC Form 1051-R-E (Airborne Operations Flash Report).
- AMC Form 168 (Airdrop/Airland/Extraction Zone Control Log). Personnel use this form to record strike report information. The DZSTL forwards the AMC Form 168 to his air operations officer, who in turn submits it through the chain of command to the USAF representative.
- Malfunction report.

DROP ZONE SURVEYS

3-98. All information concerning the DZ is placed on an AF Form 3823. This form provides the user with the essential information needed to operate the DZ. Section 4 of the form states what type of missions may be conducted on the DZ. USAF aircraft require a DZ survey for training airdrop missions involving U.S. personnel and/or equipment. Completing the DZ survey process involves physically inspecting the DZ and documenting the information on AF Form 3823. Using units may conduct the surveys. The using unit is defined as the unit whose equipment or personnel are airdropped. For exercises and joint training operations, users must ensure the survey is completed and meets the appropriate criteria for operational and safety standards. The user must conduct a physical inspection of the DZ before use to identify and evaluate potential hazards to airdropped personnel and equipment, man-made or natural structures, and ground personnel. The regional or wing tactics office or designated individual will perform the safety-of-flight review to ensure there are no obstructions prohibiting overflight. If a DZ survey is done on an existing surveyed DZ to meet new run-in axis requirements for a particular mission, only a safety-of-flight review is required. When conducting operations on a DZ that was previously surveyed by another unit, the commander of the using unit ensure the DZ meets the criteria for that operation. In all cases, the using unit must accept responsibility for all personnel injuries, parachute or load damage, and property damage that occurs on the DZ.

Host Nation Surveys

3-99. When dropping host nation (HN) military jumpers and/or equipment on a surveyed DZ in a HN, the mission can be performed by using only a safety-of-flight review of the HN survey. Users remain responsible for ground operational and safety criteria as above. However, when U.S. personnel and equipment are airdropped, HN surveys will not be used instead of a survey completed by U.S. forces IAW survey procedures. An airdrop-qualified pilot or navigator on all DZ surveys completes a safety-of-flight review. The purpose of a safety-of-flight review is to ensure an aircraft can safely ingress and egress the DZ.

Tactical Surveys

3-100. During exercises and contingencies, when time or situation do not permit completion of a full DZ survey, a tactical DZ survey may be required to support highly mobile ground forces. Though preferable, the use of an AF Form 3823 is not required for a tactical survey. Requests and surveys may be passed electronically. Enough information as practicable should be obtained and forwarded for review. Requests for tactical surveys will be forwarded to the designated exercise or contingency airlift or SO airlift component senior representative for final review. When using a tactical DZ, the airlift unit assumes responsibility for aircraft safety-of-flight and the receiving unit assumes responsibility for injury to personnel or damage to equipment and air items. The mode of delivery, load dispersal, and discussion with receiving unit regarding air item recoverability and load survivability determines the DZ size.

CONTINGENCY OR WARTIME OPERATIONS

3-101. During contingency or wartime operations and major exercises, DZSTs may need to prepare for follow-up airdrop of resupply or reinforcement. The DZST would then tactically locate, inspect, and approve a potential DZ.

Tactical Assessment

3-102. The DZST prepares a tactical DZ assessment using the—
- DZ name or intended call sign.
- Topographical map series and sheet number.
- Recommended approach axis magnetic course.
- PI location (eight-digit grid).
- Leading edge centerline coordinates (eight-digit grid).
- Air traffic restrictions and hazards.
- Name of surveyor and unit assigned.
- Recommended approval or disapproval (cite reason for disapproval).
- Remarks (include a recommendation for airdrop option: CARP, GMRS, VIRS, or blind drop).

3-103. Airdrop operations on tactically assessed DZs are made only when—
- During training events, the airdrop occurs within a military reservation or on property leased by the U.S. Government.
- The supported service accepts the responsibility for any damage that occurs because of airdrop activity.
- There is adequate time for safe, effective planning.

Drop Zone Review Process

3-104. The AZAR is a comprehensive listing of assault zones in the assault zone database. AZAR is available for use by the Department of Defense (DD). Use of the AZAR will expedite mission planning, enhance safety, and avoid duplication of surveys. Information in the AZAR does not replace the need for a completed survey before conducting assault zone operations. The appropriate agencies will forward all completed surveys to Headquarters (HQ) AMC, Director of Operations, Plans, for inclusion in the worldwide assault zone database. Completed surveys are also available via a facsimile (fax)-on-demand system located at Scott Air Force Base, Illinois. The Internet site available for military (.mil) users is located at https://amc.scott.af.mil/do/dok/zar.htm.

Note. DZ surveys now have an indefinite shelf life. If the DZ survey was current as of 1 June 1999 (having a USAF major command approval date after 31 May 1994), it qualifies for an indefinite shelf life IAW AFI 13-217 dated 10 May 2007.

3-105. All DZs must be surveyed before use. Aircrews review safety-of-flight requirements during mission planning. Survey users will redo the surveys when the user and/or airlift provider determines changes in the ground or air aspects of the DZ data. The surveyor performs the actual ground portion of the DZ survey (that is, calculating measurements, coordinates, and size; obtaining maps; and creating diagrams) and annotates results on AF Form 3823. The surveyor may be a member of the unit that intends to use the DZ. A member of another unit may perform the ground portion of a survey if requested and time permits. For example, a USAF member may perform the ground survey for an Army unit and vice versa. To facilitate future use of surveyed DZs, initial surveys will encompass the largest area available and will not be limited by specific mission requirements. The surveyor will forward the completed survey to the ground operations review authority with a transmittal letter. The surveyor will include recommended use, any deviations from DZ standards contained in Service or Army command directives, and other pertinent remarks. Throughout the review process, DZ survey packages will include all applicable maps, photos, charts, and diagrams needed to determine the safety and utility of the DZ. The ground operations review authority (AF Form 3823, item 4c) is normally the surveyor's commander or designated representative. This review ensures the survey form is complete and accurate and the DZ meets the criteria for planned airborne operations. The safety-of-flight reviewer performs the safety-of-flight review (AF Form 3823, item 4d), ensuring the DZ can be safely used from a flight perspective.

AIR OPERATIONS APPROVAL

3-106. Before a DZ is used, the appropriate operational group commander approves the surveys for air operations. This approval ensures the safety-of-flight review has been done and the DZ is considered safe for air operations. Once item 4e of the AF Form 3823 is completed, the survey is ready for use. The S-3 forwards copies of the survey to HQ, AMC, Director of Operations, Plans (DOK), 402 Scott Drive, Scott Air Force Base, IL 62225-5320, to maintain the most current data in the AZAR database. Assault zone surveys document the conditions that existed at the time the survey was done. Recommended uses may be based on minimum requirements and should not be considered all-inclusive. For example, a DZ recommended for personnel may be suitable for a single parachutist but not for 15, or it may be suitable for a C-130. The airlift and ground units involved must ensure that any DZ being considered for use meets the requirements for their specific operation.

This page intentionally left blank.

Chapter 4

Landing Zones

SF units conduct LZ operations using rotary, tilt-rotor, and fixed-wing aircraft to insert, resupply, or recover personnel or equipment into or from a JSOA. These operations normally take place at night on a preselected LZ with or without a reception committee. This chapter describes the LZ procedures and techniques that mainly apply to helicopters, tilt-rotors, and two categories of fixed-wing aircraft—light and medium weight. STOL aircraft are usually classified under the lightweight category. Because of the missions SF units perform, they may have to use aircraft that are not in the Active Army or AF inventory. Helicopter LZ criteria apply to past, present, and future models. Where differences exist between the standard LZ size requirements, each specific aircraft manual will take precedence. Because of space and time considerations, this chapter addresses the selection criteria and dimensions for night or day single aircraft LZs. FM 3-21.38, *Pathfinder Operations*, or AFI 13-217 provide information about multiship or multilift helicopter operations. This information is in agreement with Standardization Agreement (STANAG) 3601, *Criteria for Selecting and Marking of Landing Zones for Fixed Wing Transport Aircraft*. This STANAG gives the ideal criteria, but the ultimate decision will rest with the supporting aviation unit.

AUTHORIZATION

4-1. SF-qualified individuals are authorized to establish LZs for fixed- and rotary-wing aircraft without CCT support. For fixed-wing aircraft, this authorization is limited to single-ship operations under VMC and using overt Airfield Marking Pattern-2 (AMP-2) markings. The operation must not require ATC services.

FIXED-WING LANDING ZONE TRAINING OPERATIONS

4-2. The mission must be conducted on a current surveyed LZ. ATC services are normally required whenever the aircraft will be operating in class D airspace. When other aircraft, of any type, are operating in or around the LZ, air-to-air communications should be coordinated. The class D airspace restriction does not normally affect LZs located on military installations. Local installations may require that the landing zone controller have ATC capabilities or may further restrict or prohibit the ability of units to conduct LZ operations without CCT support. Units wishing to conduct LZ operations without CCT support for training should contact the local airspace controlling agency and the supporting air unit to ensure all parties concerned are in agreement.

4-3. References that govern fixed-wing LZ operations include USASOC Reg 350-2, Air Mobile Command Reg 55-60, *Assault Zone Procedures*, and AFI 13-217. Individuals performing landing zone controller duties must be thoroughly familiar with applicable sections of these references before conducting LZ operations.

AIRCRAFT CLASSIFICATIONS

4-4. Aircraft are classified into two broad categories—rotary or tilt-rotor and fixed-wing. These categories are broken down into subsequent categories based on different criteria.

Chapter 4

ROTARY-WING OR TILT-ROTOR AIRCRAFT

4-5. This classification includes helicopters and vertical takeoff and landing (VTOL) aircraft. The CV-22A Osprey, a tilt-rotor aircraft, can land vertically and horizontally.

FIXED-WING AIRCRAFT

4-6. Fixed-wing aircraft are any aircraft in which the wings and engines do not move. Fixed-wing aircraft can be further classified by their weight, number of engines, range, type of engines, and size of LZ needed. For the purpose of this manual, fixed-wing aircraft will be classified as light (weighing 12,499 pounds or less), medium (12,500 pounds to 299,999 and greater), and heavy (300,000 pounds) aircraft. STOL aircraft are aircraft in any weight class that have the ability to clear a 50-foot obstacle within 1,500 feet (500 meters) of commencing takeoff or land and stop within 1,500 feet (500 meters) after passing over a 50-foot obstacle. Not all lightweight and medium aircraft are considered STOL, and some STOL may need a longer runway because of environmental factors. An aircraft that may meet the stated STOL criteria at sea level on a cool day may not be able to have the same performance at 5,000 airport elevation in the middle of summer. For the purpose of this manual, light STOL aircraft include the UV-20A Chiricahua/Porter and UV-18A Twin Otter, and medium STOL aircraft include the Aviocar CASA-212. The light aircraft category includes the C-206, C-207, and C-208. The heavy aircraft category includes the C-130 Hercules, MC-130 Combat Talon, C-17, and C-5.

LANDING ZONE CONSIDERATIONS

4-7. The selection criteria for LZs are similar to those for DZs, but there are some factors that have less importance while other factors have greater significance. When selecting an LZ for fixed- or rotary-wing aircraft, the user or mission planners consider the following:

- Aircraft limitations.
- Mission and security.
- Identification.
- Size and features.
- Meteorological conditions.

Note. Additional criteria is outlined for each type of LZ.

AIRCRAFT LIMITATIONS

4-8. The aircraft limitations are the primary factors in site selection. Landing rules cannot arbitrarily be set, but certain specified minimums must be met. The deploying detachment and RCL must have a thorough knowledge of all requirements to provide aircrew and aircraft safety when landing on unprepared LZs.

MISSION AND SECURITY

4-9. The planning and coordination required to conduct an airland operation closely parallels DZ planning procedures. LZs will not be located near heavily defended areas, since low-flying aircraft are extremely vulnerable to ground fire. The user or mission planners designate an alternate LZ and set up prearranged signals to divert an aircraft to an alternate LZ.

IDENTIFICATION

4-10. Mission planners should select a site easily identifiable from the air. The pilot can better locate an LZ that has distinctive terrain features near it or on the approach path to it. Not all aircraft in support of SF detachments will be equipped with the latest electronic guidance measures.

SIZE AND FEATURES

4-11. The physical requirements of the LZ, such as site size, ground or water surface conditions, and approach and takeoff features, are important. The LZ size required depends on the type of aircraft used. There must be strict adherence to minimum dimensions to ensure safe operations.

Landing Zones

METEOROLOGICAL CONDITIONS

4-12. Ground personnel determine prevailing wind direction, velocity, and visibility restrictions, such as ground fog, haze, ambient light, or low cloud formations in the landing area. Prevailing weather and other meteorological conditions should favor the operation. The wind direction and speed may determine if the aircraft can land on the LZ or from which direction it has to approach the LZ.

HELICOPTER LANDING ZONES

4-13. Helicopters support a variety of SOF missions by airlanding, airdropping, or hovering to accomplish infiltration, exfiltration, and resupply. The following paragraphs discuss the capabilities of helicopter aircrews and the selection criteria for helicopter landing zones (HLZs).

AIRCREW CAPABILITIES

4-14. Deploying detachments that use special operations aviation (SOA) units should coordinate closely with the pilots to determine their exact limitations. Generally, the SO pilots of the Special Operations Aviation Regiment (SOAR) and the AF SO wings can far exceed the criteria required of conventional aviation units. Keep in mind the capabilities will vary from unit to unit.

SELECTION CRITERIA

4-15. The selection criteria for HLZs are similar to those for DZs. However, some factors have less importance while other factors have greater significance. The following selection criteria and dimensions of helicopter LZs for night and day operations are in agreement with the North Atlantic Treaty Organization STANAG 3597, Helicopter Tactical or Non-Permanent Landing Sites. This STANAG gives the ideal criteria. Reduced criteria may sometimes be accepted. The ultimate decision will rest with the supporting helicopter unit.

4-16. The SF commander must know how to select a helicopter landing zone (HLZ) that best supports his mission and meets the minimum requirements for the type of helicopter that is supporting the mission. The HLZ size depends on the type of helicopter used, the nature of the load, and the climatic conditions. The commander selects an HLZ using maps, aerial photographs, or actual ground reconnaissance. The limitations in this section pertain to conventional aviation units and may be waived by SOA units. Consider the following factors:

- Type of helicopter.
- Security.
- Clearing.
- Surfaces.
- Ground slope.
- Density altitude.
- Height and type of obstacles on the approach and departure paths.
- Approach and departure paths.
- Prevailing winds.

Type of Helicopter

4-17. Helicopters are given a number classification of 1, 2, 3, 4, or 5 to determine the minimum LZ dimensions (Figure 4-1, page 4-4). These numbers can be changed because of considerations such as helicopter type, unit proficiency, nature of loads, climatic conditions, and day versus night operations. These considerations are covered by aviation unit SOPs, or they are prearranged by the aviation unit commander in coordination with the deploying detachment. The final decision concerning minimum landing considerations rests with the aviation unit commander. If the type of helicopter that is being used is unknown, the detachment prepares a Size 5 LZ.

Chapter 4

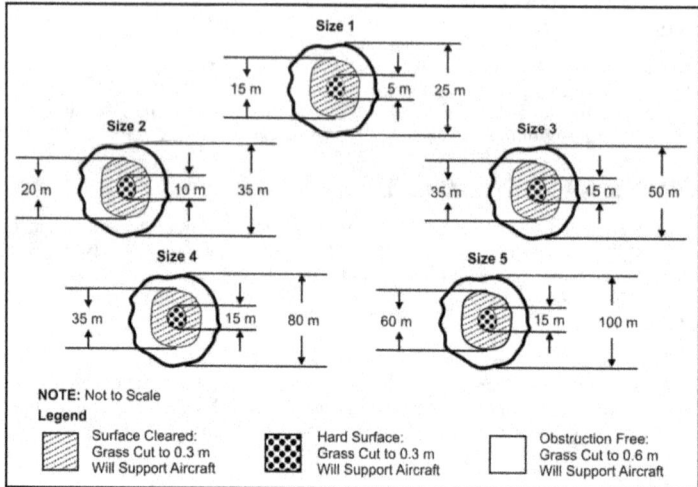

Figure 4-1. Helicopter landing point dimensions

4-18. Figure 4-2 shows the following recommended minimum distances between landing points on operations using multiple aircraft. These distances are measured from center to center of the landing points:
- Size 1: 25 meters.
- Size 2: 35 meters.
- Size 3: 50 meters.
- Size 4: 80 meters.
- Size 5: 100 meters.

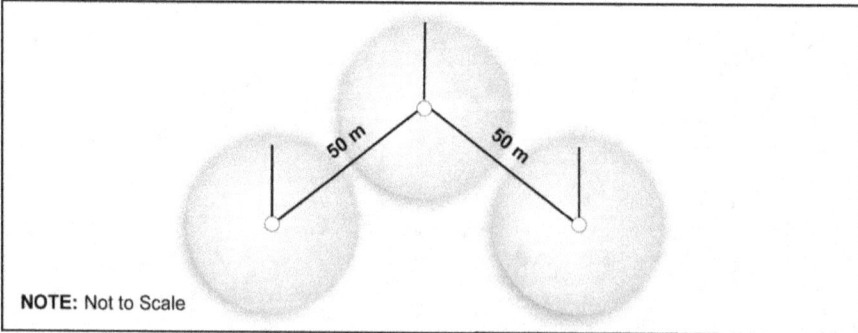

Figure 4-2. Sample landing points for multiple size 3 aircraft

Security

4-19. The LZ must facilitate helicopter operations and offer some degree of security from enemy observation and direct fire. Good LZs will allow safe helicopter operations without exposing personnel or the aircraft to unnecessary risks. The small size of SF teams will usually preclude establishing security around the entire perimeter of the LZ, but if possible security should be established. Smaller LZs are easier to reconnoiter and secure, but it is not always possible to use them. If using large LZs, the ground user locates the LZ close to a covered and concealed area and uses terrain (small hills, ridges, or rises) or vegetation (treeline or scrub brush) to mask the actual LZ site. For resupply or multilift LZs, the ground personnel should—

- Establish security in defensive positions to defend the LZ for short periods, depending on the size of the reception committee available.
- Set up fire team sectors or, if necessary, two-man positions to provide 360-degree security. Elements may be shifted after the initial flight(s) to maintain security and assist in security—plan and coordinate supporting fires with the helicopter—if possible.
- Carefully position automatic weapons to ensure maximum effectiveness. Emplace claymore mines to cover avenues of approach and/or dead space.
- Remain alert and ensure all personnel remain hidden so that the location of the LZ is not compromised.

Clearing

4-20. To ensure a safe landing, the ground user clears solid obstacles, loose materials, and flammable materials that could cause damage to the rotor blades, turbine engines, or underside of the fuselage. The term "cleared to ground level" is used to indicate this. Figure 4-1, page 4-4, shows how to determine the size of the cleared area required. It would not, for instance, be necessary to clear grass up to 0.3 meters high that might cover a level field unless a fire risk existed. If ground obstructions, such as trees and tree stumps, cannot be cleared, the ground users should mark obstructions with a red chemical light as the helicopter hovers above the LZ.

Surfaces

4-21. The LZ surface must be solid enough to bear the weight of the helicopter. The term "hard surface" is used to indicate this. Ground users must clear loose materials from the HLZ to prevent possible engine damage or personnel injury from flying debris. Since rotor wash on dusty, sandy, or snow-covered surfaces may cause loss of visual ground contact, the ground user considers stabilizing or covering these surfaces by an agreed-upon method. Snow should be packed or removed to reveal hazardous objects and to reduce the potential of blowing snow. A marker is essential to provide a visual reference for pilot depth perception and to reduce the effect of whiteout.

Ground Slope

4-22. Ideally, the LZ should be level, but if there is a slope, it should be uniform. Helicopters can touchdown hover (one skid or one wheel, but not all skids or wheels on the ground) on a sloping surface that exceeds the slope limits as long as the angle of the slope provides the necessary rotor clearance. Landing should be upslope when the ground slope is less than 7 degrees (15 percent). During daylight, the slope should not exceed 7 degrees if the helicopter must land. During night, forward and lateral slopes should not exceed 3 degrees (5 percent) and a down slope approach is not normally acceptable. Figure 4-3, page 4-6, shows the slope landing rules, and Figure 4-4, page 4-7, shows the formulas for determining ground slope.

Density Altitude

4-23. Altitude, temperature, and humidity determine air density. For planning purposes, as density altitude increases, the required size of the LZ increases proportionately. High, hot, and dry conditions at a given LZ decrease the lift capability of a helicopter using that site. Appropriate aircraft technical manuals contain detailed information. Aircrews refer to these manuals during premission planning to determine the effects of air density on aircraft performance at specific operating altitudes.

Chapter 4

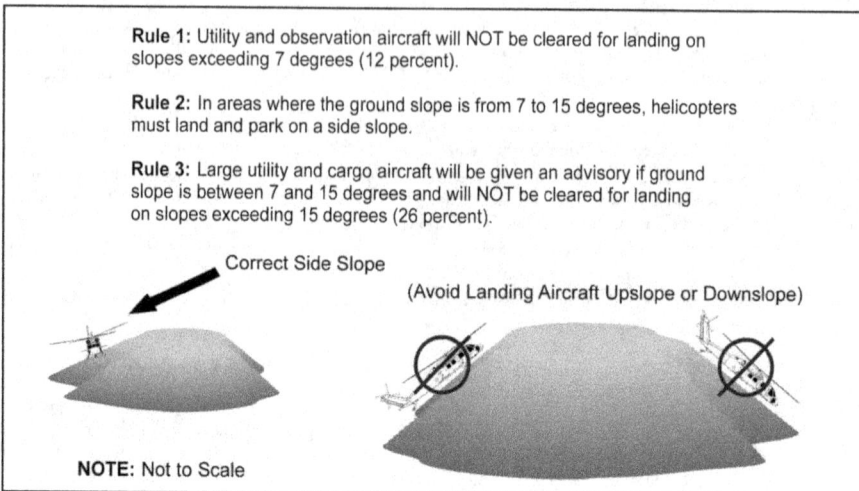

Figure 4-3. Slope landing rules

Approach and Departure Paths

4-24. Although helicopters can take off and land vertically, it is not desirable to use LZs that require this capability. Doing so requires greater power to ascend and descend vertically, reducing their allowable payload. Ideally, the approach and departure paths should be over the lowest obstacles and into the prevailing wind. However, if there is only one satisfactory approach path because of obstacles or the tactical situation, or if maximum use of available landing area is desired, most helicopters can land with a crosswind or a tailwind. A tailwind may increase the size requirement for the LZ. Specific limits should be confirmed with the supporting helicopter unit.

4-25. During daylight, within the approach and exit path, the maximum obstruction angle should not exceed 6 degrees. In Figure 4-5, page 4-8, the obstruction angle is measured at the landing point center to a distance of 500 meters. The maximum obstacle height at 500 meters is 52 meters. A ratio of 10 to 1 (1 unit of vertical clearance to 10 units of horizontal clearance) is a field-expedient way of determining this. For example, if the approach or departure path were directly over a tree 10 meters high, the LZ would require 100 meters of horizontal clearance from the tree. Greater obstruction angles may be acceptable, but this must be confirmed by the supporting helicopter unit.

4-26. During night operations, within the approach and exit path, the maximum obstruction angle should not exceed 4 degrees. In Figure 4-5, the obstruction angle is measured from the landing point center to a distance of 3,000 meters. The maximum obstacle height at 3,000 meters is 210 meters. A ratio of 14 to 1 (1 unit of vertical clearance to 14 units of horizontal clearance) is a field-expedient way of determining this. For example, if the approach or departure path were directly over a tree 20 meters high, the LZ would require 280 meters of horizontal clearance from the tree. Greater obstruction angles may be acceptable, but the supporting helicopter unit must confirm this.

Prevailing Winds

4-27. When considering the approach and departure paths and prevailing wind, the more important factor is the best approach and departure path unless the crosswind velocity exceeds 10 knots. The ability to land crosswind or downwind will vary, depending on the type helicopter. Smaller aircraft can accept less crosswind or tailwind than larger, more powerful aircraft.

Landing Zones

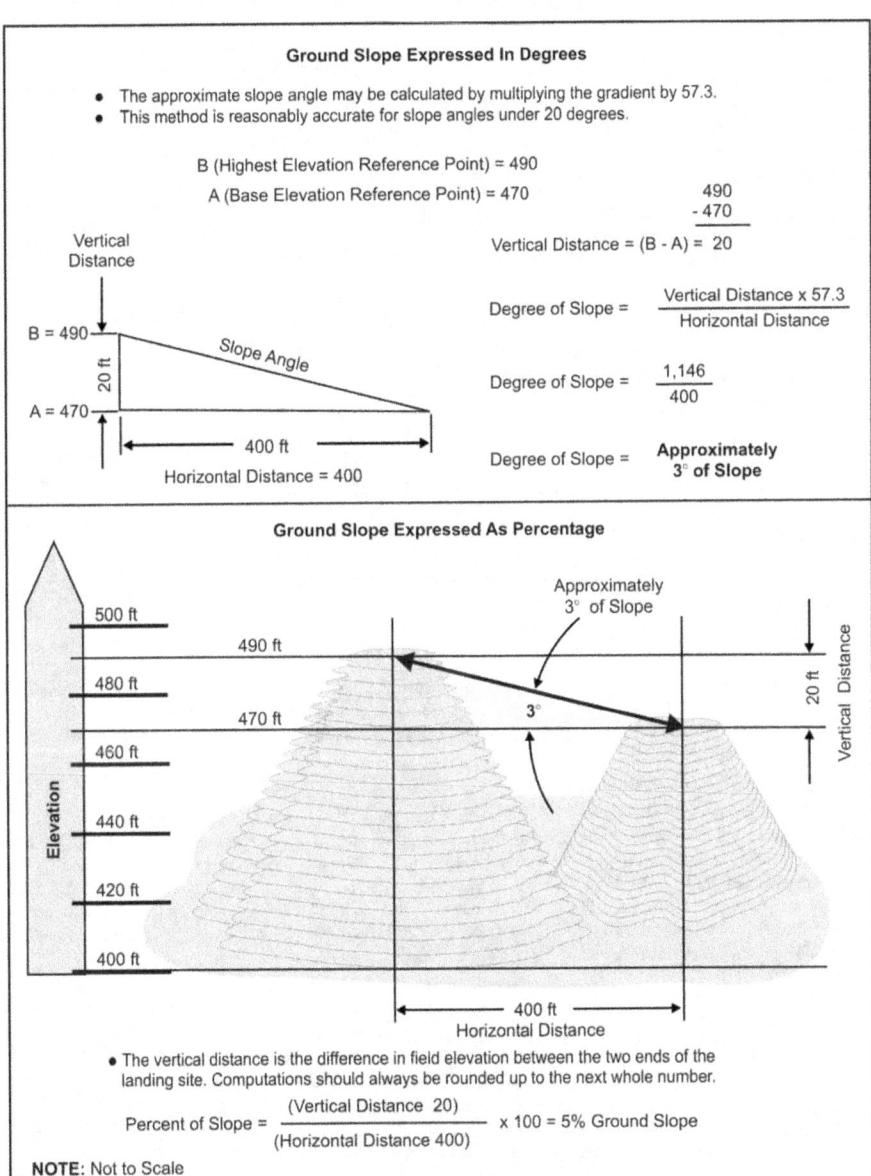

Figure 4-4. Determining ground slope

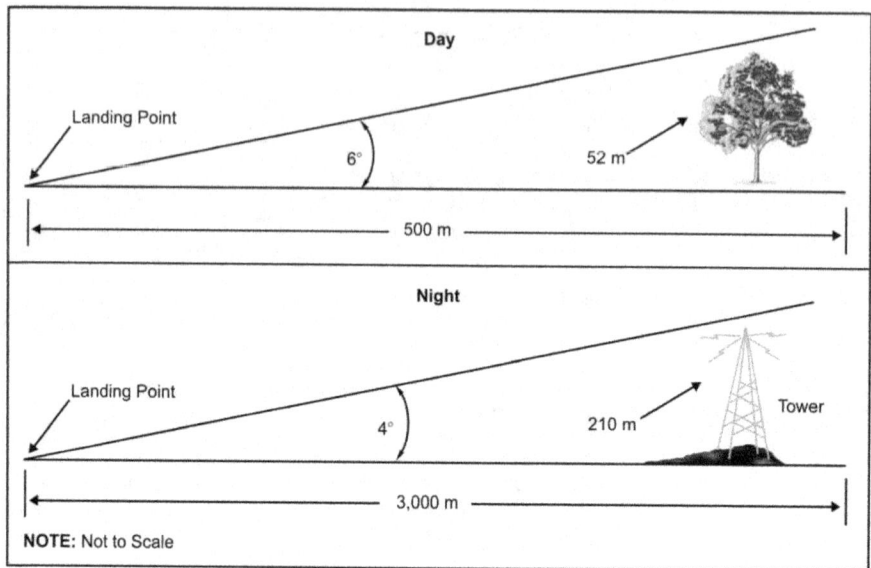

Figure 4-5. Day and night LZ obstruction angle

TYPES OF LANDING ZONES

4-28. The exact method a detachment uses to operate the LZ will depend on premission coordination, mission, situation, and unit SOP. Most SF missions are single-ship clandestine infiltrations or exfiltrations into small LZs. FM 3-21.38 contains guidance for teams to establish a DZ for multiship or multiflight missions.

INFILTRATION

4-29. Infiltration LZs should be small, unmarked, and unmanned. Detachments use them to infiltrate an AO or the vicinity of an AO. As soon as the helicopter sets down, the detachment exits as per SOP. The aircraft takes off after all personnel and equipment is off-loaded. The aircraft then continues on a predetermined route toward other possible LZs to prevent singling out the actual infiltration LZ. The aircraft may stay in the area for several minutes in the event that the detachment requires immediate exfiltration. As soon as the aircraft has departed, the detachment moves out on a predetermined azimuth to distance themselves from the LZ.

EXFILTRATION

4-30. Exfiltration LZs should be small and used for exfiltrating the detachment from the AO. The detachment plans to arrive at the LZ with enough time to determine if the LZ is usable or not. They may conduct a thorough 360-degree reconnaissance of the LZ or just the immediate vicinity of where the helicopter is going to touchdown. If the LZ is not usable because of enemy presence or natural obstacles, the detachment must have time to move to the alternate LZ, reconnoiter it, and set it up. The detachment is divided into a marking party, security party, command element, and other elements, as needed. Once the security is established, and just before the arrival time, the marking party marks the LZ IAW premission coordination, SOP, or SOI. The marking party does not visibly display or activate the markers until 2 minutes before the TOT. Once the helicopter arrives and lands, the detachment gathers up the markers and loads the aircraft as per SOP.

Resupply

4-31. Resupply LZs are used to bring in additional supplies and to exfiltrate personnel, intelligence, prisoners of war (PWs), or other items. In this case, the detachment serves as a reception committee. Resupply LZs are established in the same way as exfiltration LZs, but the actions after the aircraft lands are different. Once the aircraft lands, personnel and supplies are off-loaded first. The ground commander ensures all personnel and supplies are physically secured and moved to a holding area by members of the reception committee. Then the reception committee loads personnel and items to be exfiltrated onto the aircraft. The GUC or his representative ensures all items to be exfiltrated are loaded onto the aircraft. Once the off-loading and onloading are complete, the GUC signals the aircraft to depart. When the aircraft has departed, members of the reception committee sterilize the LZ (as best as possible), ensuring no signs are left behind. Then, the newly arrived personnel and supplies are accounted for again and loaded for movement. New supplies should be prepackaged in rucksacks or packed into rucksack-size waterproof bags to facilitate the loading into empty rucksacks brought by the reception committee. Depending on the tactical situation, resupply operations should be conducted as shortly after sunset as possible. The reception committee conducts a reconnaissance during twilight hours to have the most amount of time for moving during the hours of darkness.

MARKING LANDING ZONES

4-32. Visual ground markings for helicopter LZs provide the wind direction, identification, direction of approach, and the designated touchdown area. Ground personnel can use several types of markings, depending on the visibility and if the pilot has NVGs. Most U.S. military aviation units use NVGs during night operations. However, allied or nonmilitary pilots may not have them. The LZ may be marked by IR lights or visible light sources at night and panels in daylight, as determined by the premission coordination or the SOI. Depending on the tactical situation, smoke, flares, a signal mirror, an IR strobe light, or other means identifies the LZ for the helicopter. Pilots discourage the use of a visible strobe light to identify an LZ since it interferes with their NVGs and may be confused with antiaircraft (AA) fire. If a strobe light is to be used, it should be turned off once the pilot has identified the LZ and is making his approach.

4-33. Normally, the only marking used during daylight is a VS-17 panel at the touchdown point. The LZ may be marked with the Y (Figure 4-6, page 4-10). The Y marking depends on premission coordination and winds.

> *Note.* When using the Y, the marking team firmly secures the markers so they will not be blown away or sucked into the engine intakes. If the situation permits, designated members of the exfiltrating detachment recover the markers before boarding the aircraft. The marking team uses various techniques to ensure quick recovery. For example, they may attach markers to stakes, rucksacks, and so on.

4-34. Ground personnel mark the LZ with IR chemical lights in the "Y" marking pattern. An IR chemical light tied to a 3-foot string being swung in a vertical circle marks the RCL station. The LZ is marked in the same manner as for pilots with NVGs with one exception. Chemical lights (instead of IR chemical lights) mark the LZ.

4-35. When surface winds are a factor, 10 knots or more for infiltration under load and 15 knots or more for exfiltration under load, ground personnel position the markers to ensure the landing is made into the wind, regardless of the approach track established in the mission request and confirmation message. The pilot makes the initial approach to the LZ along the designated track and, if necessary, adjusts to the final approach track indicated by the LZ markings.

Display

4-36. The marking team does not display or activate markers until 2 minutes before the scheduled arrival time. The marking team displays markings for a total of 4 minutes—2 minutes before until 2 minutes past the scheduled arrival time—or until off-loading and onloading are completed and the helicopter departs.

Chapter 4

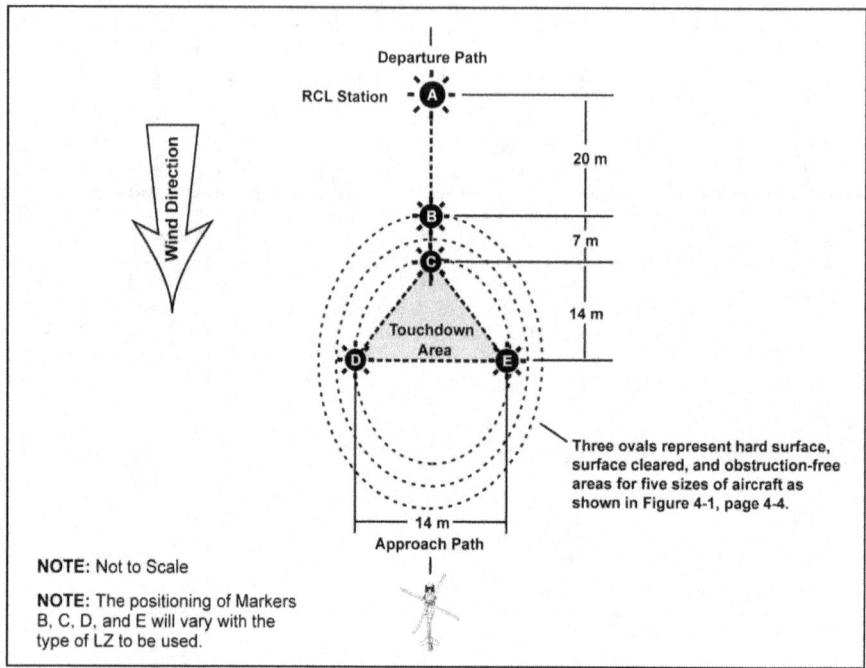

Figure 4-6. Y-marked LZ pattern

AUTHENTICATION

4-37. The SOI prescribe the authentication procedures or code signals. However, the marking team should—
- Arrive at the LZ within the specified time block on or near the designated approach track, which authenticates the mission aircraft.
- In daylight, display a distinctive panel or smoke signal. When using smoke, position it so that the prevailing wind will not cause the smoke to obscure the LZ.
- At night, display the proper IR or lighted signal. This signal may be just the swinging chemical light, the "Y," or both as per premission coordination.

IMPROVISED LANDING PLATFORMS OR PADS

4-38. Under ideal conditions and provided the necessary clearance for the rotors exists, a helicopter can land on ground slightly larger than the spread of its landing gear. Landing platforms may be built in swampy or marshy areas by using locally available materials (Figure 4-7, page 4-11). Normally, this type of LZ is used for daylight operations only. In addition to the size of the clearing and the approach and takeoff requirements already discussed for helicopter LZs, the following are additional requirements for the improvised landing platforms or pads:
- The platform should be large enough to accommodate the spread of the landing gear plus 3 meters.
- The platform should be capable of supporting the weight of the aircraft.
- The platform should be of firm construction so that it will not move when the helicopter touches down and rolls slightly forward.
- The platform should be level.
- If logs or bamboo poles are used, they should be arranged so that the top layer of poles is at right angles to the touchdown direction.

Landing Zones

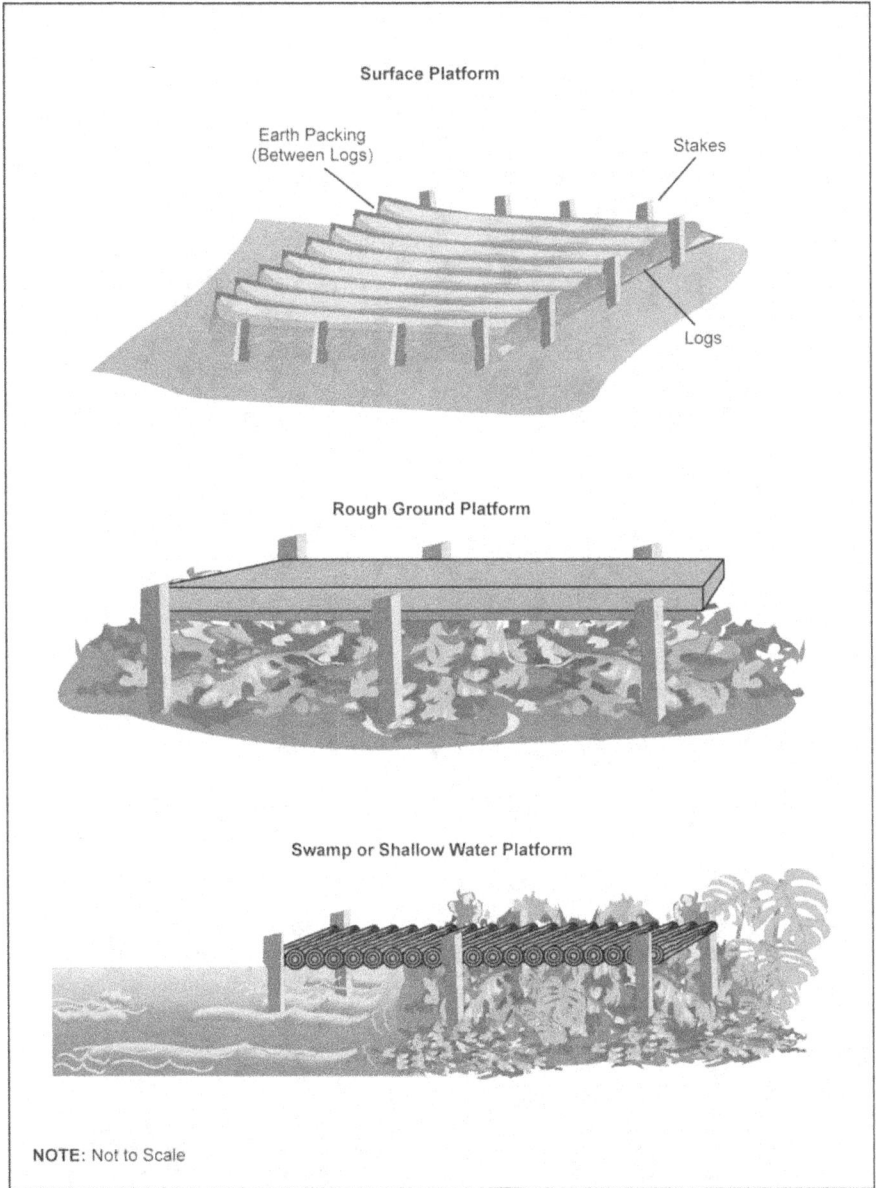

Figure 4-7. Platform LZs for rotary-wing aircraft

4-39. Landing pads can also be prepared in mountainous terrain or on hillsides by cutting and filling (Figure 4-8). Make sure there is adequate clearance for the rotors.

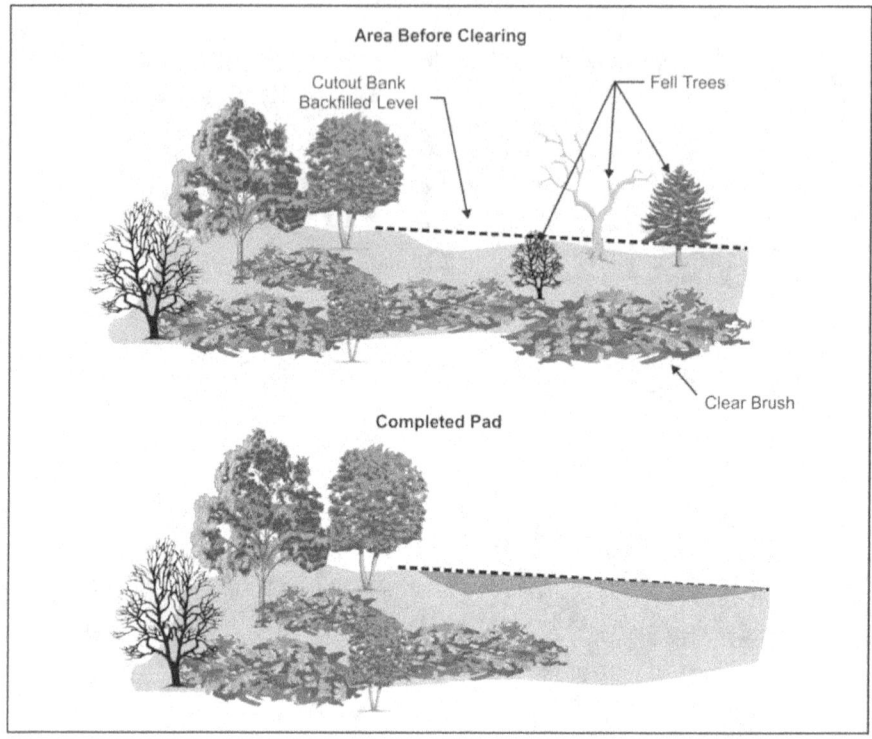

Figure 4-8. Preparing mountainous terrain landing pads for rotary-wing aircraft

4-40. Helicopters with a flotation capability present no problem in LZ preparation. They can land in water of any depth. However, helicopters without flotation capability can land in water without the use of special flotation equipment provided—
- The water depth does not exceed the height of the landing gear. Specific limits should be confirmed with the supporting unit.
- A firm bottom, such as gravel or sand, exists.

LIGHT AND MEDIUM STOL AND MEDIUM AIRCRAFT LANDING ZONES

4-41. The following information applies when working with U.S. Army and USAF aircraft in training situations, during training or operational deployments, and in combat situations. The correlation of these airfields may not be exact, and specifications are dependent upon aircraft gross weight (GW), use of aircraft arresting equipment, criteria for the particular instrument approach planned, and model and type of aircraft.

COMBAT CONTROL OR SPECIAL TACTICS TEAM

4-42. CCT has or will soon be converted to an STS. For the purpose of this manual, the terms are interchangeable. The USAF team is composed primarily of SO combat control and pararescue personnel. The team supports joint SO by—
- Selecting, surveying, and establishing assault zones.
- Providing assault zone terminal guidance and ATC.
- Conducting DA missions.
- Providing medical care and evacuation.
- Coordinating, planning, and conducting air ground and naval fire support operations.

4-43. The team is equipped with handheld pocket transits, clinometers, and levels to check approach zone clearance and airfield or dynamic cone penetrometers to check weight-bearing capability of unsurfaced LZs. The STS gathers data from the on-site survey, prepares an LZ survey package using the AF Form 3822 (Landing Zone Survey) and recommends approval or disapproval to the appropriate agency for use. They determine LZ suitability by using the general criteria in FM 5-430-00-2, *Planning and Design of Roads, Airfields, and Heliports in the Theater of Operations—Airfield and Heliport Design*, and additional criteria contained in Army command publications for the type of aircraft involved. The STS may be tasked to perform assessment of semipermanent and permanent installations, such as captured enemy airfields, for possible aircraft hazards and correct dimensions before use. The STS may be required to assist the airlift commander as a designated representative in selecting LZ sites. They are not qualified to—
- Evaluate hard-surfaced pavements for traffic cycles and weight bearing.
- Perform engineering surveys.

ENGINEERING TEAMS

4-44. A team from the AF Civil Engineering Support Agency will survey existing or proposed airfields that require precise determination of gradients. Semipermanent runways are usually surveyed by engineering units and do not require a survey by a STS.

AIRFIELD OR LANDING ZONE CLASSIFICATION

4-45. Airfields are classified as permanent or expedient. Assault LZs are either unprepared, prepared, or surfaced.

4-46. The airfield classifications are as follows:
- *Permanent.* AF airfields are usually constructed to standards that are based primarily on the expected life of the airfield. Airfields intended for extended use are generally of semipermanent construction and built to the full operational standards for the theater of operations. In most cases, the use of conventional asphalt or concrete upgrades pavement standards. The asphalt or concrete must be thick enough to meet predicted use.
- *Expedient.* The runway surface for these airfields consists of dirt, membrane, gravel, landing mat, or any combination of these.

4-47. The assault LZ classifications are as follows:
- *Unprepared.* Unprepared surfaces are natural areas. Examples of natural areas are deserts, dry lake beds, and flat valley floors.
- *Prepared.* Airstrips with prepared surfaces may or may not have an aggregate surface. Airstrips with prepared surfaces are short and have a limited use.
- *Surfaced.* Surfaced areas are paved. Examples of surfaced areas include roads and highways.

PARTS OF THE AIRSTRIP OR LANDING ZONE

4-48. Conventional airfields are composed of many parts and areas. Different aircraft require different widths and lengths of these parts. The following paragraphs describe the airfield requirements for USAF aircraft and various other aircraft (Figure 4-9, page 4-14).

Chapter 4

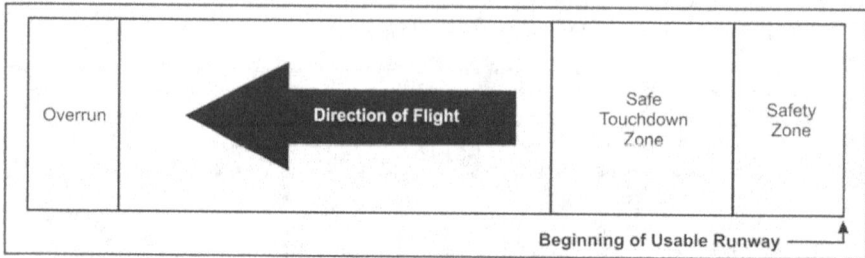

Figure 4-9. Basic airfield layout

Runway

4-49. A runway is a defined rectangular area of an airfield that has been prepared for the landing and takeoff run of aircraft along the length of the rectangular area. The length and width of the runway is determined by the type of aircraft that will be using it. Tables 4-1 and 4-2 explain the minimum length and width requirements for USAF aircraft. FMs 5-430-00-1, *Planning and Design of Roads, Airfields, and Heliports in the Theater of Operations—Road Design*, and 5-430-00-2, provide additional information.

Table 4-1. Minimum airfield criteria (standard) for USAF aircraft

Type of Aircraft	Landing Zone[1]	Width in Meters[1]		
		No Turn Required	180° Turn (Normal)	180° Turn (3 Point)
C-27	549[2]	14	14	9[4]
C-130	914[3]	18	18	15[5]
C-17	914[3]	27	40	24[5]
C-5	1,829	46	46	NA

Notes.
[1] Minimum operational criteria without a waiver during peacetime operations.
[2] The length normally used for routine training is 640 m.
[3] The length normally used for routine training is 1,067 m.
[4] Does not include any safety margin. Increase by 4 m for normal operations.
[5] Does not include any safety margin. Increase by 3 m for normal operations.

Table 4-2. Minimum airfield criteria (special use) for USAF aircraft

Type of Aircraft	Landing Zone[1]	Width in Meters[1]	
		No Turn Required	180° Turn (on Runway)
M/HC-130	914[1]	18[2]	23
C-130 NVG	914[1]	18[2]	23
C-5 SOLL II	1,829	30	46

Notes.
[1] The length normally used for routine training is 1,067 m. For NVG-equipped crews, the length is 1,219 m.
[2] Unqualified crews require a width of 23 m.

Landing Zones

Touchdown Zone

4-50. The first portion of the runway, beginning at the threshold, is the touchdown zone. It varies depending on the aircraft. For light and medium STOL, and light and medium aircraft, it is the first 100 meters.

Clear or Safety Zone

4-51. The clear or safety zone is an area located at each end of the runway. The zone width is normally equal to runway, shoulders, and clear areas. Zone length is normally 100 meters for medium aircraft and 10 percent of the runway length for light or medium STOL aircraft.

Overrun

4-52. The overrun is a graded and compacted portion of the clear zone (it may be as long as the clear zone but not as wide) located as an extension to the end of the runway. An overrun is not normally considered part of the usable runway when establishing airfield markings. Overruns are used to minimize risk to aircraft on takeoff or undershooting on landing. The width normally equals that of the runway and the type of aircraft used determines the length. The area of transition from runway to clear zones to overruns should be smooth with no "lips."

Note. The overrun distance for the C-130 may soon be reduced to 91 meters.

Shoulder

4-53. The shoulder is a graded and compacted area on either side of the runway that helps minimize the risk to aircraft of running off or landing off the runway. Shoulders should have rocks and tree stumps cut flush with the ground. Objects that could be ingested by engines or cause damage to the bottom of aircraft should be removed from the shoulders. Shoulders are normally 10 feet wide.

Graded Area

4-54. A graded area is a section located adjacent to and outside of the runway shoulders. Graded areas should not have any obstacles over 4 inches high except vegetation, runway edge markers, runway distance remaining markers, Mobile Aircraft Arresting Systems, or other visual or electronic navigational aids that must be sited in this area because of their function. Width of a graded area varies from 15 meters to 114 meters depending upon type of aircraft for which the airfield is intended. Appropriate dimensions are given in FM 5-430-00-2, Table 11-3.

Transitional Area

4-55. The normal requirements for a transitional area are 17 meters wide, extending outward and upward at a 5:1 ratio from the outer edge of the clear zone. Transitional areas should meet the criteria for the most restrictive type of aircraft using the LZ.

Approach Zone

4-56. An approach zone is a trapezoidal area extending outward from each clear zone within which no object may penetrate the glide slope angle. Approach zones should meet the criteria for the type of aircraft using the LZ. The normal clearance surface is established on a 35:1 ratio for close combat area airfields and a 50:1 ratio for support area airfields.

Taxiway

4-57. A taxiway is a specially designed or prepared path on an airfield. It is used by taxiing aircraft.

Parking Apron

4-58. A parking apron is a designated area to park aircraft. It is used for loading or unloading.

Chapter 4

> *Note.* In a UW environment, it is not required to have a complete airfield, but the following parts are required for conventional USAF aircraft: runway, touchdown zone, clear or safety zone, overrun, shoulder or clear area, and approach zones. For SOF, non-USAF, or STOL aircraft, premission coordination and aircraft requirements determine what parts are needed.

FIXED-WING LANDING ZONE CRITERIA

4-59. As a general rule, the same criteria used in selecting DZs apply when selecting fixed-wing LZs. However, ground surface, LZ size, terrain features, and approach and takeoff clearances are more important in LZ selection. Personnel must consider the slope and elevation of the runway, aircraft capability, taxiways, and loading area restrictions. Loading areas should also be able to support aircraft weight. Factors to consider in selecting an LZ are as follows:

- Size.
- Surface tolerances, drainage, and clearances.
- Terrain features.
- Approach and takeoff obstacle clearance.
- Dimensions and layout.
- Crosswinds.

4-60. The LZ minimum size requirements for fixed-wing aircraft vary according to the type of aircraft, load, LZ elevation, and climatic conditions. They will also be different for training and actual operations. Figure 4-10, page 4-17, depicts the general LZ size and marking requirements for training with light and medium STOL aircraft. The LZ size and marking requirements for actual operations are depicted in Figures 4-11 and 4-12, pages 4-18 and 4-19.

4-61. Landings at higher elevations require increased LZ dimensions because of decreased air density. If the site is above 1,220 meters altitude or in an area with a high temperature range, the minimum length should increase as follows:

- Add 10 percent for every 305 meters of altitude above 1,220 meters.
- Add 10 percent when temperatures are between 90 and 100 degrees Fahrenheit (F).
- Add 20 percent when temperatures exceed 100 degrees Fahrenheit.

Elevation changes should be made first, and then temperature changes.

LZ Surface Tolerances and Clearances

4-62. LZ terrain may be soil, dirt, sand, or another suitable surface. Tolerance of roughness will depend upon sheer strength, hardness, and size of items that cause roughness. Roughness interrupts smooth rotation of aircraft tires and interferes with marginal aerodynamic lift of flight control surfaces at slow speed. Ground personnel must minimize roughness for sustained operations. The following paragraphs are a guide for determining suitability of runway surface, shoulders, and clear areas. Exceeding these limits may result in structural failures of the aircraft. Whenever possible, ground personnel determine the weight-bearing capacity of the landing, taxiing, and parking areas.

4-63. *Surface.* The surface must be level and free of obstructions, such as ditches, deep ruts, logs, fences, hedges, rocks larger than the fist, or grass over 15 centimeters high. A surface unsuitable in summer may be ideal in winter.

4-64. *Subsoil.* The subsoil must be firm to a depth of 61 centimeters. A surface containing gravel, small stones, or thin layers of loose sand over a firm layer of subsoil is acceptable.

4-65. *Surface Gradient.* The length and width of the surface gradient should not exceed 2 percent. A gradient of more than 1 percent will adversely affect the performance of the aircraft.

4-66. *Ice and Snow.* Ice 48 centimeters thick will support light STOL aircraft. Ice 91 centimeters thick will support medium STOL aircraft. Unless the aircraft is equipped for snow landing, snow in excess of 10 centimeters must be packed firmly or removed.

Landing Zones

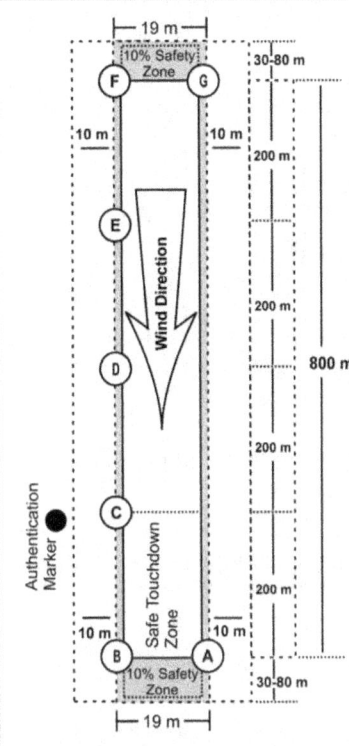

NOTE:

- The marked length of the LZ is 800 meters. Mission weight, fuel requirements, and environmental conditions may require a longer LZ. If required, additional markers are placed between the 'D' and 'E' markers at 200-meter intervals.

- The marked width of the LZ is 19 meters.

- A 10% "safety zone" is established at each end of the LZ. The safety zone may be reduced to no less than 30 meters in confined LZs. However, the 11:1 obstacle clearance descent ratio is still required (measured from the "C" marker to the highest obstacle on the approach/glide path).

- A 10-meter wide "clear area" parallels each side of the marked LZ. These areas must be clear of obstacles over 1 meter high.

- An improved field marker or improvised light marker is installed at each point - "A" through "G." LZ markers must be of the same type. LZs will not be marked using a combination of improved and improvised markers.

NOTE: The authentication marker is placed abeam the "C" marker to indicate the landing direction, mark the end of the touchdown zone, and allow earlier identification of the point at which the aircraft can begin a normal descent (critical in confined LZs).

The authentication marker may be a red improved field marker or an improvised marker. If an improvised marker is used, it must be visible from more than 2 miles and be distinct from the other LZ markers. The authentication marker must remain on during the descent and landing.

Following coordination with the supporting organization light STOL aircraft LZs may be marked IAW Figure 4-11, page 4-18.

Figure 4-10. Training LZ size and night marking requirements for light and medium STOL aircraft

4-67. ***Rocks.*** Rocks in traffic areas must be removed, embedded, or interlocked with each other.

4-68. ***Soil Balls.*** Soil balls or dry, cohesive dirt clods (clay excluded) up to 15 centimeters in diameter that will burst upon tire impact can be allowed. Hardened clay clods that have similar characteristics as rocks and exceed 10 centimeters in diameter must be pulverized or removed from the traffic areas.

4-69. ***Tree Stumps.*** All stumps must be removed. The holes must be filled with compacted soil to the firmness of the surrounding surface.

4-70. ***Ditches.*** Ditches must be eliminated from traffic areas. Filled ditches must be as firm as the surrounding area.

Chapter 4

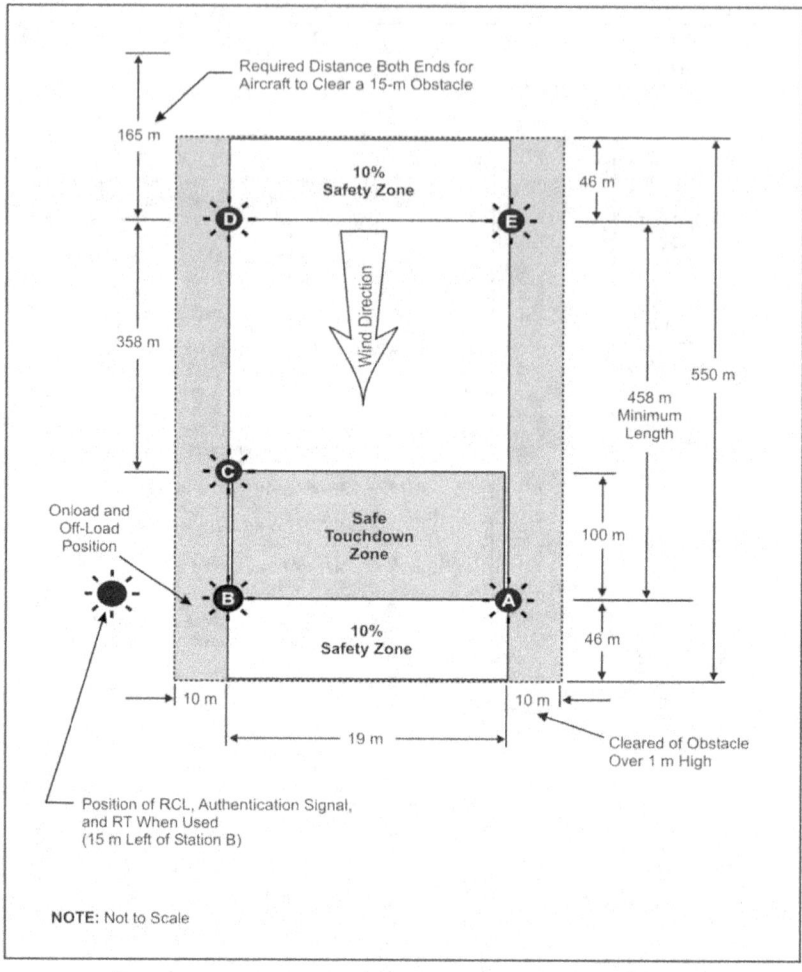

Figure 4-11. Marking land LZ for light and medium STOL aircraft

4-71. *Plowed Fields.* Plowed or planted fields should be avoided if possible. Contours of dirt patterns created by agricultural plowing to reduce erosion and water drain-off and for planting preparation usually contain a soft core. Therefore, these dirt patterns will not normally require removal. However, such dirt patterns should be examined carefully to determine the need for removal.

4-72. *Depressions and Soil Mounds.* Depressions and soil mounds cannot have sharp corners. They are recognized as oval or circular gradual downward sinks or rises. Depressions or mounds that exceed 38 centimeters across the top and 15 centimeters in depth or height should be leveled or filled until they meet grade tolerance criteria.

4-73. *Potholes.* Potholes are circular or oval in shape and are distinguished from depressions by their smaller size and sharp corners. Potholes must be filled if they exceed 38 centimeters across their widest point and 15 centimeters in depth.

Landing Zones

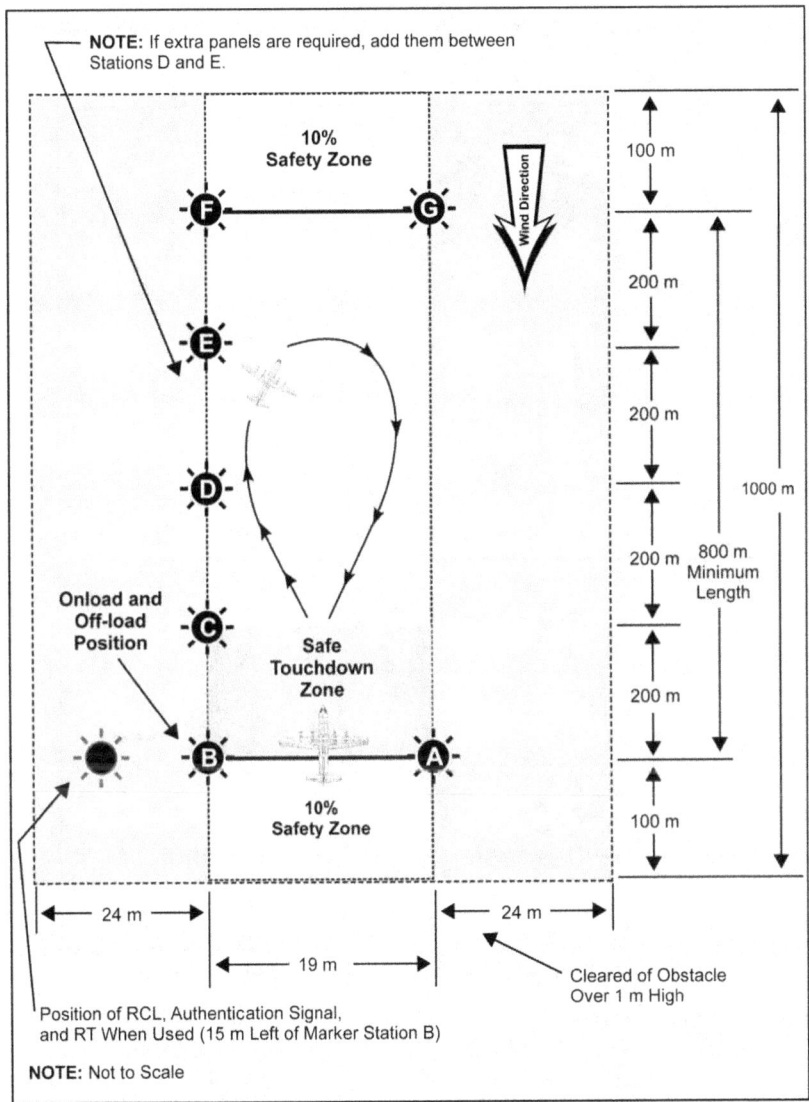

Figure 4-12. Marking land LZ for medium aircraft

Terrain Features

4-74. In mountainous or hilly country, planners should use a valley or plateau of sufficient size as an LZ. Planners must not select a pocket or small valley completely surrounded by hills for fixed-wing landing operations. If using a single-approach site, although undesirable, ground users should ensure—

- Sufficient clearance is available at both ends of the strip to permit a 180-degree turn to either side within the radius of 2 kilometers for light and medium STOL aircraft (Figure 4-13, page 4-20) and

Chapter 4

within 4 kilometers for medium aircraft (Figure 4-14) or as prescribed in appropriate aircraft operation manuals.
- All landings and takeoffs are into the wind.

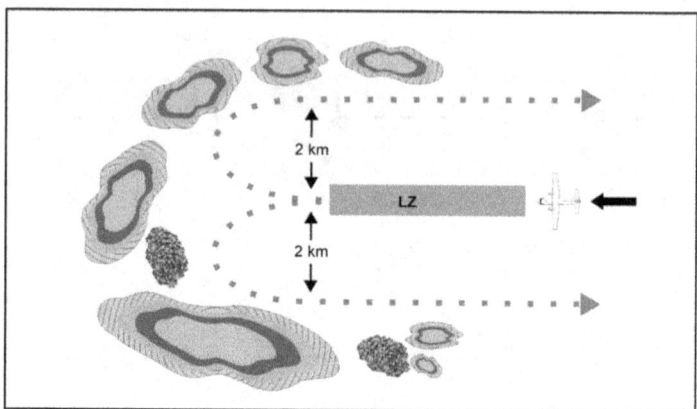

Figure 4-13. Level turning radius required for light and medium STOL aircraft

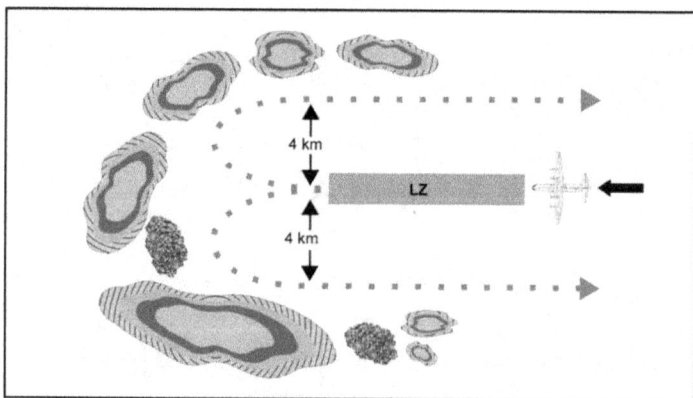

Figure 4-14. Level turning radius required for heavy aircraft

Approach and Takeoff Clearance

4-75. The approach and takeoff clearances are based on the following:
- Descent and ascent characteristics of the aircraft.
- Obstacle height.
- Proximity of obstacles to the LZ.

4-76. The descent and ascent ratio, or the so-called glide and climb ratio, is the proportion of aircraft gain or loss of altitude to distance traveled. For example, a 11:1 ratio means that for every 11 feet of distance traveled, there is a 1-foot gain or loss of altitude. The ratio for—
- Light STOL aircraft is 11:1 (U-10, AU-23A/Porter, and UV-18A).
- Light aircraft is 11:1 (C-206, C-207, and C-208).
- Medium aircraft is 40:1 (C-27 and CASA-212).

Landing Zones

> *Note.* CASA-212 has a 11:1 descent ratio and a 42:1 ascent ratio.

4-77. The descent and ascent ratios are applied as 1 unit of vertical clearance to 11 (or 42) units of horizontal clearance to determine the horizontal clearance required—
- Between obstacles and the approach and takeoff ends of an LZ.
- Between LZ markers and terrain and obstacle masks.

4-78. Natural or man-made obstacle height is measured from ground level at the center of the LZ. Where land falls away from the site, obstacles that do not cut the line of aircraft descent or ascent may be disregarded. This condition is most likely to exist in mountainous areas where plateaus are selected for a landing site (Figure 4-15, page 4-22).

Layout and Dimensions of Landing Zones

4-79. For UW operations, LZs are classified into just two categories—light and medium STOL and medium. If unsure of the classification of the incoming aircraft, establish a medium LZ.

4-80. *Runway Length.* The wartime minimum runway length for STOL aircraft is 458 meters and 800 meters for medium aircraft with no more than a 50-foot obstacle at either end of the LZ. These lengths do not include the 10 percent safe areas at each end of the runway and are insufficient for peacetime and training conditions (Figure 4-11, page 4-18, and Figure 4-12, page 4-19).

4-81. *Safe Area.* A safe area (cleared surface) must be added to each end of the LZ. This safe area is a distance equal to 10 percent of the minimum length. This area must support the weight of the aircraft and will never be less than 46 meters (151 feet) long for light STOL aircraft nor less than 100 meters for medium aircraft.

4-82. *Runway Width.* The 19-meter width, shown in Figures 4-11 and 4-12, depicts the minimum LZ width required based on aircraft wheelbase. STOL and medium aircraft wingspans may extend beyond the 19-meter width. Personnel must be cautious at LZ markers. They must be aware of the size and characteristics of the aircraft to include the following:
- Wingspan.
- Number, type, and location of engines.
- Turning radius.
- Specifics of landing gear position and width.

The marking party should use all the area available for LZ layout, which can extend the LZ width out to 25 meters for STOL and 46 meters for medium aircraft. LZ widths will not extend beyond these widths, since the pilot cannot establish good horizontal perception of the runway to ensure a safe landing.

4-83. *Shoulder or Clear Zone.* For LZs used by SOF, the shoulder and clear zone are combined and referred to as the safety strip. Safety strip dimensions are calculated as follows:
- *Light and Medium STOL Aircraft.* Add a 10-meter wide safety strip along both sides of the LZ for wing tip clearance. This strip must be clear of all obstacles over 1 meter high (Figure 4-16, page 4-23).
- *Medium Fixed-Wing Aircraft.* Add a 24-meter wide safety strip for wing tip clearance along both sides of the LZ. This safety strip must be cleared to a maximum of 0.6 meters for the first 14 meters of this safety strip. The remaining 10 meters must be cleared to a maximum height of 2 meters (Figure 4-16).

4-84. *Marker Placement.* Markers and lights shall not be manned within 20 meters of the LZ centerline on medium aircraft LZs. Under all circumstances, the markings will outline the usable area of an LZ (that area meeting weight-bearing criteria). The marking pattern outlining the limits of the LZ consists of the following:
- Wartime: five marker stations for STOL aircraft.
- Peacetime: markings as in Figure 4-10, page 4-17.

Chapter 4

- Medium aircraft: seven or more marker stations. For LZs longer than the depicted minimum length, the number of markers is increased by placing additional markers between Stations D and E at 200-meter intervals.

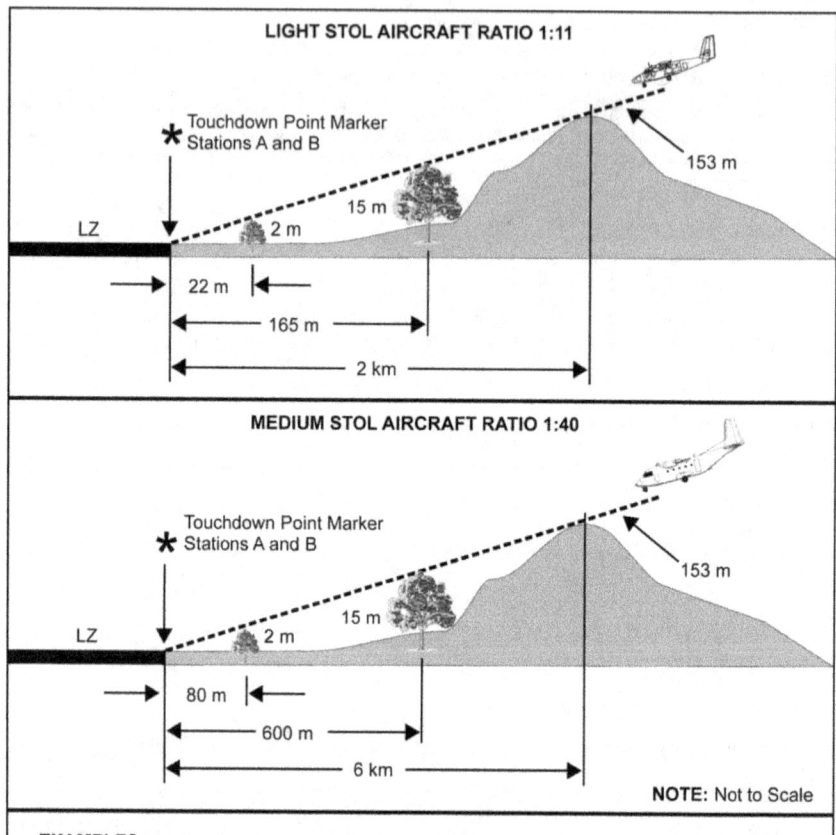

EXAMPLES:

A 15-m tree is in the approach path. A STOL aircraft has a descent and ascent ratio of 11:1. Therefore, the approach end of the LZ would be located 165 m (11 x 15) from the tree.

A 2-m obstacle is in the approach path. The approach or takeoff path must not be closer than 22 m (11 x 2) for STOL aircraft and 80 m (40 x 2) for medium aircraft.

A 15-m obstacle is in the approach path. The approach or takeoff path must not be closer than 165 m (11 x 15) for STOL aircraft and 620 m (40 x 15) for medium aircraft.

A 153-m obstacle is in the approach path. The approach or takeoff path must not be closer than 2 km (153 x 11) for STOL aircraft and 6 km (153 x 40) for medium aircraft.

Figure 4-15. Obstacle clearance

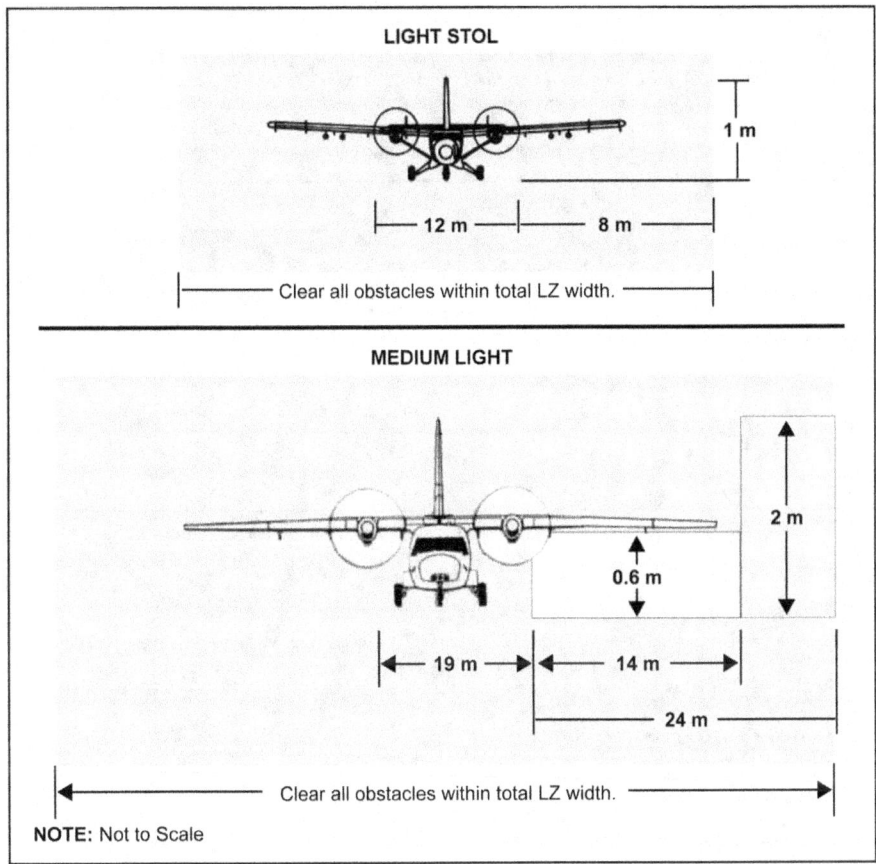

Figure 4-16. Light and medium STOL aircraft wing-tip clearance

4-85. *Marker Placement.* Markers and lights shall not be manned within 20 meters of the LZ centerline on medium aircraft LZs. Under all circumstances, the markings will outline the usable area of an LZ (that area meeting weight-bearing criteria). The marking pattern outlining the limits of the LZ consists of the following:
- Wartime: five marker stations for STOL aircraft.
- Peacetime: markings as in Figure 4-10, page 4-17.
- Medium aircraft: seven or more marker stations. For LZs longer than the depicted minimum length, the number of markers is increased by placing additional markers between Stations D and E at 200-meter intervals.

4-86. Stations A and B always mark the downwind end and provide the entrance "gate" for aircraft approach. These stations represent the first point at which the aircraft should touch the ground. Station B is the aircraft off-loading or onloading position. The RCL station should be placed 15 meters (49 feet) to the left of Station B as viewed from the landing aircraft. Station C marks the very last point at which the aircraft can touch down and still complete a safe landing. Stations D and E (Figure 4-11, page 4-18) or F and G (Figure 4-12, page 4-19) mark the upwind extreme of the landing area.

Chapter 4

4-87. *Extended STOL LZ*. If a STOL LZ is long enough, the pilot may stop at the D and E panels, off-load and onload personnel or equipment, and continue takeoff in the same direction. An additional 550 meters plus the safe area (10 percent) must be available beyond the D and E panels. This additional length will be marked at the end of the LZ with two additional panels, F and G (Figure 4-17). The minimum length of the addition is equal to the minimum length of the original LZ to include temperature and elevation changes and safety area. This added area must meet the same criteria as the original LZ. This technique is especially useful when using highways or roadways as LZs.

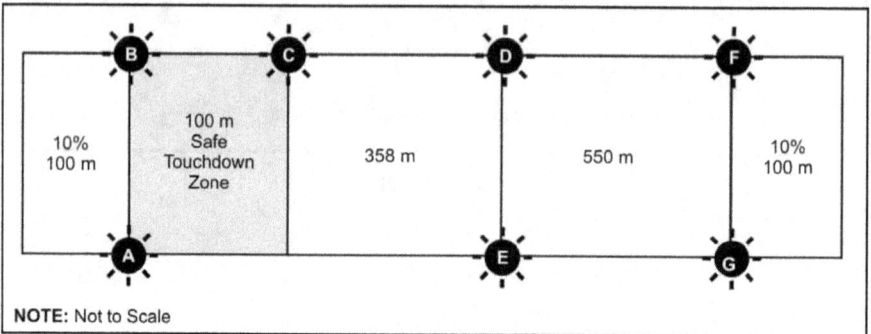

Figure 4-17. Extended STOL LZ for one-direction landing and takeoff of light and STOL aircraft

Crosswinds

4-88. Ground crosswind velocities are difficult to predict. Accurate reconnaissance prevents selecting LZs that are not usable because of prevailing winds. The pilot of each mission aircraft is the final authority on the crosswind limits of his aircraft and on what level of crosswinds he can manage.

SETUP OF LANDING ZONES

4-89. Securing and establishing an LZ is accomplished in much the same way as it is for a DZ in the UW environment. A reception committee is required for an LZ but is an option on the DZ. The reception committee is organized and functions the same as discussed in Chapter 3. The only significant changes are in marking the LZ.

LANDING ZONE MARKERS

4-90. Ground personnel use visual markers to outline the limits of the landing strip, to indicate landing direction, and to identify the RCL station. During darkness or periods of limited visibility, ground personnel use visible light sources such as L32 field markers, IR lights, improvised light sources (flashlights and fuel pots), or beacons. During daylight, ground personnel use standard or improvised panels or beacons. LZ markers are the same as those used for DZs. Ground personnel position the markers so they can be seen from the approaching aircraft. Flashlights, when used, must be handheld to ensure directional control. The landing direction is always indicated by the row of marker stations aligned along the left edge of the strip and by the RCL signal station that is always on the approach or downwind end. Each marker should be manned so that it can be extinguished once the aircraft passes it. Figures 4-10, 4-11, 4-12, 4-17, and Figure 4-18, page 4-25, explain the placement of markers.

Display

4-91. The marking team displays LZ markings for 4 minutes. They begin displaying the LZ markings 2 minutes before until 2 minutes past the scheduled aircraft arrival time or until the aircraft completes touchdown and landing roll.

Landing Zones

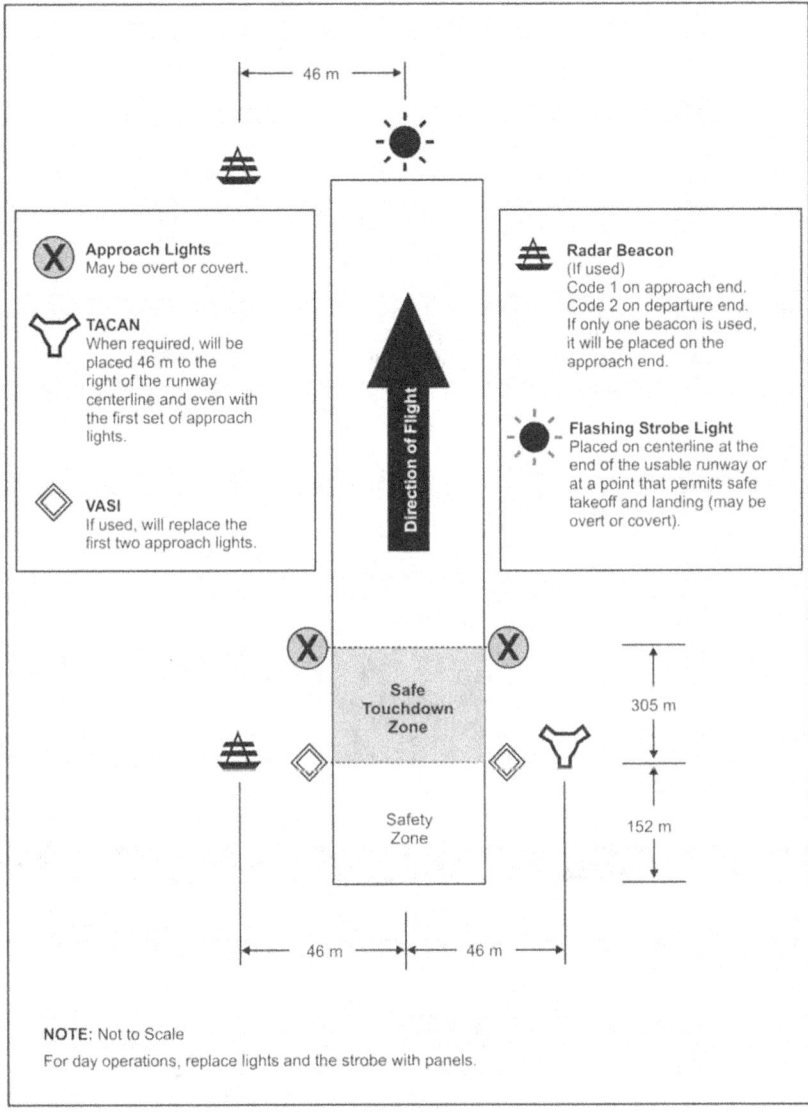

Figure 4-18. Land LZ marked with beacons

Identification

4-92. Identification is how the aircrew and the reception committee identify themselves to one another. Identification means may include time, approach, and marking pattern.

Chapter 4

4-93. *Air-to-Ground.* The aircraft is identified to the reception committee by—
- Arriving in the objective area within the specified time limit, usually 2 minutes before to 2 minutes after scheduled drop time.
- Approaching at designated drop altitude and track.

4-94. *Ground-to-Air.* The reception committee is identified to the aircraft by—
- Displaying the correct marking pattern within the specified time limit.
- Using the proper authentication code signal.
- Setting the proper code on the RT or LTD.

Authentication

4-95. There is no standard authentication system for UW reception operations. The authentication system to be used is agreed upon by the commanders concerned during mission planning. Authentication between the aircraft pilot or navigator and the RCL may be achieved by using a coded light source, colored light, North Atlantic Treaty Organization code letter, signal panel, radio contact, RT, or combinations thereof. These may be employed individually or with the marking pattern. There should always be a primary and alternate means of authentication.

4-96. *Morse Code.* When using international Morse code light signals, personnel will not use code letters identified by all dots or dashes—I, E, M, O, S, T, and H. The following time intervals will be used to assist pilot or navigator recognition:
- 2 seconds for dots.
- 4 seconds for dashes.
- 2 seconds for intervals between dots and dashes.
- 5 seconds for intervals between repetitions.

4-97. *RTs or LTDs.* During mission planning and when RTs or LTDs are used for authentication, the ground and air commanders concerned decide how, when, and where to use the RTs or LTDs. They jointly agree upon positioning, codes, and turn-on or turn-off times during mission planning.

4-98. *Standard Aircraft Signals.* Table 4-3 lists some of the standard air traffic control light signals that may be used to communicate with the pilot. Use of these signals should be coordinated with the pilot before the mission.

Table 4-3. Standard ATC light signals

Signal	Aircraft on the Ground	Aircraft in the Air
Steady Green	Cleared for takeoff	Cleared to land
Flashing Green	Cleared for taxi	Return for landing
Steady Red	Stop	Give way to other aircraft, and continue circling
Flashing Red	Taxi clear of runway	Field unsafe; do not land
Flashing White	Return to starting point	NA
Alternate Red and Green	Use extreme caution	Use extreme caution

Actions on the LZ

4-99. Depending on the form of signaling used between the LZ and the aircraft, authentication may occur before identification, but normally identification occurs first and then authentication occurs. The RCL directs marking teams to display all marker stations 2 minutes before the scheduled arrival time. The RCL displays the appropriate authentication signal into the direction of expected approach of the aircraft. If planned, the aircrew may acknowledge the code light. When the RCL determines the aircraft is on its final approach (within 15 degrees to either side of the approach track and below 1,000 feet), he ceases the code signal. The pilot tries to make a straight-in landing on the initial approach. If he cannot do so because of a sudden change in wind direction or conditions, he flies a modified landing pattern at minimum altitude for security purposes.

4-100. *Aircraft Flying With NVGs.* If the pilot is using NVGs, once the code signal is ceased, only the marking lights for the LZ remain lit. The pilot will land by following the marking lights and using NVGs. No lights are aimed at a plane whose aircrew is using NVGs.

4-101. *Aircraft Flying Without NVGs.* If the pilot does not have NVGs, once the RCL ceases the code signal, the RCL aims his light in the direction of the landing aircraft. If the aircraft must make a "go-around," the light follows the aircraft. The light continues to follow the aircraft during touchdown and landing roll. The reception committee extinguishes the marker lights as the aircraft passes each successive marker station.

4-102. The pilot will not land his aircraft when—
- There is a lack of proper identification or authentication received from the LZ.
- The RCL gives an abort signal; for example, extinguishing the LZ markings.

4-103. After touchdown and landing roll, the pilot executes a right turn if possible. Stations A and B shine a solid light toward the aircraft to guide the pilot in taxiing the aircraft to the takeoff position. The pilot keeps the engines running during the entire operation.

4-104. Incoming personnel and materiel are off-loaded first to preclude confusion and ensure rapid handling. To ensure safety, all off-loading and onloading is done behind the running engine(s). The aircraft should always be approached from the rear.

4-105. The pilot prepares the aircraft for immediate takeoff after the off-loading and onloading are completed. The RCL moves to a vantage point clear of the aircraft, directs the LZ be illuminated, and signals the pilot to take off by flashing his light toward ground level in front of the aircraft. The reception committee extinguishes LZ illumination as soon as the aircraft is airborne.

WATER LANDING ZONE (SINGLE- AND TWIN-ENGINE AIRCRAFT)

4-106. During an operation using a water LZ, a seaplane will land on the designated body of water and taxi to the reception party located on or near the shore, onload or off-load supplies and personnel, and take off. Aircraft that can land on water are divided according to several criteria. For the purpose of this manual, these aircraft are classified as single-engine or twin-engine seaplanes. Following is guidance on how to select, lay out, mark, and operate water LZs when working with single- or twin-engine seaplanes.

DIMENSIONS AND LAYOUT

4-107. Ground personnel select the water LZ on the characteristics and limitations of the aircraft. As a general rule, any body of water that is 1,000 meters long, after considering all the other factors such as climb ratios, depth, surface conditions, and so on, is acceptable. The landing and takeoff distance for water aircraft varies depending on the type of aircraft. Most single-engine seaplanes can land in 200 to 300 meters, but they require a minimum of 300 meters for takeoff. The twin-engine seaplane can land in 300 to 400 meters but may require more than 600 meters for takeoff.

4-108. If the distance is shorter than 1,000 meters, ground personnel notify higher HQ of the distance available. The detachment requests the distances required for the aircraft so they can compute the LZ length. The detachment must consider that high altitudes and extreme temperatures require lengthening a water LZ. The detachment determines the additional length in the same manner as described for land LZs. Additionally; the detachment adds a safe area to each end of the LZ, a distance equal to 10 percent of the minimum length. These areas will never be less than 30 meters long.

LOCATION OF THE RECEPTION COMMITTEE

4-109. Ground personnel or mission planners plan the location of the reception committee when determining the location and layout of the LZ. The ideal situation would be for the plane to land parallel to

Chapter 4

the shore, turn into a small cove with a beach, onload or off-load, and then depart by continuing into the wind and taking off. When deciding on the location of the reception committee, mission planners or ground personnel consider the following:

- The pilot can see the signal on his approach.
- The pilot can bring his plane into or close to the shore. A sandy beach is ideal.
- The location is secluded and secure. Preferably, the location is unobservable from most of the shoreline.
- Movement to and from the site is not restricted so that the reception committee can withdraw from the area quickly.
- The prevailing wind direction does not force the pilot to do an excessive amount of taxiing.

WATER SURFACE AND DEPTH

4-110. The water surface must be free of obstructions such as boulders, rock ledges, shoals, sunken pilings, logs, moored craft, floating debris, or seaweed. The minimum safe water depth for single-engine aircraft is 1 meter. The minimum safe water depth for twin-engine aircraft is 3 meters.

WEATHER CONDITIONS

4-111. Surface wind conditions are critical for water landings. Crosswinds are difficult to predict, but accurate reconnaissance can preclude choosing an LZ that is inconsistent with the prevailing winds. The pilot of each mission aircraft is the final authority on the crosswind limits of his aircraft and his ability to manage his aircraft in crosswinds.

4-112. Any direction may be used for landing or takeoff in winds less than 8 knots. The landing may vary up to 15 degrees from the wind direction when surface winds do not exceed 8 knots and it is impossible to land directly into the wind. The landing must be made into the wind when surface winds exceed 8 knots on open water. Landings will not be made in winds exceeding 20 knots. If a downwind landing or takeoff is absolutely required, it will be made directly downwind.

WAVE HEIGHT

4-113. The maximum wave height is 0.3 meter for single-engine aircraft. Maximum wave height for a twin-engine aircraft is 1 meter. The height of surface swells must not exceed 0.3 meter, and the wind wave must not be more than 1 meter for twin-engine aircraft when all swells and wind waves are in phase. The state of the tide should have no bearing on the suitability of the landing area. However, the low-tide depth must exceed the minimum safe water depth required for the mission aircraft.

WATER AND AIR TEMPERATURE

4-114. Water and air temperatures vary with the type of water. Because of the danger of icing, water and air temperatures must conform to the minimums indicated in Table 4-4.

Table 4-4. Water and air temperature

Water Type	Water Temperature (Degrees in Fahrenheit)	Air Temperature (Degrees in Fahrenheit)
Salt Water	+18	+26
Fresh Water	+35	+35
Brackish Water	+30	+35

Landing Zones

STATE OF THE SEA

4-115. The state of the sea is the state of agitation of the sea resulting from various factors such as wind, swell, currents, angle between swell and wind, and so on. Table 4-5 is an extract from the 1974 Edition of the *World Meteorological Organization Manual on Codes*, Number 306, Volume I, International Codes.

Table 4-5. State of sea (Code Table 3700) extract

Code Figure	Descriptive Terms	Height in Meters
0	Calm (Glassy)	0
1	Calm (Rippled)	0 to 0.1
2	Smooth (Waveless)	0.1 to 0.5
3	Slight	0.5 to 1.25
4	Moderate	1.25 to 2.5
5	Rough	2.5 to 4
6	Very Rough	4 to 6
7	High	6 to 9
8	Very High	9 to 14
9	Phenomenal	Over 14

Notes.
1. The values for height in meters refer to well-developed wind waves of the open sea. While priority shall be given to the descriptive terms, these height values may be used for guidance by the observer when reporting the total state of agitation of the sea resulting from various factors such as wind, swell, currents, angle between swell and wind, and so on.
2. The exact bounding height shall be assigned for the lower code figure. For example, a height of 4 m is coded 5.

APPROACH AND TAKEOFF CLEARANCE

4-116. Water LZs require approach and takeoff clearances the same as land LZs. The ratio for single-engine aircraft is 11:1, and the ratio for twin-engine aircraft is 40:1.

MARKING

4-117. The reception committee uses visible light sources during darkness or panels in daylight to mark their location and the LZ location. Position the marker on the shore so that it is visible to the pilot as he flies his approach, or the marker may be placed in a boat or secured to a flotation device just offshore. If positioned in a boat, the marker must be handheld and the boat must maintain position. Two persons are usually required in the boat—one to maintain station position and the other to signal. If the marker is attached to flotation devices, anchor the flotation device to prevent drifting. In deep or rough water, use improvised sea anchors. Light sources will be at least 0.3 meter above the surface of the water to prevent waves from causing a blackout.

4-118. Any hazards in the water should be marked by using a different color light or panel. The pilot should be notified in advance which markers to avoid.

DISPLAY

4-119. The marking teams display DZ markings for 4 minutes. They begin displaying the DZ markings 2 minutes before and until 2 minutes past the scheduled aircraft arrival time or until the aircraft completes touchdown and landing roll.

Chapter 4

AUTHENTICATION AND IDENTIFICATION

4-120. Authentication and identification procedures can be the same as those described for land LZs or a variation.

ACTIONS ON THE WATER LANDING ZONE

4-121. Actions on the water DZ are dependant on whether or not the seaplane can come to the shore or must stay offshore. The following procedures should be used if the seaplane is going to come to the shore:
- After the plane lands, the pilot will taxi to the shore.
- If any additional authentication is planned, the RCL and pilot will exchange authentication before the plane touches shore.
- After the plane beaches, personnel and equipment are off-loaded.
- After everyone and everything is off-loaded, anything to be exfiltrated is loaded onto the aircraft.
- During the off-loading and onloading, the RCL briefs the pilot of any hazards or additional information that he may need.
- The plane then departs.

The following procedures should be used if the seaplane is unable to beach and the off-loading and onloading must be done by boat:
- The RCL stations a boat containing himself and personnel and equipment to be exfiltrated just offshore from the remainder of the reception committee.
- After the plane has landed, the pilot turns the aircraft toward the RCL boat.
- The RCL guides the aircraft to his position by shining a continuous light beam in the direction of the taxiing aircraft or displaying a light on his boat. He must take care not to blind the pilot with this light.

4-122. Depending on prior coordination and the plan, the boat may come to the aircraft or the aircraft may come to the boat. It is critical that the RCL and pilot coordinate this before the landing.

Stationary Aircraft

4-123. The pilot holds the aircraft into the wind at minimum speed. The pilot then lets the RCL boat maneuver into position alongside the left door.

Stationary RCL Boat and Single-Engine Aircraft

4-124. The RCL boat remains stationary while the pilot taxis the aircraft 15 to 30 meters from the boat where the aircrew releases a buoyant dragline from the left door. The dragline is about 19 meters long and has a flotation device attached to its end. The flotation device has a small marker light for night operations. The pilot taxis the aircraft to the left around the RCL boat to position the dragline close enough to be secured to the boat. The RCL maintains his light either on his position or the aircraft at all times to permit the pilot to position the dragline and keep a safe distance between the RCL boat and the propeller. If the pilot loses sight of the RCL light, he may turn on a landing light immediately and keep it directed downward toward the water. The pilot continues a left turn until the dragline has been secured to the RCL boat. Personnel in the boat pull and secure the boat alongside the left float and begin off-loading or onloading personnel and cargo.

> **CAUTION**
> Do not allow the boat to drift forward of the aircraft door where it could be struck by the propeller. Upon completing off-loading and onloading, back the boat and move it away from the aircraft.

Stationary RCL Boat and Twin-Engine Aircraft

4-125. The RCL boat remains stationary during the operation. The pilot taxis the aircraft 15 to 30 meters from the boat. At this point, the aircrew releases a buoyant dragline from the left door. The dragline is about 45 meters long and has three flotation devices attached as follows: one about 15 meters from the aircraft, a second at midpoint, and a third on the extreme end of the line. When used for night operations, the flotation devices have small marker lights. The pilot taxis the aircraft to the left around the RCL boat to position the dragline close enough to be secured. Once the line is secured to the boat, personnel in the boat will not try to pull on the line because of the danger of swamping the boat. The aircrew then pulls the boat to the door of the aircraft.

Note. Should the boat drift by the aircraft door toward the running engine, all personnel must immediately abandon the boat when it passes under the trailing edge of the wing.

4-126. Personnel and cargo are loaded on the aircraft first and then the incoming personnel and supplies are loaded into the RCL boat. The aircrew receives any information that will aid in the takeoff after completing the off-loading and onloading. The RCL boat then moves backwards to a safe vantage point. The RCL signals the pilot all clear for takeoff by flashing his light toward the waterline in front of the aircraft. The reception committee extinguishes all markers as soon as the aircraft is airborne.

4-127. The pilot will not land the aircraft when—
- There is a lack of proper identification or authentication received from the LZ.
- The RCL gives an abort signal; for example, extinguishing the LZ markings.

SNOW LANDING ZONES

4-128. The procedures for ski plane operations described in the following paragraphs apply only to STOL-type aircraft equipped with wheel or spring skis (Figure 4-19, page 4-32).

4-129. Almost any snow-covered field or frozen lake of the proper size makes an acceptable ski LZ. The minimum ice thickness for a ski landing is 48 centimeters. The minimum snow depth is 3 centimeters on ice and 10 centimeters on a hard surface. The RCL is responsible for determining that these minimums exist. If these minimums do not exist, the RCL aborts the operation.

4-130. The ski plane approach and takeoff clearances are identical to those of land LZs. They are based on the same descent and ascent ratios as for land LZs. The standard marking, display, and authentication procedures used are the same as those described for land LZs (light and medium STOL and medium aircraft).

4-131. Depth perception is usually very poor during landings on a large snow-covered surface. Therefore, the reception committee uses small bushes to augment the marking pattern and to assist pilot depth perception.

4-132. The maximum crosswind velocity for a landing or takeoff on skis is 10 knots. More power is required to start the aircraft moving when it is on skis. If the skis are slightly frozen to the snow, the maximum power of the aircraft may be needed to move the aircraft.

4-133. The takeoff and landing distances for a wheel and ski plane operating on snow will vary according to snow conditions that are difficult to define. Figure 4-19 shows the minimum dimensions required for a typical light and medium STOL snow LZ. These minimum dimensions depend on—
- Snow and surface conditions.
- Weather.
- Weight of aircraft.

Chapter 4

Note. The pilot should leave the engines running to enhance aircraft security. However, all ground personnel must take extra safety precautions. The RCL plans for the dispersal or withdrawal of personnel or cargo in case of enemy interference. All elements involved in the mission should carefully coordinate these plans and conduct practice withdrawals or dispersions, if necessary.

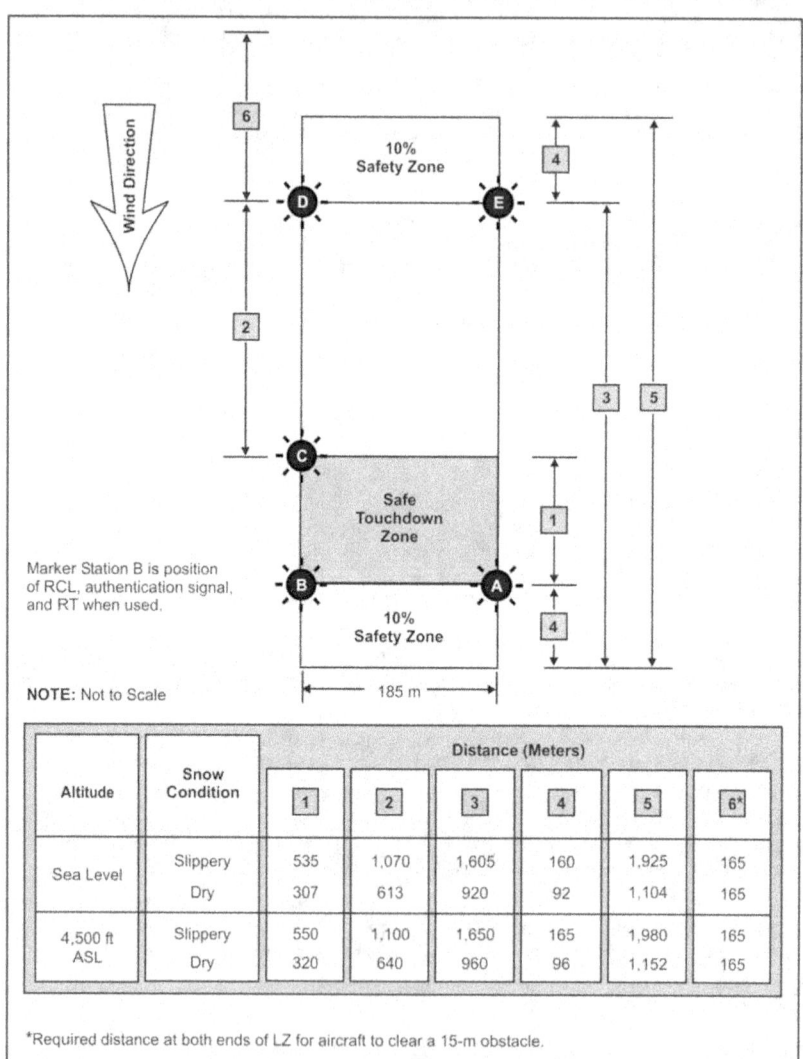

Altitude	Snow Condition	Distance (Meters)					
		1	2	3	4	5	6*
Sea Level	Slippery	535	1,070	1,605	160	1,925	165
	Dry	307	613	920	92	1,104	165
4,500 ft ASL	Slippery	550	1,100	1,650	165	1,980	165
	Dry	320	640	960	96	1,152	165

*Required distance at both ends of LZ for aircraft to clear a 15-m obstacle.

Figure 4-19. Snow LZ (STOL aircraft)

Chapter 5

Army Special Operations Aviation Units and Aircraft

The SOAR has SO rotary-wing aircraft, including the AH/MH-6M Little Bird, the MH-60L/K Blackhawk, and the MH-47E/G Chinook. SOAR units can conduct and support SO missions for the ARSOF commander or for the theater special operations command (TSOC). The SOAR can be task-organized based on expected missions, the requirements of the units being supported, the environmental conditions in the theater of operations, and sustainment requirements. The SOAR task-organizes around one of the SOA battalions. With proper personnel and equipment augmentation, the SOAR battalion commander and his staff could also serve as a joint special operations air component commander. When two or more battalions are required in the theater, the regimental commander could serve as the joint SO air commander.

UNIT ORGANIZATION

5-1. The SOAR supports other SOF units by conducting special air operations in all operational environments. The specially organized, trained, and equipped aviation units give the joint force special operations component commander (JFSOCC) the capability to infiltrate, resupply, and exfiltrate SOF elements engaged in all core tasks, missions, and environments.

MISSION

5-2. The SOAR mission is to conduct and support special air operations by clandestinely penetrating hostile and denied airspace. SOAR units can operate in harsh environments and across the full spectrum of operations. They also support SOF in conducting joint, multinational, interagency, liaison, and coordination activities in support of the USSOCOM commander and the GCC's concept of operations. The participation of the SOAR in the ARSOF core tasks varies based upon the type of conflict, the environment, and the scope of the operation. As a direct reporting unit to the United States Army Special Operations Command (USASOC), the SOAR organizes, equips, trains, validates, sustains, and employs assigned aviation units for USASOC missions.

MH-6M HELICOPTER

5-3. The primary mission of the MH-6M helicopter is to conduct overt and covert infiltration, exfiltration, and combat assaults over a wide variety of terrain and environmental conditions. The MH-6M also performs command and control (C2) and reconnaissance missions.

DESCRIPTION

5-4. The MH-6M is a light assault helicopter (Figure 5-1, page 5-2). It is a single-engine, light utility helicopter modified to transport up to six combat troops and their equipment externally. Its small size allows for rapid deployability in C-130, C-17, and C-5 transport aircraft. Aircraft modifications and aircrew training allow for extremely rapid upload and download times.

Chapter 5

Figure 5-1. MH-6M helicopter

AIRCRAFT SURVIVABILITY EQUIPMENT

5-5. Each aircraft has the APR-39 Radar Warning Receiver (RWR). This passive omnidirectional warning set detects and identifies hostile search and acquisition and fire control radar. It provides audio and visual alerts to the flight crew.

STANDARD MISSION EQUIPMENT

5-6. Some aircraft have the forward-looking infrared (FLIR), a passive radar system that provides an infrared image of terrain features and ground or airborne objects of interest. A standard videocassette recorder can play back recorded images.

5-7. The MH-6M can have two Goliath tanks installed as an internal auxiliary fuel system (IAFS). The tanks provide 62 additional gallons of fuel each. Each tank adds approximately 90 minutes of flight time.

ARMAMENT

5-8. The MH-6M has no standard armament.

SPECIAL MISSION EQUIPMENT

5-9. Personnel can rapidly configure the aircraft for fast-rope operations. Motorcycle racks provide the capability to insert and extract up to two motorcycles.

TRANSPORTABILITY OF MH-6M AIRCRAFT

5-10. AC-130 can carry 3 MH-6Ms, a C-17 can carry 9, and a C-5 can carry 21. In each case, tactical uploading and downloading of the aircraft can take place in an extremely short time.

PLANNING CONSIDERATIONS

5-11. The following paragraphs discuss considerations that must be taken into account when planning to use MH-6M aircraft in a mission.

Weather Minimums

5-12. A minimum 500-foot ceiling and a 2-mile visibility capability must exist for day and night flying over all types of terrain. The unit commander may reduce weather minimums on a case-by-case basis.

5-13. A visible horizon must exist in two of the four horizontal quadrants at all times. All MH-6M missions must take place under VMC. Instrument flight rule flights are unauthorized.

Winds

5-14. The maximum wind allowed to start the aircraft is 40 knots, with a 20-knot gust spread.

Flight Altitudes

5-15. For training missions, the minimum en route altitude for routes not reconnoitered is 300 feet AGL. The minimum overwater altitude is 50 feet. For operational missions, the minimum en route altitude is dependent upon METT-TC.

Landing Areas

5-16. The MH-6M can land on any structure that allows clearance for the rotor systems (30 feet) and meets stress requirements. Single-aircraft confined landing areas require a minimum size of 50 feet by 50 feet.

Shipboard Operations

5-17. The MH-6M can operate day or night from any ship with at least a one-spot helicopter-landing capability.

Aircrew Composition

5-18. The normal aircrew for most training exercises and operational or contingency missions consists of a pilot and a copilot. All overwater flights require a pilot and copilot current and qualified in overwater flight. All aircrews can conduct NVG infiltration and exfiltration, stabilized body operations, FRIES, and aerial suppression operations to urban, mountainous, desert, and jungle objectives, as well as to ships and offshore drilling platforms. Aircrews have training in long-range precision navigation and formation flight over land and water to arrive at objectives at a prearranged time (+30 seconds).

Aircraft Capabilities

5-19. Table 5-1, pages 5-3 and 5-4, lists the capabilities of the MH-6M aircraft. Figures 5-2 and 5-3, page 5-4, illustrate specific dimensions of the aircraft.

Table 5-1. MH-6M aircraft capabilities

Aircraft Weight	
Basic weight	1,925 pounds
Mission weight	3,100 pounds
Maximum gross weight	3,750 pounds
Aircraft Dimensions	
Length blades folded	22 feet 6 inches
Blades unfolded	32 feet 1 inch
Width blades folded	6 feet 5 inches
Blades unfolded	27 feet 4 inches
Height	8 feet 11 inches
Diameter of main rotor	27 feet 4 inches
Aircraft turning radius	36 feet 9 inches

Chapter 5

Table 5-1. MH-6M aircraft capabilities (continued)

Fuel Tank	Range and Endurance at 240 Pounds Per Hour			
	Endurance (Hours + Minutes)	Aircrew	Passenger	Fuel Range (Nautical Miles)
Main	1+20	2	3	110
Main plus one auxiliary	3+00	2	2	240
Main plus two auxiliaries	3+50	2	1	310
Airspeed. The cruise airspeed for the MH-6M helicopter is 80 knots.				

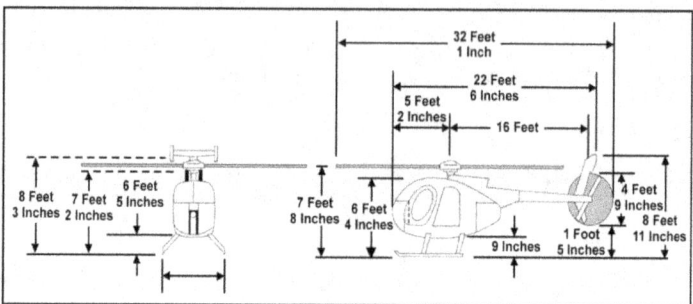

Figure 5-2. MH-6M and AH-6M aircraft dimensions

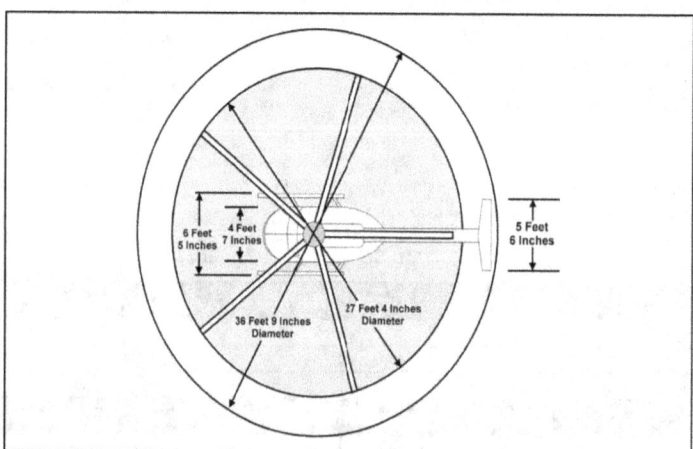

Figure 5-3. MH-6M and AH-6M aircraft dimensions and turning radius

SAFETY

5-20. The MH-6M has no seat belts installed for passengers. Each passenger must provide his own means of securing himself. A short length of rope wrapped and knotted around the waist with a snap link attached to one end allows each passenger to secure himself to hard points on the aircraft.

AH-6M HELICOPTER

5-21. The mission of the AH-6M helicopter is to provide a rapidly deployable light attack helicopter to meet the need for precise, small-area target destruction or neutralization, with provisions for close-air fire support for ground assault operations.

DESCRIPTION

5-22. The AH-6M is a highly modified version of the McDonnell Douglas 530-series commercial helicopter (Figure 5-4). The aircraft is a single turbine-engine, dual-flight-control, light attack helicopter. Its primary employment is CAS of ground troops, target destruction raids, and armed escort of other aircraft.

Figure 5-4. AH-6M helicopter

AIRCRAFT SURVIVABILITY EQUIPMENT

5-23. Each aircraft has the APR-39 RWR. This passive omnidirectional warning set detects and identifies hostile search and acquisition and fire control radar. It provides audio and visual alerts to the flight crew. Additional survivability equipment includes fire extinguishers, underwater beacons, first-aid kits, and survival kits.

STANDARD MISSION EQUIPMENT

5-24. Some aircraft have FLIR, a passive radar system that provides an infrared image of terrain features and ground or airborne objects of interest. A standard videocassette recorder can play back recorded images.

5-25. The AH-6M can have two Goliath tanks installed as an IAFS. The tanks provide 62 additional gallons of fuel each. Each tank adds approximately 90 minutes of flight time.

SPECIAL MISSION EQUIPMENT

5-26. Personnel may configure the MH/AH-6M to employ the family of loudspeakers—aircraft configuration (FOL-AC) for PSYOP missions.

Chapter 5

WEAPONS SYSTEMS

5-27. The AH-6M uses the plank system (Figure 5-5), which features detachable, foldable outboard store stations. The system permits simplified aircraft transportability. Because of the flexibility of the plank system, numerous configurations of weapons systems are possible. The M-27 system is a single minigun mounted on the left side of the AH-6M. It has a maximum of 1,500 rounds loaded in the ammunition can. The HGS-17 system is a single rocket pod (7- or 19-shot) mounted on the right side. The standard plank configuration for an AH-6M aircraft is two miniguns and two seven-shot rocket pods.

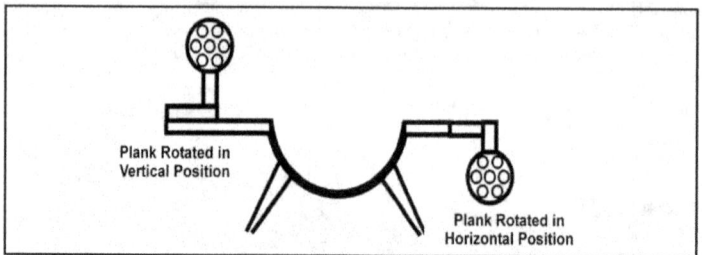

Figure 5-5. AH-6M plank system for aircraft weapons configurations

5-28. Provisions are available on the AH-6M plank system to mount and fire the following systems:
- M134, 7.62-millimeter (mm) minigun, 2,000- or 4,000-round-per-minute rate of fire (Figure 5-6). The ammunition can holds a maximum of 2,625 rounds of ball, tracer, low-light tracer, or Sabot-launched armor-piercing (SLAP) ammunition. Each aircraft uses two ammunition cans. The normal load is 1,500 to 2,000 rounds per gun. The mounting site for miniguns is normally on the inboard stores.

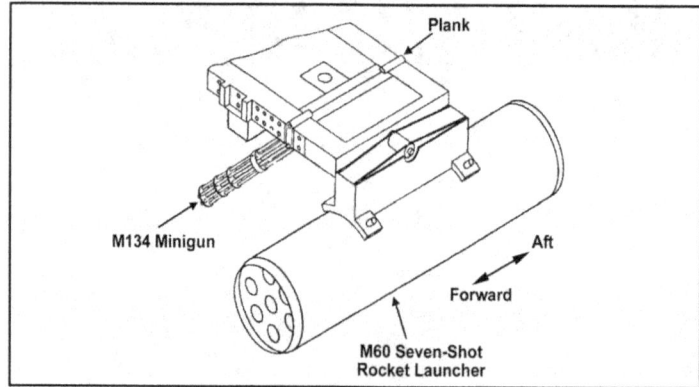

Figure 5-6. AH-6M weapons variation

- M260 rocket launcher, seven-shot 2.75-inch folding-fin aerial rocket (FFAR) pod. This system can fire both Mark 40 and Mark 66 rocket motors and numerous warheads, including flare, infrared flare, chlorobenzaimalononitrile, flechette, 17-pound high-explosive dual-purpose (HEDP), 17-pound high-explosive (HE) proximity, white phosphorus, 10-pound HEDP, smoke, and inert. The mounting site for rocket pods is normally on the outboard stores.

- M261 rocket launcher, 19-shot 2.75-inch FFAR rocket pod. All other data on the M261 matches the M260 rocket launcher.
- Hellfire missile launchers attach to the plank system in pairs. The mounting site is on the outboard stores. Each launcher can hold two missiles, for a total of four missiles. The AH-6M has a Hellfire missile laser designator mounted on the aircraft.

5-29. Some of the optional configurations are as follows:
- Two seven-shot rocket pods.
- One minigun, with two seven-shot rocket pods.
- Two 19-shot rocket pods.
- One minigun and two Hellfire missiles.
- Four Hellfire missiles.
- Four Hellfire missiles and one minigun.
- One minigun and one seven-shot rocket pod.

WEAPONS EMPLOYMENT

5-30. The pilot may fire weapons systems from either pilot station. He may fire rockets in singles (one at a time), pairs (two rockets, one from each rocket pod), or multiple rockets (depressing and holding down the firing button). The pilot can select, while in flight, the rocket he wishes to fire next on a 7-shot rocket pod or on a 19-shot pod. The pilot can select a zone he wishes to fire upon next (two to three rockets per zone, zones loaded with the same type of warhead). This arrangement allows the pilot to select the type of warhead to use on the target.

5-31. During a call for fire, the radiotelephone operator can request a type of round for firing (for example, a minigun only or flechettes), but normally the pilot selects the type of rounds fired during an engagement. Normal engagement ranges are as follows:
- Minigun, 10 meters to 750 meters.
- 2.75-inch FFAR, 100 meters to 600 meters.
- Hellfire missiles, 800 meters to 8,000 meters.

5-32. Minigun and 2.75-inch FFAR targets include ground troops, buildings, small boats, aircraft, and thin-skinned vehicles (SLAP rounds can penetrate 3/4-inch homogeneous rolled steel). Hellfire missile targets include tanks and other hard-skinned vehicles, bunkers to some degree, larger boats, and buildings (the shaped warhead causes very localized damage).

TRANSPORTABILITY OF AH-6M AIRCRAFT

5-33. The C-130 can carry 3 AH-6Ms, a C-17 can carry 9, and a C-5 can carry 21. In each case, tactical uploading and downloading of the aircraft can take place in an extremely short time. Off-load times vary, based upon numerous factors, such as ramp space, ramp condition, ramp type, off-load area, aircraft configuration, and mission configuration. General planning times for off-load from ramp down to takeoff (except C-5 deployment) are as follows:
- With the plank system, approximately 10 minutes.
- With the "T" tail removed, approximately 15 minutes.

PLANNING CONSIDERATIONS

5-34. The following paragraphs discuss considerations that must be taken into account when planning to use the AH-6M aircraft in a mission.

Weather Minimums

5-35. A minimum 500-foot ceiling and 2-mile visibility capability must exist for day and night flying over flat or mountainous terrain or over water. The unit commander may reduce weather minimums on a mission-essential, case-by-case basis.

Chapter 5

5-36. A visible horizon must exist in two of the four horizontal quadrants at all times. All AH-6M missions must take place under VMC rules.

Winds

5-37. The maximum wind allowed to start the aircraft is 40 knots, with a 20-knot gust spread.

Flight Altitudes

5-38. For training missions, the minimum en route altitude for routes not reconnoitered is 300 feet AGL. The minimum overwater altitude is 50 feet. For operational missions, the minimum en route altitude is dependent upon METT-TC.

Landing Areas

5-39. The AH-6M is capable of landing on any structure that allows clearance for the rotor systems and meets stress requirements. Single-aircraft confined landing areas require a minimum size of 50 feet by 50 feet.

Shipboard Operations

5-40. The AH-6M can operate day and night from any ship having at least a one-spot helicopter-landing capability. Because of the high radio and radar electromagnetic interface (EMI) signature onboard U.S. Navy vessels, only Mark 66 MOD-3 rocket motors are compatible with shipboard operations without waiver approval.

Aircrew Composition

5-41. The aircrew of an AH-6M consists of two pilots—a pilot in command (PIC) and a copilot. The PIC is responsible for the employment and actions of his aircraft. The copilot assists the PIC in accomplishing the mission. Both aircrew members have extensive training in navigation, gunnery, shipboard operations, overwater training, mountain flying, urban operations, and desert flying. The lead aircraft has a flight-lead–qualified pilot during all operations. The flight-lead pilot is responsible for mission accomplishment and is the primary mission planner.

Aircraft Capabilities

5-42. Table 5-2, page 5-9, lists the capabilities of the AH-6M. The aircraft dimensions illustrated in Figures 5-2 and 5-3, page 5-4, also pertain to the AH-6M.

5-43. The cruise airspeed of the AH-6M is 90 knots indicated airspeed (KIAS). The maximum airspeed is 108 KIAS. All speeds are dependent on mission configuration and load.

SAFETY

5-44. Personnel must observe the following safety precautions:
- Never walk in front of armed aircraft.
- Wear protective headgear at all times when working around the turning rotor blades of the low rotor and tail rotor system of the AH-6M.
- Wear hearing and eye protection when working around operating aircraft.
- Be aware that the aircraft exhaust can start ground fires in extremely dry conditions with combustible material present (for example, dry grass or straw).
- Approach operating AH-6M aircraft as depicted in Figure 5-7, page 5-9.

Table 5-2. AH-6M aircraft capabilities

Aircraft Weight	
Basic weight	2,196 pounds
Mission weight	3,100 pounds (fully fueled, dual-pilot)
Maximum gross weight	3,950 pounds
Aircraft Dimensions	
Length blades folded	22 feet 6 inches
Blades unfolded	32 feet 1 inch
Width blades folded	6 feet 5 inches
Blades unfolded	27 feet 4 inches
Height	8 feet 11 inches
Diameter of main rotor	27 feet 4 inches
Aircraft turning radius	36 feet 9 inches

Range and Endurance at 240 Pounds Per Hour (With Optional Fuel Tank)		
Fuel Tank	Endurance (Hours + Minutes)	Fuel Range (Nautical Miles)
Main	1+17	116
Main plus one auxiliary	2+57	266

Note. Because of weight restrictions, the use of the optional fuel tank prevents the installation of a minigun and ammunition cans, and requires a reduced rocket load.

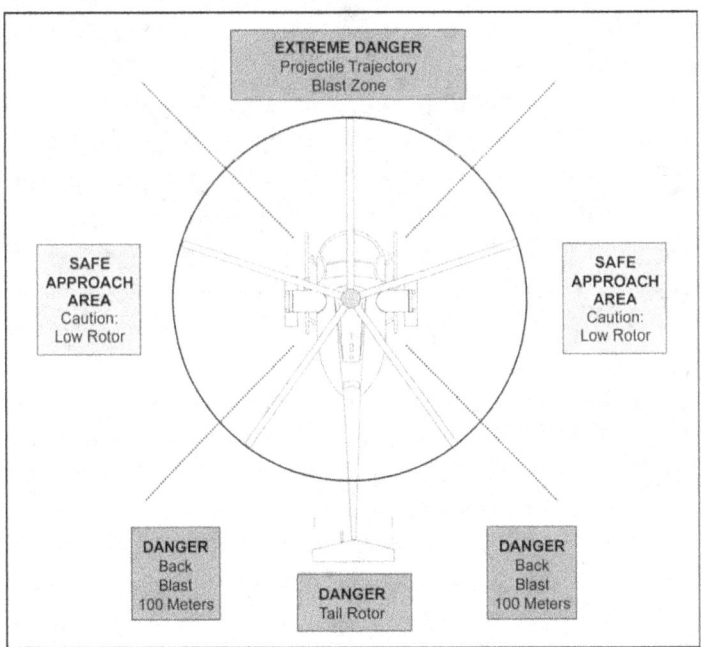

Figure 5-7. AH-6M safety approach areas

Chapter 5

MH-60L HELICOPTER

5-45. The primary mission of the MH-60L is to conduct overt and covert infiltration, exfiltration, and resupply of SOF across a wide range of environmental conditions and across the spectrum of conflict. Other missions of the MH-60L include C2, external load, CSAR, and MEDEVAC operations. The MH-60L can operate from fixed-base facilities, remote sites, or oceangoing vessels.

DESCRIPTION

5-46. The MH-60L (Blackhawk) is a highly modified twin-engine utility helicopter (Figure 5-8). Its configuration may include a number of auxiliary fuel systems to allow for operational times of as much as 5.5 hours with a range of 640 nautical miles (NM). The MH-60L has secure Selective Adaptive Communications Processor (SELSCAN) high frequency (HF), FM, ultrahigh frequency (UHF), very high frequency (VHF), satellite communications (SATCOM), and Sabre communications. The FRIES allows for rapid insertion and extraction of personnel in areas blocked from air-land maneuvers. The aircraft has two M134 7.62-mm Gatling guns (miniguns), a ballistic armor subsystem (BASS), and aircraft survivability equipment (ASE) to increase aircrew survivability in all threat environments. Mission-selective systems include a cargo hook for external load operations, a personnel locator system (PLS) for CSAR, and a four-place C2 console for airborne C2 operations. An armed version of the MH-60L—the DAP—is capable of mounting two M134 7.62-mm miniguns, two 30-mm chain guns, two 2.75 rocket pods, Hellfire missiles, or combinations of these systems for armed escort and fire support operations.

Figure 5-8. MH-60L helicopter

AIRCRAFT SURVIVABILITY EQUIPMENT

5-47. The following explain each survival element found on the MH-60L helicopter.

- The AN/APR-39A (V) 1 RWR identifies threat pulse radar in the C/D and H through J bands. It provides audio and visual alerts to the flight crew.
- The AN/APR-44 (V) 3 RWR detects continuous wave (CW) SAM threat radar emissions. It provides audio and visual warnings to the flight crew.
- The AN/ALQ-144 Infrared Countermeasures Set provides false infrared signals to defeat the threat of infrared-sensing missiles.
- The M-130 Chaff Dispenser dispenses decoy chaff as an effective countermeasure against radar-guided missiles.
- The AN/APX-100 (V) I IFF Transponder provides automatic identification of the helicopter to suitably equipped ground and airborne interrogators.

Army Special Operations Aviation Units and Aircraft

- The emergency locator transmitter transmits a distress signal on UHF and VHF guard frequencies. It goes on the avionics rack on the left side of the right pilot's seat. Impact activates the transmitter, or personnel may turn it on manually.
- The underwater acoustic beacon radiates a pulsed acoustic signal of 37.5 kilohertz (kHz) detectable by hydrophone-equipped vessels. Water activates the beacon.
- Fire extinguishers are hand-operated. One is mounted on the cabin wall forward of the right gunner's window and one is mounted on the right side of the left pilot's seat.
- First-aid kits are located on the back of the left and right pilots' seats.
- Survival kits are one or two environment-specific kits attached to the internal auxiliary fuel tanks.

STANDARD MISSION EQUIPMENT

5-48. The following paragraphs discuss standard equipment found on the MH-60L helicopter.

Armament

5-49. The standard armament is the M134 (7.62-mm minigun). The M134 is a six-barrel, air-cooled, electrically operated Gatling gun, with maximum effective fire of 1,000 meters. The gun fires A165 (7.62-mm balls), A257 (7.62-mm low-light balls), and SL66 (armor-piercing sabots). One gun each is on the outside of the left and the right gunners' windows. The crew chiefs normally operate the guns, using open-steel, aim point, or aim-1 sights.

Ballistic Armor Subsystem

5-50. This fabric-covered steel plating provides increased ballistics protection in the cockpit and cabin.

Guardian Auxiliary Fuel Tanks

5-51. Two 172-gallon tanks, mounted in the cabin area at the aft bulkhead, provide range extension of approximately 2 hours (mains plus two auxiliary tanks, 4 hours total). Each tank occupies approximately 18 square feet of usable cabin floor space. Normal operational time without the guardian tanks is approximately 2 hours.

Fast-Rope Insertion and Extraction System Bar

5-52. Each side of the FRIES bar (Figure 5-9, page 5-12) can support a maximum weight of 1,500 pounds.

MISSION-SELECTIVE SYSTEMS

5-53. The following items are mountable on the MH-60L to support a primary mission or to enhance the capabilities of aircraft performing assault or DAP missions:
- *Cargo Hook.* This item is mountable in the belly of the aircraft, below the main rotor. The hook can support external loads up to 9,000 pounds.
- *External Rescue Hoist System.* This system is a hydraulic hoist capable of lifting 600 pounds. It contains 200 feet of usable cable. The crew chief or the hoist operator maneuvers the hoist using a handheld pendant.
- *Internal Auxiliary Fuel System.* The MH-60L has wiring provisions for four additional 150-gallon fuel cells, mountable in the cargo area. Each fuel cell provides approximately 50 minutes flight endurance. Ambient conditions and weight restrictions limit the maximum number of additional fuel cells. The use of all four IAFS cells reduces usable cargo area space to near zero.
- *External Extended Range Fuel System (ERFS).* This system consists of two 230-gallon, or two 230- and two 450-gallon, or four 230-gallon jettisonable fuel tanks mountable on the external stores support system (ESSS) for long-range deployment of the aircraft. The use of the ERFS restricts the employment of the M134 miniguns. Center-of-gravity or maximum-gross-weight restrictions and ambient conditions may limit the specific configuration of the ERFS.

Chapter 5

- *C2 Console.* This system provides four operator positions with access to the four AN/ARC-182 (V) multiband transceivers and FLIR display. Personnel may configure the MH-60L with an ESSS to employ the FOL-AC with the supporting amplifier array frame on the cabin floor for PSYOP missions.

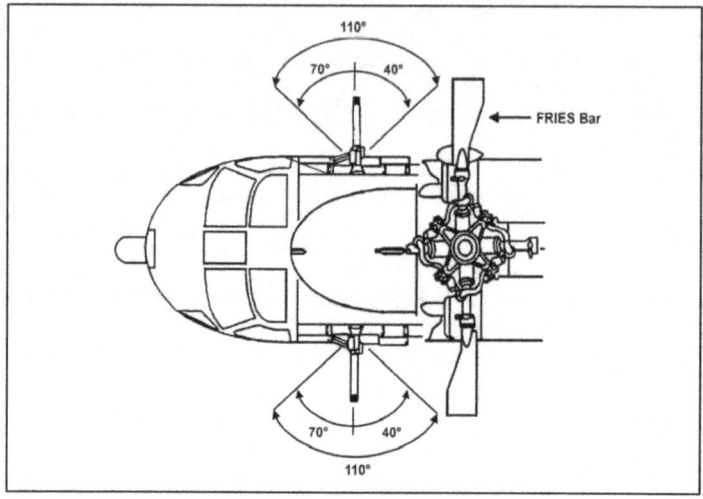

Figure 5-9. MH-60L FRIES bar

DEFENSIVE ARMED PENETRATOR (DAP) SPECIAL CONFIGURATION

5-54. The mission of the armed MH-60L DAP (Figure 5-10) is to conduct attack helicopter operations using area fire or precision-guided munitions and armed infiltration or exfiltration of small units. The DAP is a multimission aircraft capable of deploying on short notice and of conducting DA missions. It is also capable of reconfiguring for troop assault operations. The DAP is capable of conducting all missions during day or night, or in adverse weather.

Figure 5-10. MH-60L defensive armed penetrator

Army Special Operations Aviation Units and Aircraft

5-55. The DAP can provide armed escort for employment against threats to a vertical-lift formation. Using team tactics, the DAP is capable of providing CCA suppression or CAS for formations and teams on the ground. In the defensive armed role, the DAP is not a primary transport for troops or supplies because of high GWs. The DAP conducting deep attacks has a combat radius of 225 nautical miles (takeoff, fly 225 nautical miles, no loiter, and return).

MH-60L DAP Weapon System and Employment

5-56. The CMS-80 of the MH-60L DAP has integrated fire control systems. The integration gives the pilot a reduced cockpit workload and an increased weapons-selection capability through cockpit control driver and hands-on collective and stick weapons selection.

AN/AVQ-34 Monocular Head-Up Display (MONOHUD)

5-57. The MONOHUD provides a lightweight, infinity focus, optical sight that allows the pilot to deliver rockets and gunfire effectively at targets. It gives the pilot cues for accurately launching missiles. It also provides aircraft flight symbology. The symbology is concise and provides all pertinent information in a manner that accommodates the pilot operating in daytime or with NVGs.

5-58. The standard armament configuration of the DAP is one rocket pod, one 30-mm cannon, and two miniguns. The configuration changes based on METT-TC (Figure 5-11).

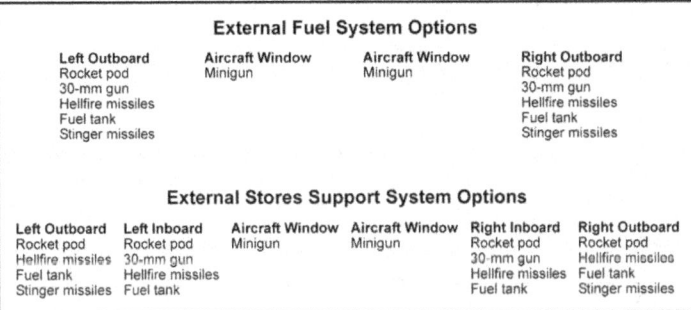

Figure 5-11. Armament options for the MH-60L DAP

Note. To avoid exceeding maximum GW limitations, reconfiguration of the ammunition or fuel mix may be necessary to achieve the desired insertion ranges for personnel when the MH-60L is in the DAP configuration.

M134 7.62-mm Minigun

5-59. The M134 is a six-barrel, air-cooled, link-fed, electrically driven Gatling gun, with a 1,000-meter maximum effective range and a tracer burnout at 900 meters. The weapon has a rate of fire of 2,000 or 4,000 rounds per minute. The weapon is mountable in the fixed position on the left and right sides of the aircraft. The minigun fires a variety of 7.62-mm rounds. Nighttime operations use a 7.62-mm ball with a special low-light tracer, which prevents the shutting down of NVGs. The weapon also fires 7.62 SLAP ammunition for light-armor penetration. The DAP normally carries 3,000 rounds of 7.62-mm ammunition.

M261 19-Shot Rocket Launcher

5-60. The M261 fires a 2.75-inch FFAR with a variety of special-purpose warheads. It has a 10-pound and a 17-pound high-explosive warhead for light armor and bunker penetration. The bursting radius for the 10-pound warhead is 8 to 10 meters, and for the 17-pound warhead it is 12 to 15 meters. The antipersonnel flechette warhead contains 2,200 flechettes. Its minimum launch distance is 800 meters, and its optimum range is 1,100 meters. Another warhead is white phosphorous used for smoke. The illumination warheads come in two types. One provides a bright light; the other, a bright infrared light. Firing of the warheads is within 3,000 meters of the target area. After deploying, the warheads provide 120 seconds of overt light or 180 seconds of infrared light. The multipurpose submunition warhead contains nine submunitions that are effective against light armor and personnel. The multipurpose round has a fuse that can be preset and that deploys the submunitions at the desired distance. The 2.75-inch FFAR is useful as a point target weapon at ranges from 100 to 750 meters and an area fire weapon at ranges up to 7,000 meters. The DAP can also fire chlorobenzaimalononitrile, HE-proximity, and inert rockets. The aircraft can carry an additional load of rockets internally, allowing the aircrew to reload the rocket pod. The aircrew can accomplish the reload within 15 minutes.

M230 30-mm Chain Gun

5-61. The M230 has its own magazine capable of carrying 1,100 rounds. The M230 has a cyclic rate of fire of 625 + 25 rounds per minute. The M230 is capable of firing the HEDP, target practice, and target practice tracer. The HEDP is effective against light armor and personnel at ranges of 4,000 meters. With the use of the MONOHUD as a sighting system, the 30-mm cannon is a point-target weapon at a range of 1,500 meters or less. It is also an area fire weapon at ranges up to 4,000 meters.

AGM-114 Hellfire

5-62. The AGM-114 is a 100-pound semiactive laser-guided missile, capable of defeating any known armor. The M272 launchers are able to hold four Hellfire missiles each. The minimum engagement range is 0.8 kilometer; the maximum is 8 kilometers. Any ground or air North Atlantic Treaty Organization standard laser designator can designate the missile.

AN/AAQ-16D Airborne Electronic Special Operations Payload FLIR

5-63. The AN/AAQ-16D is a FLIR with a laser range finder or designator. The AN/AAQ-16D allows the DAP to detect, acquire, identify, and engage targets at extended ranges with laser-guided munitions. The FLIR is a controllable, infrared surveillance system that provides a television-video-type infrared image of terrain features and ground or airborne objects of interest. The FLIR is a passive system and detects long-wavelength radiant infrared energy emitted, naturally or artificially, by any object in daylight or darkness.

Air-to-Air Stinger

5-64. The DAP can fire the infrared seeking, fire-and-forget missile.

MH-60L DAP RECONFIGURATION

5-65. The MH-60L DAP has the capability to perform utility and armed missions. The time to reconfigure the aircraft from the armed to the utility or vice versa is minimal. The 7.62 miniguns remain with the aircraft regardless of the mission.

TRANSPORTABILITY

5-66. C-5A/B and C-17 aircraft can deploy the MH-60L, including the DAP configuration. The C-5A/B can carry a maximum of six MH-60Ls. The helicopters need a short time to prepare for on-load and again for rebuild upon arrival at the destination. The C-17 can carry three MH-60Ls.

PLANNING CONSIDERATIONS

5-67. Successful mission accomplishment is largely a function of adequate premission planning time. Mission notification should occur in time to have an adequate mission-planning session and briefing, followed by a period of rest before mission execution.

Army Special Operations Aviation Units and Aircraft

Weather Minimums

5-68. For training missions, forecast and actual weather requirements are a 500-foot ceiling and a 2-mile visibility. For contingency missions, as directed by the commander, a 500-foot ceiling and a 2-mile visibility work well for planning purposes. This type of forecast allows for en route cruise speed of the standard 120 knots and ample opportunity to adjust mission execution in the event of lower weather.

Winds

5-69. The MH-60L rotor has the capability to start and stop in actual winds no greater than 45 knots.

Flight Altitudes

5-70. For training missions, the minimum altitude is 300 feet AGL for routes not reconnoitered and 150 feet AGL for reconnoitered routes. For contingency missions, the minimum altitude is dependent upon METT-TC.

Landing Areas

5-71. The minimum landing area for the MH-60L is 100 feet by 100 feet.

Shipboard Operations

5-72. The MH-60L, including the DAP, can operate day and night from U.S. Navy ships with Level II, Class II helicopter-landing pads. For DAP, because of the high radio or radar EMI signature onboard Navy vessels, only Mark 66 MOD-3 rocket motors are compatible with shipboard operations without waiver approval.

Aircrew Composition

5-73. Most training flights and all NVG operations require four aircrew members—a PIC, a pilot, and two aircrew chiefs or gunners. One aircrew chief is at the right gunner's position. He scans for hazards, operates the hoist, conducts FRIES operations, operates the minigun, and conducts external load operations. The other aircrew chief is at the left gunner's position and scans for hazards, conducts FRIES operations, operates the minigun, and assists in external load operations.

Aircrew Qualifications

5-74. All aircrews are qualified to support flight operations for the missions stated in FM 3-05 and JP 3-05, *Doctrine for Joint Special Operations*. Aircrew qualifications include multiship NVG infiltration, exfiltration, and live-fire operations in urban, overwater, mountain, desert, jungle, and chemical, biological, radiological, and nuclear (CBRN) environments to LZs, buildings, ships, and oil rigs. Aircrews are trained in NVG long-range overland and overwater navigation, with an arrival standard of +30 seconds.

Aircraft Capabilities

5-75. Table 5-3, pages 5-15 and 5-16, lists the capabilities of the MH-60L.

Table 5-3. MH-60L aircraft capabilities

Maximum Gross Weight (Ferry)	
Ferry configuration	23,500 pounds
Assault configuration	22,000 pounds
Aircraft Dimensions	
Length	64 feet 10 inches
Width	53 feet 8 inches
Height	16 feet 10 inches
Diameter of main rotor	53 feet 8 inches
Diameter of main rotor	53 feet 8 inches

Chapter 5

Table 5-3. MH-60L aircraft capabilities (continued)

Range and Endurance		
Fuel Tank	Endurance (Hrs + Min)	Fuel Range (Nautical Miles)
Main	1+45	212
Main plus one auxiliary	3+02	364
Main plus two auxiliaries	4+10	496
Main plus three auxiliaries	5+00	600
Airspeed		
Cruise		120 KIAS
Maximum		165 KIAS

MH-60K HELICOPTER

5-76. The primary mission of the MH-60K (Blackhawk) is to conduct overt or covert infiltration, exfiltration, and resupply of SOF over a wide range of environmental conditions. The MH-60K (Figure 5-12) can operate from fixed-base facilities, remote sites, or oceangoing vessels.

Figure 5-12. MH-60K helicopter

DESCRIPTION

5-77. The MH-60K is a highly modified twin-engine utility helicopter. The aircraft can be configured with a number of auxiliary fuel systems to allow for operational times of as much as 5.5 hours with a range of 634 nautical miles. The MH-60K is equipped with secure HF, single-channel ground and airborne radio station (SINCGARS), FM, UHF, VHF, SATCOM, and Sabre communications. The FRIES allows for rapid insertion and extraction of personnel in areas occluded from AirLand maneuvers. The aircraft has two M134 7.62-mm Gatling guns, a BASS, and ASE to increase aircrew survivability in all threat environments. The GPS, area navigation unit, inertial navigation unit, attitude and heading reference system, and multimode radar systems allow pinpoint navigation. Mission-selective systems include the cargo hook for external load operations and the PLS for CSAR.

AIRCRAFT SURVIVABILITY EQUIPMENT

5-78. The following explain each survival element found on the MH-60K helicopter:
- The AN/APR-39A (V) 1 RWR identifies threat pulse radar in the C or D and the H through J bands and provides audio and video alerts to the flight crew.
- The AN/APR-44 (V) 3 RWR detects CW SAM threat radar emissions and provides audio and visual warnings to the flight crew.
- The AN/AVR-2 Laser Warning Receiver detects laser emissions directed toward the helicopter.
- The AN/AAR-47 Missile Warning Receiver detects plume emissions from a missile's exhaust.
- The AN/ALQ-162 (V) 2 CW Radar Jammer detects and jams CW radar emitters.
- The AN/ALQ-136 (V) 2 Pulse Radar Jammer detects and jams pulse radar emitters.
- The AN/ALQ-144 Infrared Countermeasures Set provides false infrared signals to defeat infrared-sensing missile threats.
- The M-130 Chaff and Flare Dispenser dispenses decoy chaff and flare as an effective countermeasure against radar-guided and infrared missile threats.
- The AN/APX-100 (V) 1 IFF Transponder provides automatic identification of the helicopter to suitably equipped ground and airborne interrogators.
- The emergency locator transmitter is mounted on the left side of the right pilot's seat avionics rack. It transmits a distress signal on UHF and VHF guard frequencies. The transmitter is impact-activated but may be turned to the ON state manually.
- The underwater acoustic beacon is activated by contact with water. It radiates a pulsed acoustic signal of 37.5 kHz detectable by hydrophone-equipped vessels.
- Fire extinguishers are hand-operated—one mounted on the cabin wall forward of the right gunner's window and one mounted on the right side of the left pilot's seat.
- First-aid kits are located two each on the back of the left pilot's seat and one each mounted on the back of the right pilot's seat.
- Survival kits are environment-specific kits attached to the internal auxiliary fuel tanks.

STANDARD MISSION EQUIPMENT

5-79. The following are systems and equipment always on board the aircraft during tactical missions. This list does not include avionics, ASE, and sensors, as they are considered part of the basic aircraft.

Armament

5-80. The standard armament for the MH-60K is the M134 (7.62-mm minigun), 6-barrel, air-cooled, and electrically operated Gatling gun (Figure 5-13, page 5-18). The maximum effective fire is 1,000 meters. The M134 fires A165 (7.62 ball), A257 (7.62 low light ball), and SL66 (armor-piercing sabot) ammunition. One gun is mounted outside both the left and right gunner's windows. Aircrew chiefs normally operate the weapon system. Weapon sighting is by open steel sights.

Ballistic Armor Subsystem

5-81. The system consists of fabric-covered steel plating, which provides increased ballistic protection in the cockpit and cabin areas.

Guardian Auxiliary Fuel Tanks

5-82. Two 172-gallon fuel tanks provide a range extension of approximately 2 hours (4 hours total). The tanks are mounted in the cabin area at the aft bulkhead. They occupy approximately 18 square feet of usable cabin floor space. The normal operational time without the Guardian tanks is approximately 2 hours 10 minutes.

Chapter 5

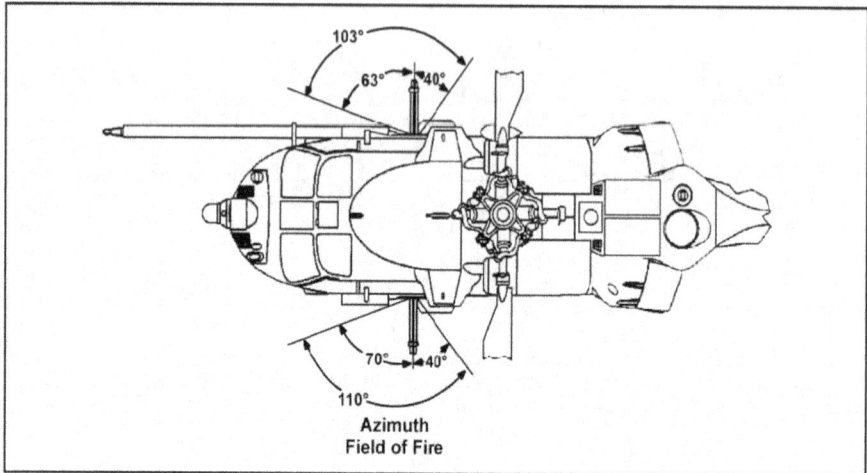

Figure 5-13. MH-60K with M134 minigun window-mounted field of fire

Fast-Rope Insertion and Extraction System Bar

5-83. Each side of the FRIES bar can support a maximum weight of 1,500 pounds.

MISSION-SELECTIVE SYSTEMS

5-84. The following systems are mountable on the MH-60K to support a primary mission or to enhance the capabilities of aircraft performing assault missions:
- *Cargo Hook.* This system is mountable in the belly of the aircraft below the main rotor. It can support external loads up to 8,000 pounds.
- *External Rescue Hoist.* This hydraulic hoist is capable of lifting 600 pounds with 200 feet of usable cable. The aircrew chief or hoist operator uses a handheld pendant to control the system.
- *External Tank System (ETS).* This system has two 230-gallon external fuel tanks that may be jettisoned during emergencies. The fuel tanks are mounted on the ETS for long-range deployment of the aircraft. Installation of the ETS restricts the use of weapons systems. The aircraft's center of gravity, maximum gross-weight limitations, or ambient conditions may limit specific configurations.
- *Aerial Refueling System.* This system is an aerial refueling probe that allows extended range and endurance by refueling from HC/MC/KC-130 tanker aircraft.
- *Aerial Loudspeaker System.* The MH-60K with ESSS can be configured to employ the 2,700-watt FOL-AC with the supporting amplifier array frame on the cabin floor for PSYOP aerial loudspeaker missions.

TRANSPORTABILITY

5-85. The MH-60K may be deployed by C-5A, C-5B, and C-17 aircraft. A maximum of six MH-60Ks can be loaded on a C-5A or a C-5B. A short time is needed to prepare the helicopters for on-load and again for rebuilding on arrival at the destination. Three MH-60Ks can be loaded onto a C-17. Ammunition for the weapon systems is palletized and loaded on the same aircraft for distribution at the destination.

Planning Considerations

5-86. Successful mission accomplishment is largely a function of adequate premission planning. Mission notification should occur in time for an adequate mission-planning session and briefing, followed by a period of rest before execution.

Weather Minimums

5-87. Training missions require forecast and actual weather parameters of a 500-foot ceiling and a 2-mile visibility. For contingency missions, as directed by the commander, a 500-foot ceiling and a 2-mile visibility work well for planning purposes. This type of forecast allows for en route cruise speed of the standard 110 knots and ample opportunity to adjust mission execution depending upon weather conditions.

Flight Altitudes

5-88. For training missions, the minimum altitude is 300 feet AGL for routes not reconnoitered and 150 feet AGL for reconnoitered routes. Contingency missions are dependent upon METT-TC.

Landing Areas

5-89. The minimum landing area for the MH-60K is 100 feet by 100 feet. For shipboard operations, the MH-60K can operate day and night from Navy ships that have Level II, Class II helicopter-landing pads.

Aircrew Composition

5-90. Most training flights and all NVG operations require four aircrew members. These members include a PIC, a pilot, and two aircrew chiefs or gunners. One aircrew chief—stationed at the right gunner's position—scans for hazards, operates the hoist, conducts FRIES operations, operates the minigun, and conducts external load operations. The other aircrew chief—stationed at the left gunner's position—conducts FRIES operations, operates the minigun, and assists in external load operations.

Aircrew Qualifications

5-91. Aircrews can perform all mission tasks in all environments. They can perform NVG long-range overland and overwater navigation, with an arrival standard of ± 30 seconds.

Aircraft Capabilities

5-92. Table 5-4, pages 5-19 and 5-20, lists the capabilities of the MH-60K aircraft. Figures 5-14 through 5-16, pages 5-20 through 5-22, illustrate specific dimensions and capabilities of the aircraft.

Table 5-4. MH-60K aircraft capabilities

Aircraft Weight	
Basic weight	15,600 pounds
Maximum gross weight	24,500 pounds
Aircraft Dimensions	
Length	64 feet 10 inches (folded–60 feet 7 inches)
Width	53 feet 8 inches (folded–9 feet 9 inches)
Height	16 feet 10 inches
Diameter of main rotor	53 feet 8 inches
Aircraft turning radius	41 feet 8 inches

Chapter 5

Table 5-4. MH-60K aircraft capabilities (continued)

Range and Endurance		
Fuel Tank	Endurance (Hrs + Min)	Fuel Range (Nautical Miles)
Main	1+30	165
Main plus one auxiliary	2+30	275
Main plus two auxiliaries	3+30	385
Airspeed		
Cruise		115 knots indicated airspeed
Maximum		145 knots indicated airspeed

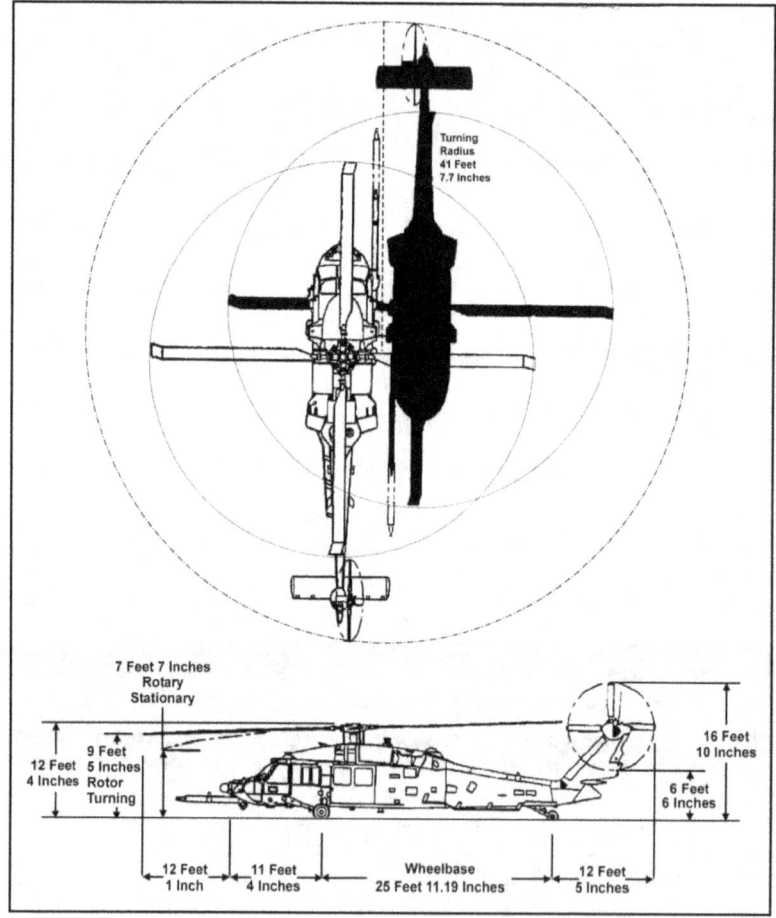

Figure 5-14. MH-60K dimensions and turning radius

Army Special Operations Aviation Units and Aircraft

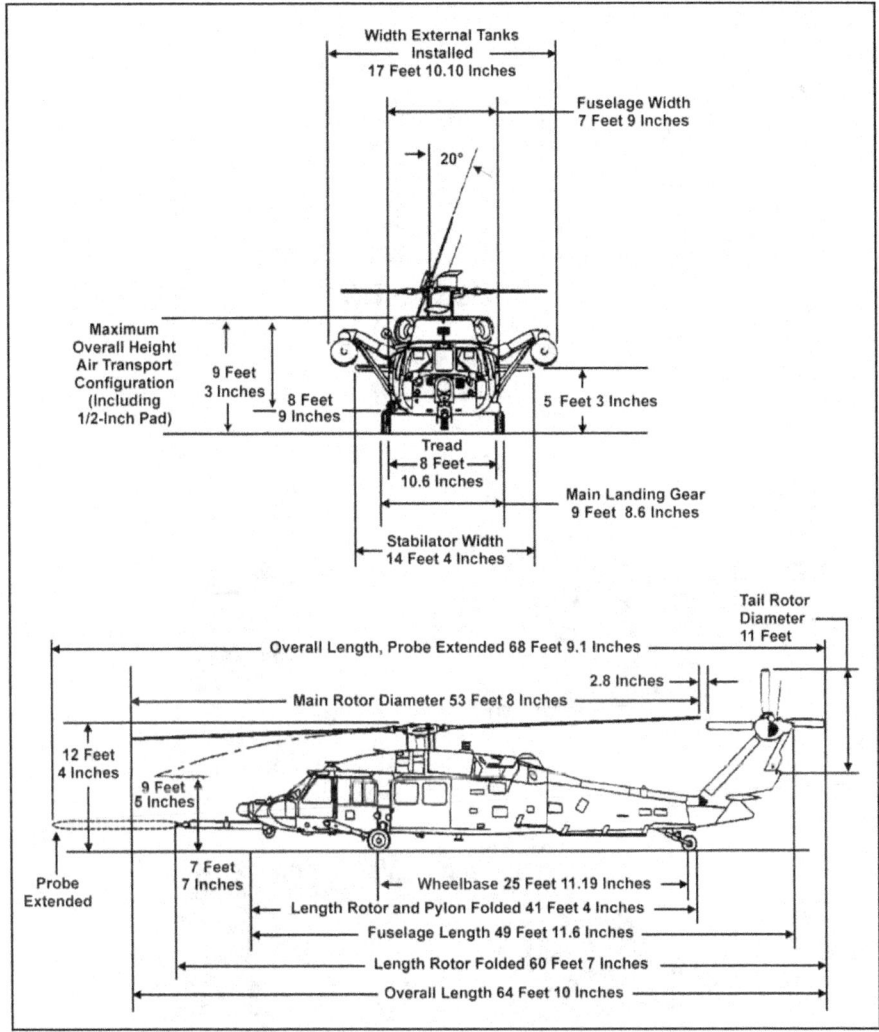

Figure 5-15. MH-60K dimensions for strategic airlift preparation

Chapter 5

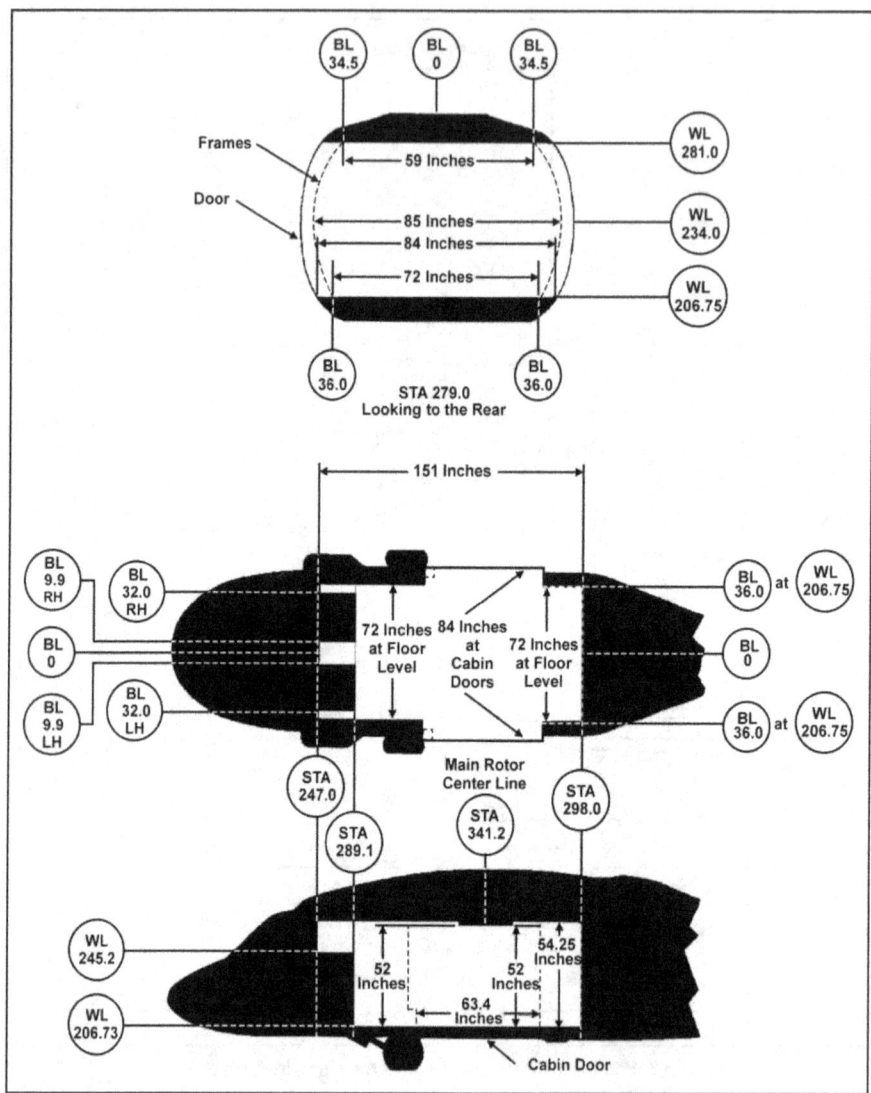

Figure 5-16. MH-60K aircraft capabilities

MH-47G HELICOPTER

5-93. The primary mission of the MH-47G is to conduct overt and covert infiltration, exfiltration, air assault, resupply, and external-sling operations under a wide range of environmental conditions. The aircraft can perform a variety of other missions, including shipboard, platform, urban, water, forward arming and refuel point (FARP), mass-casualty, and CSAR operations.

DESCRIPTION

5-94. The MH-47G (Chinook) is a twin-engine, tandem-rotor, heavy assault helicopter specifically modified for long-range SO flights (Figure 5-17). It has secure voice communications on FM, UHF with HaveQuick II, VHF, HF and SELSCAN, Sabre, and SATCOM radios. Other features include FRIES, limited aircraft survivability equipment, a defensive armament system of two M134 machine guns (one located in the left forward cabin window and one at the right cabin door), one M-60D machine gun (located on the ramp), and an internal rescue hoist with a 600-pound capacity.

Figure 5-17. MH-47G helicopter

5-95. The MH-47G adverse-weather cockpit is equipped with weather-avoidance and search radar, an aerial refueling probe for in-flight refueling, a PLS used with the PRC 112 for finding downed aircrews, FLIR, and a navigation system consisting of a mission computer using GPS, inertial navigation system (INS), or Doppler navigation sources for increased accuracy. The MH-47G helicopter is instrument-capable with an automatic direction finder (ADF), very high frequency omnidirectional range (VOR), distance measuring equipment (DME), instrument landing system (ILS), and TACAN with the ability to do fully coupled approaches. Mission-computer-generated approaches can be used when normal approaches are unavailable.

5-96. The MH-47G helicopter is capable of operating at night during marginal weather conditions. With the use of special mission equipment and night vision devices (NVDs), the aircrew can operate in hostile mission environments over all types of terrain. The aircrew can operate at low altitudes during periods of low visibility and low ambient lighting conditions with pinpoint navigation accuracy +30 seconds on target.

AIRCRAFT SURVIVABILITY EQUIPMENT

5-97. The following explain each survival element found on the MH-47G helicopter:
- The AN/APR 39A RWR identifies hostile pulse fire control radar and provides audio and video alerts to the flight crew when the system detects threat radar emissions.

Chapter 5

- The AN/ALQ 156 (V) 47 is an active airborne Doppler radar system that detects the approach of antiaircraft missiles. When detecting a missile, the system automatically triggers the M-130 flare system and ejects a decoy flare.
- The AN/ALE-47 (V) Countermeasures Dispensing System consists of five components used to provide preemptive and terminal threat protection. The pilots control the system by using the cockpit control unit mounted in the center console. The AN/ALE-47 replaces the M-130 system and enhances aircraft survivability by—
  - Integrating with avionics and EW systems.
  - Providing threat-adaptive programmable dispensing routines.
  - Providing data links for advanced expendables.
  - Using available threat sensors.
- The AN/AAR-47 (V) 3 Radar Warning System is a passive system that provides visual and aural alert indications of CW radar signals from SAM and airborne interceptor (AI) threats. The APR-44 interfaces with the APR-39 to provide visual and aural indications. The system consists of two SAM antennas, an AI receiver, a SAM receiver, and a low-pass filter.
- The AN/ALQ 44 RWR identifies hostile AI and SAM CW fire control radar, and provides audio and visual alerts to the flight crew when the system detects threat radar emissions.
- The M-130 Infrared Countermeasures Dispenser System is a chaff and flare dispenser system designed to deceive radar guidance and infrared missiles by using chaff or flares as required.

STANDARD MISSION EQUIPMENT

5-98. The following paragraphs discuss standard equipment found on the MH-47G helicopter.

Armament

5-99. The MH-47G has three weapon stations—left forward window, right cabin door, and at the ramp. The forward station mounts a 7.62-mm minigun, and the ramp station mounts an M60D 7.62-mm machine gun. An aircrew member at each station manually operates the weapon. The primary use of the weapon is self-defense and enemy suppression. The minigun is normally used for soft targets and troop suppression, which requires a high rate of fire. The minigun is air-cooled and link-fed. It has a maximum effective range of 1,500 meters, with a tracer burnout at 900 meters. The weapon has an adjustable rate of fire of 2,000 or 4,000 rounds per minute. The aircrew members currently fire ball or SLAP ammunition with a mix of four ball rounds to one tracer round (4:1) or a 9:1 mix to prevent NVD shutdown on low-illumination nights. The ammunition complement without reloading is 8,000 rounds per weapon.

Ballistic Armor Subsystem

5-100. This subsystem is a fabric-covered steel plating that provides increased ballistic protection in the cockpit and cabin.

Fast-Rope Insertion and Extraction System Bar

5-101. The FRIES is used for insertion and extraction of personnel. Applied loads at the rear ramp for insertions will not exceed nine persons per rope at the same time. Applied loads at the rear ramp for extractions will not exceed six persons per rope at the same time.

MISSION SELECTIVE SYSTEMS

5-102. The external cargo hook system facilitates greater load stability and ensures faster airspeeds during flight. Each hook (Table 5-5, page 5-25) may be used separately or with each other. All loads should be planned as a tandem-rigged load.

5-103. The external rescue hoist system is configured for use at the center cargo hook and rescue hatch. It has a 600-pound capacity and approximately 150 feet of usable cable.

Table 5-5. MH-47G external cargo hooks

Type	Capacity
Forward hook	17,000 pounds
Center hook	26,000 pounds
Aft hook	17,000 pounds
Tandem hook	25,000 pounds

Note. These are maximum hook-rated loads and may not accurately reflect the true capability of the aircraft because of external conditions, such as pressure, altitude, and temperature.

5-104. The AN/AAQ-16 FLIR is a controllable, infrared surveillance system that provides a television video-type infrared image of terrain features and ground or airborne objects of interest. The FLIR is a passive system and detects long-wavelength radiant infrared energy emitted, naturally or artificially, by any object in daylight or darkness.

5-105. The cargo compartment expanded range fuel system (CCERFS) consists of one and up to three ballistic-tolerant, self-sealing tanks. Each tank has the capacity of holding 800 gallons of fuel but normally is filled to 780 gallons. They are refillable using aerial refueling operations.

5-106. The forward area refueling equipment (FARE) consists of fueling pumps, hoses, nozzles, and additional refueling equipment to set up a two-point refueling site. Gallons of fuel dispensed are dependent upon the range of operation required of the tanker aircraft.

TRANSPORTABILITY

5-107. The SOAR has modified and validated procedures to load two MH-47Gs on a C-5 and one MH-47G on a C-17, with all support equipment. Compared to all other SO helicopters, the MH-47G requires extensive time for aircraft disassembly and assembly to become combat-ready.

PLANNING CONSIDERATIONS

5-108. Successful mission accomplishment is largely a function of adequate premission planning time. Mission notification should occur in time to have an adequate mission-planning session and briefing, followed by a period of rest before mission execution.

Weather Minimums

5-109. The weather minimum for the MH-47G is a 500-foot ceiling, with a visibility of 2 miles.

Winds

5-110. The MH-47G has no specified minimum wind; however, the maximum wind for starting and stopping the rotor system is 30 knots.

Flight Altitudes

5-111. For training missions, the minimum en route altitude for reconnoitered routes is 150 feet AGL or AHO. For routes not reconnoitered, the minimum en route altitude is 300 feet AGL or AHO. For operational missions, the minimum en route altitude is dictated by threat systems.

Landing Areas

5-112. The minimum landing area for the MH-47G is 150 feet by 100 feet.

Shipboard Operations

5-113. The MH-47G can operate day and night from Navy ships that have Level II, Class III helicopter-landing pads.

Aircrew Composition

5-114. Most training, exercises, and operational or contingency missions require five aircrew members, including a pilot, a copilot, a flight engineer, and two aircrew chiefs. The flight engineer, usually positioned at the ramp station, scans for other aircraft, targets, and obstacles. He also operates the hoist (when required), assists in FRIES operations, operates the machine gun, and conducts sling-load operations. The aircrew chiefs, positioned at the left and right forward gunners' stations, scan for other aircraft, targets, and obstacles. They also operate the miniguns and assist in sling-load and FRIES operations.

Aircrew Qualifications

5-115. MH-47G aircrews can perform all mission tasks in all environments. They can perform NVG infiltration and exfiltration operations, arriving at the target time of ± 30 seconds. MH-47G aircrews can also perform aerial refueling operations.

Aircraft Capabilities

5-116. Table 5-6, pages 5-26 and 5-27, lists the capabilities of the MH-47G aircraft.

Table 5-6. MH-47G aircraft capabilities

Aircraft Weight	
Maximum gross weight	50,000 pounds
Empty gross weight	25,000 pounds
Maximum altitude	20,000 feet

Fuel Flow	
Normal fuel consumption	2,750 pounds per hour (hr)
Maximum fuel consumption	3,300 pounds per hr

Range and Endurance		
Fuel Tank	Endurance (Hrs + Min)	Fuel Range (Nautical Miles)
Integral	2+08	256
Note. The range is limited only by aircrew and aerial refueling.		

Cruise Speed	
Normal cruise	120 knots indicated airspeed
Maximum	170 knots indicated airspeed
Note. Actual figures are dependent upon temperature, aircraft GW, and density altitude.	

Aircraft Dimensions	
Length of fuselage	50 feet 9 inches
Length of blades	98 feet 10 inches
Width of fuselage	11 feet 11 inches
Width of blades	52 feet
Height	18 feet 11 inches

Table 5-6. MH-47G aircraft capabilities (continued)

Aircraft Dimensions (continued)	
Diameter of main rotor	60 feet
Aircraft turning radius	122 feet
Cargo Area (Unobstructed)	
Height	78 inches
Width	90 inches
Depth	366 inches
Troop Capacity	
With seats	33 troops
Floor loading	65 troops
Litters	24 troops

Note. Actual amounts are dependent upon infiltration and exfiltration distances flown and the number of internal auxiliary fuel tanks installed or the availability of aerial refueling.

Typical Mission Composition

5-117. A variety of mission scenarios may employ the MH-47G. A typical mission profile for a low-to-medium threat infiltration and exfiltration sortie could take the following form:

- Departing by night, single-ship, VMC from a forward operating location to a target 260 nautical miles (range dependent upon fuel configuration and availability of prestaged FARP locations). MH-47G aircraft are capable of aerial refueling and are range limited by availability of MC-130 support.
- Navigating to an initial point using the best option of three navigational modes. GPS is the primary mode when the required number of satellites is available.
- Using guidance cues, following navigation steering to a landing site and accomplishing the approach to landing to a remote site of not less than 150 feet by 100 feet.
- On-loading an exfiltration party of up to 65 passengers from conditions not worse than 50 degrees Celsius (123 degrees Fahrenheit) at 500-foot pressure altitude.
- Reversing the route and returning to a recovery base or location in friendly territory.
- Using low en route altitudes (down to 50 feet) given favorable conditions of ambient light, visibility, and the use of infrared searchlight with NVDs.

MH-47E HELICOPTER

5-118. The primary mission of the MH-47E (Figure 5-18, page 5-28) is to conduct overt and covert infiltration, exfiltration, air assault, resupply, and sling operations over a wide range of environmental conditions. The aircraft can perform a variety of other missions, including shipboard, platform, urban, water, FARP, mass-casualty, and CSAR operations.

DESCRIPTION

5-119. The MH-47E (Chinook) is a twin-engine, tandem-rotor, heavy assault helicopter specifically designed and built for the SOA mission. It has a totally integrated avionics subsystem that combines the following:

- Redundant avionics architecture with dual mission processors.
- Remote terminal units.
- Multifunction displays and display generators to improve combat survivability and mission reliability.
- Aerial refueling probe for in-flight refueling.

Chapter 5

- External rescue hoist.
- Two L714 turbine engines with full-authority digital electronic control, which provides more power during hot or high environmental conditions.
- Two integral aircraft fuel tanks providing 2,068 gallons of fuel.
- Stormscope for thunderstorm avoidance.

Figure 5-18. MH-47E helicopter

AIRCRAFT SURVIVABILITY EQUIPMENT

5-120. The following explain each survival element found on the MH-47E helicopter:
- The AN/APR 39A (V) 1 RWR identifies hostile pulse fire control radar and provides audio and video alerts to the flight crew when the system detects threat radar emissions.
- The AN/APR 44 RWR identifies hostile AI and SAM CW fire control radar. It also provides audio and visual alerts to the flight crew when the system detects threat radar emissions.
- The M-130 Infrared Countermeasures Dispenser System deceives radar guidance and infrared missiles by using chaff or flares as required.
- The AN/AAR-47 Missile Warning System is a passive electronic warfare system that detects in-band infrared and ultraviolet radiation emanating from a missile plume.

STANDARD MISSION EQUIPMENT

5-121. The following paragraphs discuss standard equipment found on the MH-47E helicopter.

Armament

5-122. The MH-47E has three weapon stations: left forward window, right cabin door, and the ramp. The forward stations mount a 7.62-mm minigun and the ramp station mounts an M60D 7.62-mm machine gun. An aircrew member at each station manually operates the weapons. The weapons are primarily for self-defense and enemy suppression.

Note. When using the right cabin door weapon system, the forward fast-rope station cannot be used.

5-123. The minigun is normally for soft targets and troop suppression, which require a high rate of fire. The minigun is air cooled and link fed. It has a maximum effective range of 1,500 meters with tracer burnout at 900 meters. The weapon has an adjustable rate of fire of 2,000 or 4,000 rounds per minute. The aircrew members currently fire ball or SLAP ammunition with a mix of four balls to one tracer, 4:1, or a 9:1 mix to prevent NVD shutdown on low-illumination nights. The ammunition complement without reloading is 8,000 rounds per weapon.

Fast-Rope Insertion and Extraction System Bar

5-124. The FRIES is a system for inserting and extracting personnel. Applied loads for the FRIES are as follows:

- Applied loads at the rear ramp for insertions must not exceed nine persons per rope at the same time.
- Applied loads at the rear ramp for extractions must not exceed six persons per rope at the same time.

FLIR, AN/AAQ-16

5-125. The AN/AAQ-16 is a controllable, infrared surveillance system that provides a television video-type infrared image of terrain features and ground or airborne objects of interest. The FLIR is a passive system that detects long-wavelength radiant infrared energy emitted, naturally or artificially, by any object in daylight or darkness. In the MH-47E, the pilots or the onboard computer may control the FLIR, and the infrared video may be saved on VHS tape for later mission debriefings.

Map Display Generator

5-126. When used with the data transfer module, the map display generator displays aeronautical charts, photos, or digitized maps in the mission planning and 3D modes of operation based on digital terrain elevation data and digital feature analysis data.

MISSION-SELECTIVE SYSTEMS

5-127. The external cargo hook system may be used separately or with others (Table 5-7). All loads should be planned as a tandem-rigged load to facilitate greater load stability and to ensure faster airspeeds during flight.

Table 5-7. MH-47E external cargo hooks

Type	Capacity
Forward hook	17,000 pounds
Center hook	26,000 pounds
Aft hook	17,000 pounds
Tandem hook	25,000 pounds
Note. These are maximum hook-rated loads and may not accurately reflect the true capability of the aircraft because of external conditions, such as pressure, altitude, and temperature.	

5-128. The external rescue hoist is for use at the right front cabin door. It has a 600-pound capacity, with 200 feet of usable cable. Fast-rope operations can still be conducted out the front cabin door with the hoist installed.

5-129. The internal rescue hoist is for use at the center cargo hook and rescue hatch. It has a 600-pound capacity, with approximately 150 feet of usable cable.

5-130. The CCERFS consists of one and up to three ballistic-tolerant, self-sealing tanks. Each tank has the capacity of holding 800 gallons of fuel but normally is filled to 780 gallons. Filling may occur during ground or aerial refueling operations.

5-131. The FARE consists of fueling pumps, hoses, nozzles, and additional refueling equipment to set up a two-point refueling site. Gallons of fuel dispensed are dependent upon the range of operation required of the tanker aircraft.

TRANSPORTABILITY

5-132. The SOAR has modified and validated procedures to load two MH-47Es on a C-5 and one MH-47E on a C-17, with all support equipment. The time required to disassemble and assemble the MH-47E is greater than the time required of other SO helicopters.

PLANNING CONSIDERATIONS

5-133. Successful mission accomplishment is largely a function of adequate premission planning time. Mission notification should occur in time to have an adequate mission-planning session and briefing, followed by a period of rest before mission execution.

Weather Minimums

5-134. The weather minimum for the MH-47E is a 500-foot ceiling, with a visibility of 2 miles.

Winds

5-135. No wind minimums are specified for training, operational, and support missions; however, 45 knots is the maximum wind for starting and stopping the rotor system in the MH-47E.

Flight Altitudes

5-136. For training missions, the minimum en-route altitude is 150 feet AGL or AHO for reconnoitered routes and 300 feet AGL or AHO for unreconnoitered routes. For operational missions, the minimum en route altitude is dictated by threat systems.

Landing Areas

5-137. The minimum landing area for the MH-47E is 150 feet by 100 feet.

Shipboard Operations

5-138. The MH-47E can operate day and night from Navy ships possessing Level II, Class III helicopter-landing pads.

Aircrew Composition

5-139. Most training, exercises, or operational or contingency missions require five aircrew members. Aircrew members include a pilot, copilot, flight engineer, and two aircrew chiefs. The flight engineer is usually at the ramp station. He scans for other aircraft (targets or obstacles) and operates the hoist (when required), assists in FRIES operations, operates the machine gun, and conducts sling-load operations. The aircrew chiefs are at the left and right forward gunners' stations. They scan for other aircraft (targets or obstacles), operate the miniguns, and assist in sling-load and FRIES operations.

Aircrew Qualifications

5-140. All aircrews can support flight operations for the missions stated in FM 3-05 and JP 3-05. Aircrew qualifications include NVG infiltration and exfiltration operations to urban, overwater (ship, oil rigs), mountainous, desert, and jungle objectives arriving at the target at a prearranged time +30 seconds.

Aircrews are trained in formation live-fire, long-range NVD operations over land and water. MH-47E aircrews can also perform aerial-refueling operations.

Aircraft Capabilities

5-141. Table 5-8 lists the capabilities of the MH-47E aircraft. Figures 5-19 through 5-26, pages 5-32 through 5-39, illustrate specific dimensions and capabilities of the aircraft.

Table 5-8. MH-47E aircraft capabilities

Aircraft Weight	
Maximum gross weight	54,000 pounds
Empty gross weight	26,918 pounds
Aircraft Dimensions	
Length of fuselage	51 feet 1 inch
Length of probe	68 feet 5 inches
Length of blades	99 feet
Width of fuselage	15 feet 8 inches
Width of blades	52 feet
Height	18 feet 7 inches
Diameter of main rotor	60 feet
Aircraft turning radius	122 feet
Cargo Area (Unobstructed)	
Height	78 inches
Width	90 inches
Depth	366 inches
Troop Capacity	
With seats	44 troops
Floor loading	65 troops
Litters	24 troops
Airspeed	
Normal cruise	120 knots indicated airspeed
Maximum dash	170 knots indicated airspeed
Maximum altitude	20,000 feet
Fuel Flow	
Normal fuel consumption	2,750 pounds per hr
Maximum fuel consumption	3,300 pounds per hr
Range and Endurance	

Fuel Tank	Endurance (Hrs + Min)	Fuel Range (Nautical Miles)
Integral	4+30	540
Aircrew endurance and aerial refueling support may limit the range.		
Note. Actual figures depend upon temperature, aircraft GW, and density altitude.		

Chapter 5

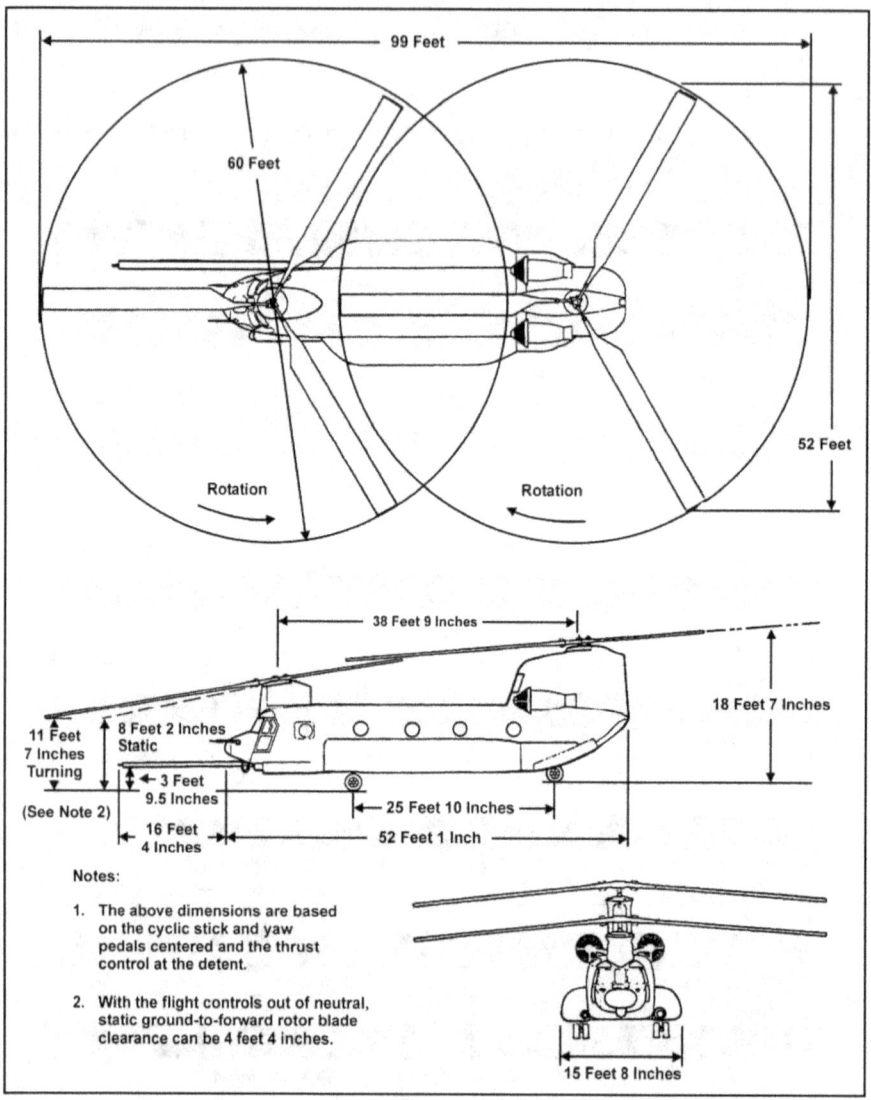

Figure 5-19. MH-47E aircraft capabilities and dimensions

Army Special Operations Aviation Units and Aircraft

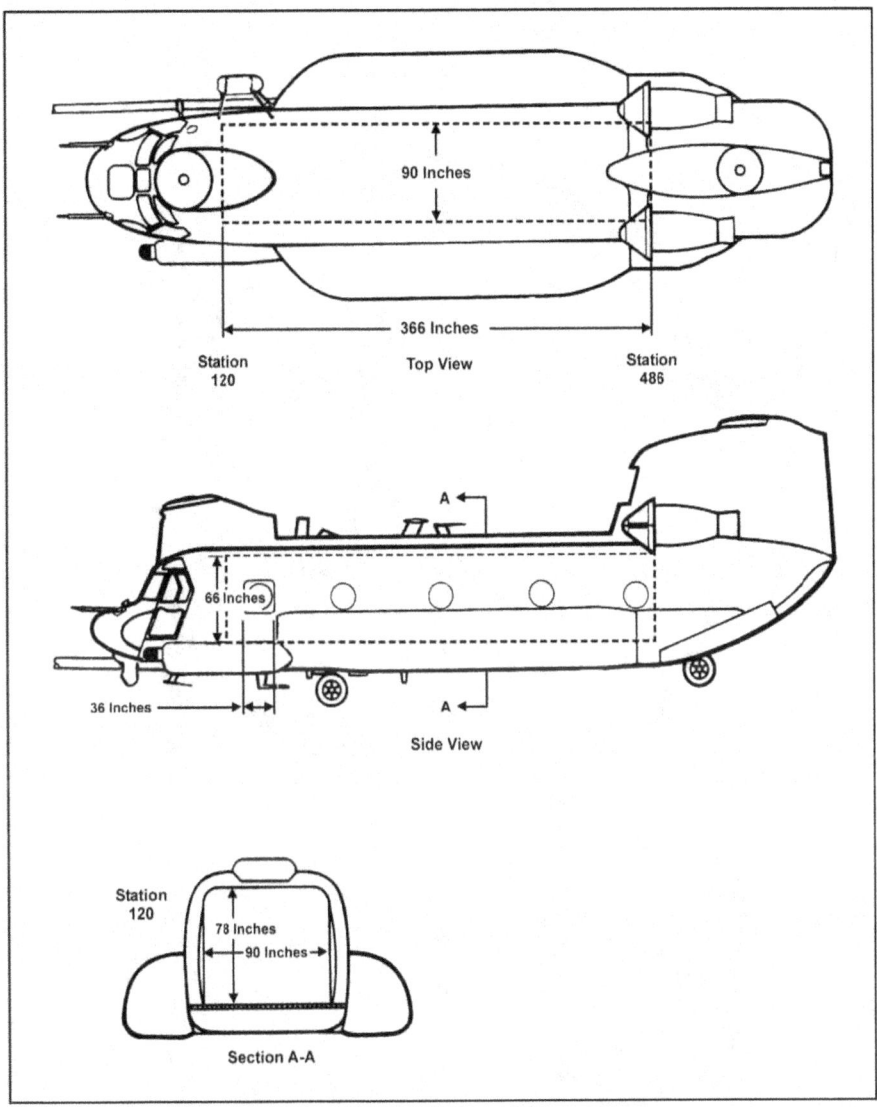

Figure 5-20. MH-47E capabilities of cargo areas

Chapter 5

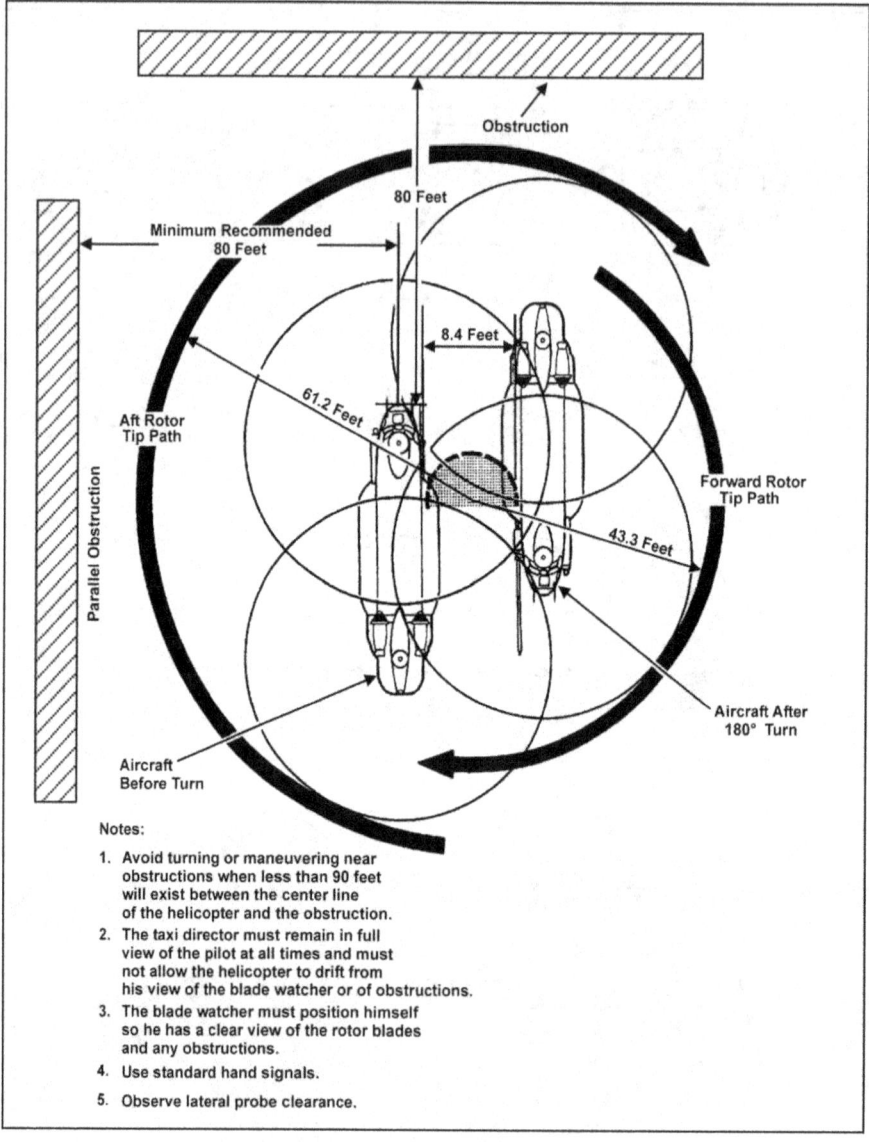

Figure 5-21. MH-47E capabilities of turning radius

Army Special Operations Aviation Units and Aircraft

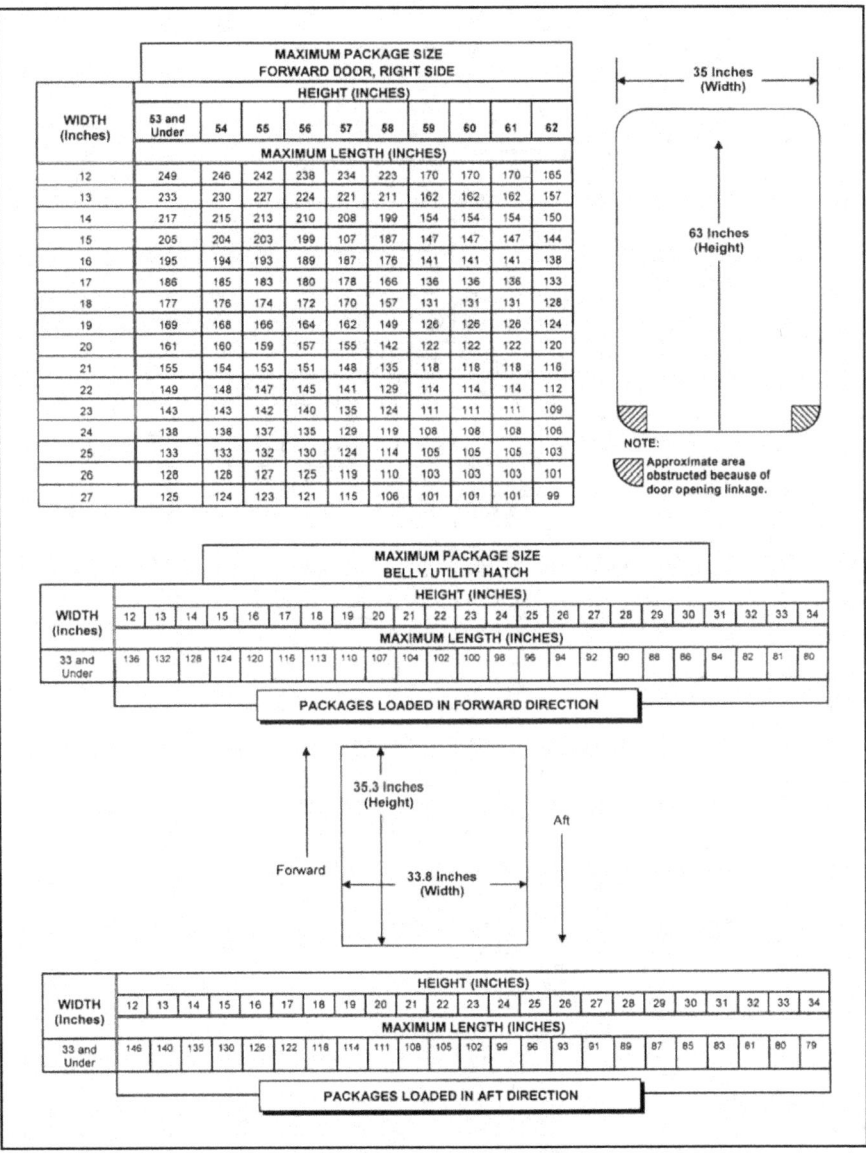

Figure 5-22. MH-47E maximum package size

Chapter 5

Figure 5-23. MH-47E ramp door maximum package size

WIDTH (INCHES)	MAXIMUM PACKAGE SIZE RAMP DOOR															
	HEIGHT (INCHES)															
	62 and Under	63	64	65	66	67	68	69	70	71	72	73	74	75	76	77
	MAXIMUM LENGTH (INCHES)															
62 and Under	362	362	362	362	362	362	362	330	282	230	180	135	100	67	30	
63	362	362	362	362	362	362	362	328	280	228	178	133	98	66		
64	362	362	362	362	362	362	362	326	278	226	176	130	96	64		
65	362	362	362	362	362	362	362	322	274	222	173	127	93			
66	362	362	362	362	362	362	362	318	270	218	169	123	90			
67	362	362	362	362	362	362	362	313	266	214	165	119	86			
68	362	362	362	362	362	362	357	307	260	208	160	114	81			
69	362	362	362	362	362	362	348	299	252	201	154	107	75			
70	362	362	362	362	362	362	339	290	243	193	146	99				
71	362	362	362	362	362	362	330	281	234	185	139	91				
72	362	362	362	362	362	362	321	272	226	177	131	83				
73	362	362	362	362	362	352	312	263	216	167	122	75				
74	362	362	362	362	362	339	298	250	203	156	112					
75	362	362	362	362	362	325	284	237	190	144	101					
76	362	362	362	362	348	311	270	223	177	132	90					
77	362	362	362	362	334	297	256	209	164	119						
78	362	362	362	346	316	278	237	191	147	104						
79	362	362	362	329	298	258	218	173	129	85						
80	362	362	362	310	276	236	195	151	108							
81	362	362	362	289	253	213	172	128	85							
82	362	362	362	267	230	188	148	105								
83	362	362	362	241	202	161	121									
84	362	362	362	213	174	133	93									
85	362	362	362	182	142	100										
86	362	362	362	146	105											
87	362	362	362	105												
88	362	362	362	362												
89	362	362	362	362												
90	362															

78 Inches (Height)

90 Inches (Width)

Figure 5-23. MH-47E ramp door maximum package size

Army Special Operations Aviation Units and Aircraft

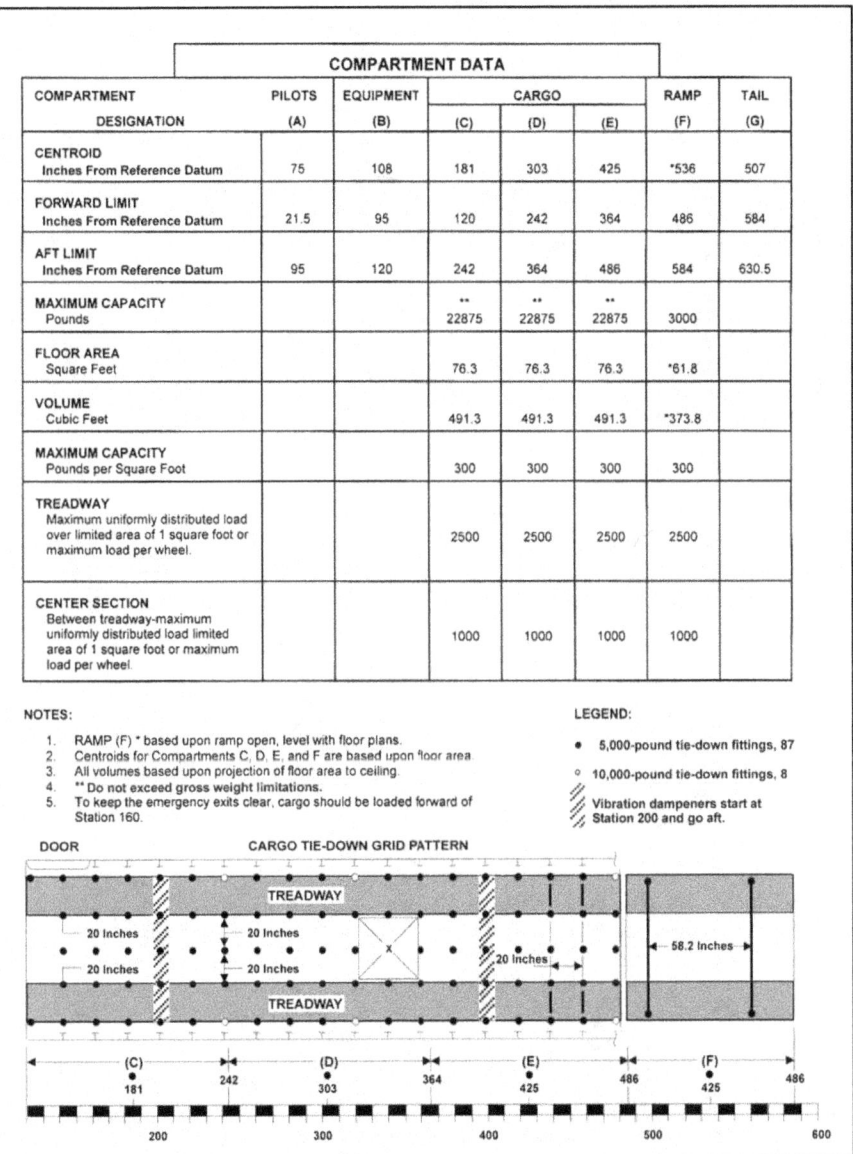

Figure 5-24. MH-47E compartment data

Chapter 5

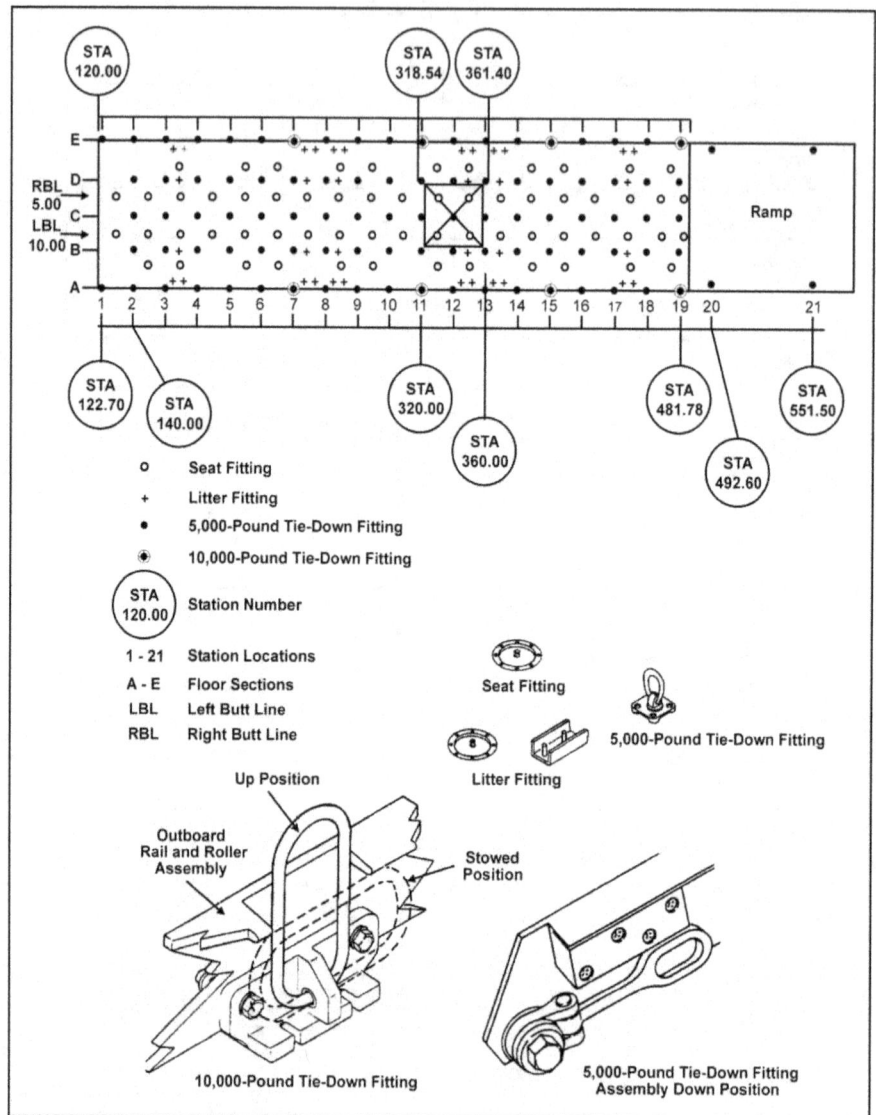

Figure 5-25. MH-47E fitting capabilities

Army Special Operations Aviation Units and Aircraft

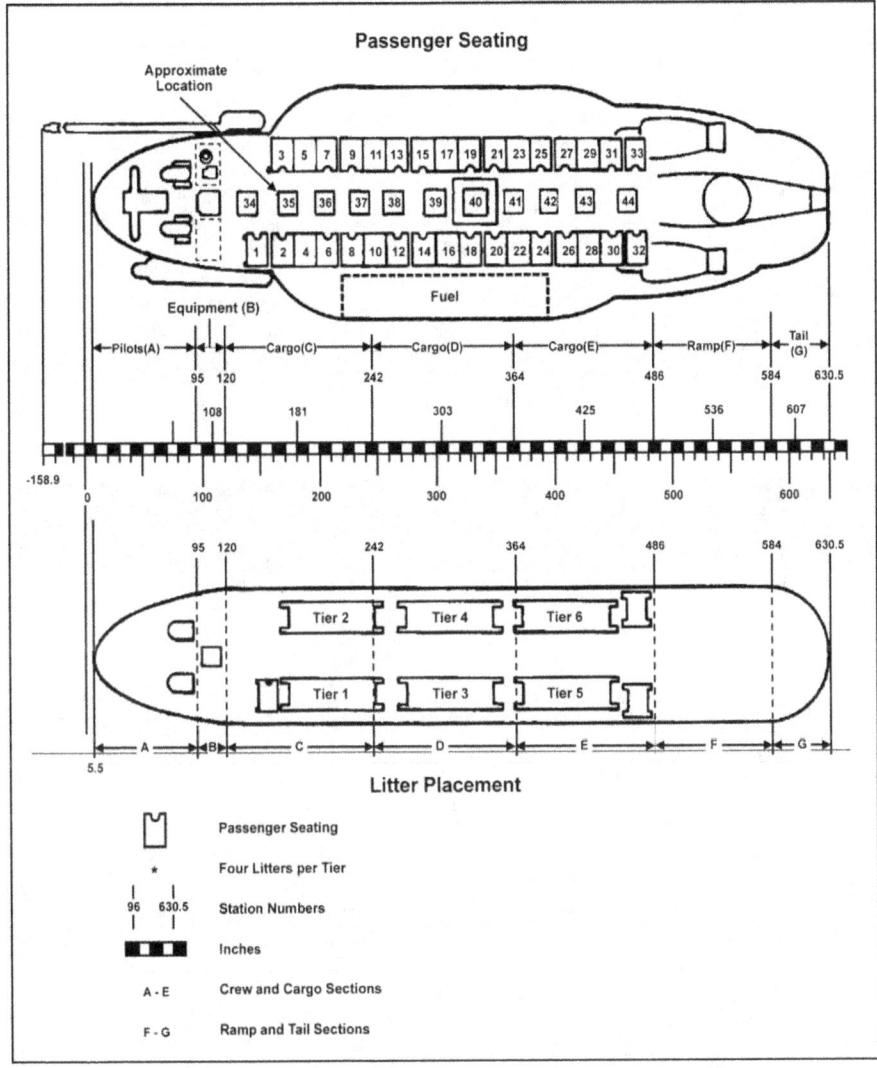

Figure 5-26. MH-47E passenger seating and litter placement

Typical Mission Composition

5-142. A variety of mission scenarios may employ the MH-47E. A typical mission profile for a low-to-medium threat infiltration or exfiltration sortie could take the following form:
- Departing by night, single-ship, VMC from a forward operating location to a target 520 nautical miles (range dependent upon fuel configuration and availability of prestaged FARP locations).

Chapter 5

The MH-47E aircraft is capable of aerial refueling. Availability of KC-, HC-, or MC-130 support limits the range.
- Navigating to an initial point using the best option of eight navigational modes available. GPS is the primary mode when the required number of satellites is available.
- Using guidance cues, following navigation steering to a landing site, and accomplishing the approach to landing to a remote site of not less than 150 feet by 100 feet.
- On-loading an exfiltration party of up to 65 passengers from conditions not worse than 50 degrees Celsius (123 degrees Fahrenheit) at 500-foot pressure altitude.
- Reversing the route and returning to a recovery base or location in friendly territory.
- Using low en route altitudes (down to 50 feet), given favorable conditions of ambient light, visibility, and use of infrared searchlight with NVDs or during adverse weather conditions.

Capabilities Matrix

5-143. The SOAR aircraft capabilities matrix (Table 5-9, pages 5-40 and 5-41) is a ready reference for mission planners. Its purpose is to reduce mission-planning time. The matrix provides instant information, without time-consuming research on the part of mission planners. The current aircraft capabilities matrix encompasses all aircraft systems. SOAR units must update the matrix periodically as technology changes occur.

Table 5-9. SOAR aircraft capabilities matrix

Types of Aircraft	MH-6M	AH-6M	MH-60L	MH-60L (DAP)	MH-60K	MH-47G	MH-47E
Aircraft Capabilities							
Cruise Speed (Knots)	80	90	120	120	110	120	120
Flight Time (Standard Tanks)	1+20	1+17	1+45	1+40	1+30	4+30	4+30
Range (Nautical Miles) (Standard Tanks)	110	116	212	200	165	540	540
Air Refuelable	no	no	no	yes	yes	yes	yes
Passengers	6	0	12	0	14	37	37
Maximum Passengers (No Seats for Rucksacks)	6	0	17	0	23	65	65
Landing Area (Feet)	50x50	50x50	100x100	100x100	100x100	150x100	150x100
Communications							
UHF	yes	yes	yes	yes	yes	yes	yes
HF	no	no	yes	yes	yes	yes	yes
VHF (FM/AM)	yes	yes	yes	yes	yes	yes	yes
SATCOM	yes	yes	yes	yes	yes	yes	yes
Single-Channel Ground and Airborne Radio Station	yes	yes	yes	yes	yes	yes	yes
Sabre	yes	yes	yes	yes	yes	yes	yes
Airborne Target Handover System	no	no	yes	yes	no	yes	yes
IMC Certified	no	no	yes	yes	yes	yes	yes
GPS	yes	yes	yes	yes	yes	yes	yes
ADF	no	no	yes	yes	yes	yes	yes
VOR/DME	yes	yes	yes	yes	yes	yes	yes

Table 5-9. SOAR aircraft capabilities matrix (continued)

Types of Aircraft	MH-6M	AH-6M	MH-60L	MH-60L (DAP)	MH-60K	MH-47G	MH-47E
Navigation							
ILS	no	yes	yes	yes	yes	yes	yes
Doppler	no	no	yes	yes	yes	yes	yes
INS	no	no	no	no	yes	yes	yes
Attitude Heading Reference System	no	no	no	no	yes	yes	yes
TACAN	yes	yes	yes	yes	yes	yes	yes
Air Data Computer	no	no	no	no	no	yes	yes
PLS	no	no	yes	yes	yes	yes	yes
Mission Computer Unit	no	no	no	no	yes	yes	yes
Long-Range Navigation (LORAN)	yes	yes	no	no	no	no	no
Armament							
M134 7.62-mm Minigun (Maximum)	no	yes	yes (2)	yes (2)	yes (2)	yes (2)	yes (2)
M230 30-mm Chain Gun	no	no	no	yes	no	no	no
M260 7-Shot Rocket	no	yes	no	yes	no	no	no
M261 19-Shot Rocket	no	yes	no	yes	no	no	no
AGM-114 Hellfire (Maximum)	no	yes (4)	no	yes (16)	no	no	no
M2 Caliber .50 Machine Gun (Maximum)	no	yes (2)	no	no	no	no	no
Standard or Special Equipment							
Ballistic Armor Subsystem	yes	yes	yes	yes	yes	yes	yes
Guardian Auxiliary Fuel Tank	no	no	2+00	2+00	2+00	no	no
FRIES	yes	no	yes	yes	yes	yes	yes
FLIR	yes	yes	yes	yes	yes	yes	yes
External Cargo Hook	no	no	9,000 pounds	no	8,000 pounds	26,000 pounds	26,000 pounds
Rescue Hoist	no	no	600 pounds	no	600 pounds	600 pounds	600 pounds
Auxiliary Fuel System (Time/Nautical Miles)	3+50/ 310	2+57/ 266	5+00/ 600	4+45/ 570	3+30/ 385	8+20/ 1000	8+20/ 1000
C2 Console	no	no	yes	no	no	yes	yes

This page intentionally left blank.

Chapter 6

Air Force Special Operations Organization and Aircraft

Air Force special operations forces (AFSOF) are America's special operations air power. AFSOF fall under the command of the Air Force Special Operations Command (AFSOC), which is a component of USSOCOM and a major command in the USAF. AFSOC organizes, equips, trains, validates, and employs air assets for worldwide deployment and assignment to geographic combatant commands. The highly trained AFSOF aircrews (active and reserve) and their uniquely equipped fixed- and rotary-wing or tilt-rotor aircraft provide precise multitarget firepower. AFSOF engages in a variety of missions, such as infiltration and exfiltration of SOF operational elements, armed escorts, reconnaissance, interdiction, resupply and refueling, as well as CSAR. AFSOF performs PSYOP by using video and radio broadcasting and conducting literature drops. AFSOF can support all core tasks and support activities of USSOCOM.

UNIT ORGANIZATION

6-1. AFSOC consists of one active group (1st Special Operations Group), one Air Force Reserve (AFR) wing (919th Special Operations Wing [SOW]), one Air National Guard (ANG) wing (193d SOW), two forward-deployed active special operations groups (SOGs) (352d and 353d), and one active special tactics group (STG) (720th).

- The 1st Special Operations Group is located at Hurlburt Field, Florida. The 1st SOG has 12 squadrons: 1st Special Operations Support Squadron, 3rd Special Operations Squadron (SOS) MQ-1 Predator/MQ-9 Reaper, (4th SOS) AC-130U Spooky Gunship, (6th SOS) UH-IN C-47 MI-8, C130, AN-26, (8th SOS) CV-22, (9th SOS) MC-130P Combat Shadow, (15th SOS) MC-130H Combat Talon II, (16th SOS) AC-130H Spectre Gunship, (19th SOS) Formal Training, (73rd SOS) MC-130W, and (319 SOS) U-28A.
- The 919th SOW (AFR) is located at Duke Field, Florida. The 919th SOW is equipped with MC-130E (711th SOS) and MC-130P (5th SOS) aircraft.
- The 193d SOW (ANG) is located at Harrisburg International Airport, Pennsylvania. The 193d SOW is equipped with the EC-130E/J COMMANDO SOLO aircraft (193d SOS).
- The 352d SOG is located at Mildenhall, England. The 352d SOG is equipped with MC-130H (7th SOS) and MC-130P (67th SOS) aircraft.
- The 353d SOG is located at Kadena Air Base, Okinawa. The 353d SOG is equipped with MC-130H (1st SOS) and MC-130P (17th SOS) aircraft.
- The 720th STG is located at Hurlburt Field, Florida, with liaison officers at HQ, AMC, Scott Air Force Base (AFB), California. The 720th STG consists of combat controllers, pararescuemen, combat weathermen, and support personnel. Under the 720th STG are the following units:
    - 21st STS at Pope AFB, North Carolina.
    - 22d STS at McChord AFB, Washington.
    - 23d STS at Hurlburt Field, Florida.
    - 24th STS at Fort Bragg, North Carolina.(JSOC)
    - 320th STS at Kadena, Japan (operational control to the 353d SOG).

Chapter 6

- 321st STS at Mildenhall, England (operational control to 352d SOG) with elements located in Germany.
- 10th Combat Weather Squadron (CWS) located at Hurlburt Field with detachments at Fort Lewis, Fort Campbell, Fort Carson, Fort Benning, and Fort Bragg.
- 123d Special Tactics Flight located at Louisville, Kentucky (ANG).

CONCEPT OF OPERATIONS

6-2. SO missions normally require undetected operations in hostile or denied areas and may require precise fire support. Most missions involve deep penetration (100 NMs or more) of hostile airspace to reach the target area. When within rotary-wing aircraft range and load limits, vertical lift insertion is generally more desirable than fixed-wing airdrop. Vertical lift insertion minimizes paradrop injuries and post-infiltration reassembly problems. On those missions where vertical lift is used, USAF rotary-wing aircraft are currently employed, with en route refueling being the essential factor to extend range. In-flight refueling is a necessary design capability to extend range and avoid security and logistics problems associated with forward area ground refueling.

ENVIRONMENT

6-3. AFSOF aircraft tasked in an SO environment must be capable of operating in hostile airspace at low altitudes under conditions of minimum visibility (darkness or adverse weather) while navigating precisely within narrow time and course parameters to arrive at specifically defined air refueling points; DZs, PZs or LZs; infiltration points; or targets.

6-4. In addition to aircraft capabilities, minimum lighting or lights out, radio discipline, deceptive course changes, and preplanned avoidance of enemy radar and air defenses and populated areas can enhance mission success. The use of deceptive ECM and passive warning systems prevent detection by enemy air defenses.

6-5. Per the tactics manual accompanying each type of aircraft, AFSOF aircraft have been type-coded based on the maximum threat level and weather conditions within which the weapon system can be expected to successfully operate. Table 6-1 depicts the type of classifications for specific aircraft:

- Type I aircraft are capable of operating at night, in adverse weather, and at low levels (below 500 feet AGL) in a medium- to high-threat environment. They are in-flight refuelable.
- Type II aircraft are capable of operating at night, in adverse weather, and at low levels (below 500 feet AGL) in a low- to medium-threat environment. They may be in-flight refuelable.
- Type III aircraft are capable of night visual flight at low levels, 500 feet AGL (fixed wing) or below (vertical lift), in a low- to medium-threat environment. They may be in-flight refuelable.
- Type IV aircraft are less capable than Types I, II, or III.

Table 6-1. Aircraft type classification

Type I	Type II	Type III	Type IV
MC-130E	None	MC-130P	EC-130E (CS)
MC-130H		C-17 SOLL II	
MV-22A		AC-130H	
		AC-130U	
		MH-60G	

CAPABILITIES

6-6. SO aircraft (transport or cargo and vertical-lift airframes) provide clandestine or covert penetration of hostile, sensitive, or politically denied airspace to infiltrate, resupply, and exfiltrate ground and maritime

forces in support of UW, DA, SR, counterterrorism (CT), and counterproliferation operations. They are equipped as follows:
- Fixed- and rotary-wing aircraft can be equipped for adverse weather, all terrain, long-range infiltration and exfiltration, suppressive fire support, personnel recovery, MEDEVAC, and PSYOP.
- Highly specialized fixed- and rotary-wing aircraft are equipped for deep penetration of hostile areas through use of electronic defense systems and terrain following, terrain avoidance, and NVG tactics. They are capable of airlanding or airdropping personnel, equipment, and psychological warfare materials.

6-7. AFSOF units may be tasked to advise, train, or assist the air forces of other nations in support of their internal defense and development strategy. These activities require language skills and detailed area orientation. Involvement in such activities may include operations, maintenance, and logistics support to host nation counterinsurgency, counternarcotics operations, or nation-building activities.

AERIAL TANKERS

6-8. SOF aerial tankers are equipped to refuel rotary-wing aircraft in flight, thereby significantly extending the range of the rotary-wing aircraft. Strategic fixed-wing tankers are capable of refueling certain SOF aerial tankers.

GUNSHIPS

6-9. AC-130 gunships are equipped with a variety of sensors and weapons to acquire and engage static and moving surface targets in an interdiction and CAS role or to provide armed escort. Aircrew training and avionics capabilities permit operations at night and in adverse weather.

PSYOP SUPPORT AIRCRAFT

6-10. Reserve component USAF PSYOP aircraft principally conduct a range of PSYOP activities in support of theater operational objectives or specific contingencies. In addition, select core SOF aircraft may be employed in PSYOP roles supporting either special or conventional operations.

AFSOF LIMITATIONS

6-11. AFSOF assets operate most effectively in particular environments and under specific methods of employment. Generally, AFSOF assets become less effective when employed outside of their intended operational environments. For instance, infiltration and exfiltration platforms may be less effective in conditions other than darkness and adverse weather. Planners must be familiar with the specific capabilities and limitations of each AFSOF platform as spelled out in appropriate technical manuals and orders. Broad-based limitations that apply generically to all AFSOF, because of their nature, include the following:
- Limited self-deployment and sustainment capability.
- Mission effectiveness degraded with increasing sophistication of enemy defenses.
- Dependence on established support and logistics packages that must accompany employment aircraft. Operations may be sustained from a bare base, but the technological sophistication of most AFSOF resources limits their "bed down" flexibility.
- High-technology avionics equipment requires extensive maintenance support.
- Long-range deployment and employment require aerial tanker support.
- Extremely limited defensive air-to-air capabilities.

AUGMENTING USAF FORCES

6-12. The Secretary of Defense has not designated AFSOF as core. However, certain general-purpose forces may receive enhanced training and may be specially equipped and organized to conduct missions in support of SO. These improvements enhance the primary combat capabilities of the conventional force and

Chapter 6

support SO on a nondedicated, mission-specific basis. These general-purpose forces do not conduct unrestricted, unilateral activities across the full range of SO missions.

SPECIAL OPERATIONS LOW LEVEL

6-13. These are basic fixed-wing strategic and theater (tactical) airlift forces. Because of special aircrew training or aircraft modification, SOLL can quickly augment AFSOF for the conduct and support of selected SO.

STRATEGIC TANKERS

6-14. USAF maintains a limited number of strategic tanker crews. These tanker crews are trained to support the often-unique refueling requirements of AFSOF fixed-wing aircraft.

AIR COMBAT COMMAND

6-15. AFSOC is employing the A-10 in SOF operations. The A-10 provides AFSOC with a hard kill and antiaircraft artillery suppression capability.

OTHER

6-16. Depending upon the mission and capabilities required, the Secretary of Defense or combatant commander allocates any Department of Defense assets to support SO. This support is usually mission-specific and of short duration. Such capabilities may include strategic or tactical bombing or airlift, airborne warning and control, electronic warfare, reconnaissance, deception, or space-based support. If the SOC commander requires these aircraft or support systems, the special operations liaison element or special operations command and control element (SOCCE) directly coordinates their use with the theater conventional joint force air component commander (JFACC) or theater service component commanders. If this support cannot be provided to the SOC because of other commitments and the SOC commander views his own requirements as critical, the theater commander may establish priorities for the requested conventional resources. Similarly, the theater JFACC, or other theater component commander, can coordinate with the SOC commander and the special operations liaison element for the use of SO aircraft to support conventional operations.

AIRCRAFT CAPABILITIES

6-17. SO aircraft are distinctly modified for specific missions. Although each type of aircraft may be able to perform several missions, some are more suited for certain missions than others. Tables 6-2 through 6-5, pages 6-4 through 6-9, list the capabilities and equipment of each type of aircraft.

Table 6-2. Aircraft capabilities

	MC-130E	MC-130H	AC-130H	AC-130U	MC-130P*	EC-130E
Airdrop With Seats Rigged	Varies[1]	Varies[1]	NA	NA	26	NA
Airdrop Without Seats Rigged	Varies[1]	Varies[1]	NA	NA	NA	NA

Table 6-2. Aircraft capabilities (continued)

	MC-130E	MC-130H	AC-130H	AC-130U	MC-130P*	EC-130E
Airdrop Maximum HE Weight	35,000 lb	35,000 lb	NA	NA	NA	NA
Airdrop Maximum CDS Bundles/ Weight	12 Bundles 34,000 lb Total Weight	16 Bundles 34,000 lb Total Weight	NA	NA	26 Bundles 5,000 lb Total Weight	NA
Airdrop Maximum HSLLADS/ CRS	4 Bundles Total Weight of 6,667 lb Not to Exceed 2,200 lb Dropped Per Pass	4 Bundles Total Weight of 6,667 lb Not to Exceed 2,200 lb Dropped Per Pass	NA	NA	NA	NA
Airdrop With Seats Rigged	Varies[1]	Varies[1]	NA	NA	26	NA
Airdrop Without Seats Rigged	Varies[1]	Varies[1]	NA	NA	NA	NA
Airdrop Maximum HE Weight	35,000 lb	35,000 lb	NA	NA	NA	NA
Airdrop Maximum CDS Bundles/ Weight	12 Bundles 34,000 lb Total Weight	16 Bundles 34,000 lb Total Weight	NA	NA	2 Bundles 5,000 lb Total Weight	NA
Airdrop Maximum HSLLADS/ CRS	4 Bundles Total Weight of 6,667 lb Not to Exceed 2,200 lb Dropped Per Pass	4 Bundles Total Weight of 6,667 lb Not to Exceed 2,200 lb Dropped Per Pass	NA	NA	NA	NA
Airland With Seats Rigged	59	77	NA	NA	34	NA
Airland (Floor Loaded)	Varies	Varies	NA	NA	Varies	NA
Litters	30	57	NA	NA	16	NA
Pallets	5	6	NA	NA	1	NA

Table 6-2. Aircraft capabilities (continued)

	MC-130E	MC-130H	AC-130H	AC-130U	MC-130P*	EC-130E
CRRC	4 on 2 Platforms and 20 Parachutists[2]	4 on 2 Platforms and 20 Parachutists[2]	NA	NA	NA	NA
Door Bundles	2 Bundles Per Pass Not to Exceed 500 lb	2 Bundles Per Pass Not to Exceed 500 lb	NA	NA	2 Bundles Per Pass Not to Exceed 500 lb	NA
Maximum Airspeed	250 KIAS	271 KIAS	250 KIAS	271 KIAS	250 KIAS	250 KIAS
Planning Airspeed for Deployment (High Level)	270 knots true airspeed (KTAS)	290 KTAS	240 KTAS	245 KTAS	300 KTAS	270 KTAS
Planning Airspeed for Deployment (Low Level)	230 GS	230 GS	NA	NA	230 GS	NA
Maximum GW[3]	155,000 lb	155,000 lb	155,000 lb	155,000 lb	155,000 lb	155,000 lb
Aircraft Weight Empty	100,000 lb	95,000 lb	115,000 lb	111,000 lb	87,000 lb	NA
Maximum Effort Takeoff GW (Using a 3,000-ft Runway) (Dry/Wet)[4]	SL, 15°C 154,000 Dry 150,000 Wet 4,000 ft, 15°C 133,000 Dry 129,000 Wet	SL, 35°C 140,000 Dry 137,000 Wet 4,000 ft, 35°C 120,000 Dry 117,000 Wet	NA	NA	SL, 15°C 154,000 Dry 150,000 Wet 4,000 ft, 35°C 120,000 Dry 117,000 Wet	NA
Normal Takeoff Runway Distance Required (Maximum Weight) (Dry/Wet)[4]	SL, 15°C 4,950 Dry 5,200 Wet SL, 15°C 6,150 Dry 6,500 Wet	SL, 15°C 4,950 Dry 5,200 Wet SL, 35°C 6,150 Dry 6,500 Wet	SL, 15°C 5,150 Dry 5,400 Wet SL, 35°C 6,450 Dry 6,800 Wet	SL, 15°C 5,150 Dry 5,400 Wet SL, 35°C 6,450 Dry 6,800 Wet	SL, 15°C 4,950 Dry 5,200 Wet SL, 35°C 6,150 Dry 6,500 Wet	SL, 15°C 5,150 Dry 5,400 Wet SL, 35°C 6,450 Dry 6,800 Wet
Normal Landing Runway Distance Required (Maximum Weight) (Dry/Wet)[4,5]	SL, 15°C 6,700 Dry 7,850 Wet SL, 35°C 7,450 Dry 8,600 Wet	SL, 15°C 6,700 Dry 7,850 Wet SL, 35°C 7,450 Dry 8,600 Wet	SL, 15°C 6,450 Dry 7,400 Wet SL, 35°C 7,200 Dry 8,350 Wet	SL, 15°C 6,450 Dry 7,400 Wet SL, 35°C 7,200 Dry 8,350 Wet	SL, 15°C 6,700 Dry 7,850 Wet SL, 35°C 7,450 Dry 8,600 Wet	SL, 15°C 6,450 Dry 7,400 Wet SL, 35°C 7,200 Dry 8,350 Wet

Air Force Special Operations Organization and Aircraft

Table 6-2. Aircraft capabilities (continued)

	MC-130E	MC-130H	AC-130H	AC-130U	MC-130P*	EC-130E
Maximum Effort Landing Runway Distance Required (130,000 lb) (Dry/Wet)[4]	SL, 15°C 1,900 Dry 2,550 Wet SL, 35°C 2,050 Dry 2,700 Wet	SL, 15°C 1,900 Dry 2,550 Wet SL, 35°C 2,050 Dry 2,700 Wet	NA	NA	SL, 15°C 1,900 Dry 2,550 Wet 4,000, 35°C 2,250 Dry 2,900 Wet	NA
Endurance (hr)/Range (NM) +20-Min Reserve. Main tanks Only, Maximum GW	NA	NA	4 hr 1,000 NM With 1 hr Reserve	3 hr 790 NM	NA	NA
Endurance (hr)/Range (NM) +20-Min Reserve. Main and Internal Aux Tanks Maximum GW	NA	NA	5 hr 45 min 1,265 NM Average	5 hr 1,224 NM	NA	NA
Endurance (hr)/Range (NM) +20-Min Reserve. Maximum Onboard Fuel, Maximum GW	NA	NA	8 hr 30 min 1,705 NM With 1 hr Reserve	11 hr 2,607 NM	NA	5 hr 30 Min 1,485 NM Average

*Information is based on normal configuration of one Benson Tank.

Notes.

[1] Personnel numbers will vary with size of troops and actual equipment carried. Floor loading may be used when standard seating configuration cannot be used because of mission requirements.

[2] Applies only if breakaway static lines are used, otherwise reduce number of parachutists by one for each platform to be dropped.

[3] Mission planners use 155,000 ft for planning purposes. Emergency war planning weight is 175,000 lb, but needs Commander, AFSOC or HQ, AFSOC waiver.

[4] Distances will increase with less than optimum conditions (wet, ice, high crosswinds, and so on), and NVG require an additional 1,000 ft above the distance required when not using NVG.

[5] Distances are for aircraft using 0% flaps, which requires the longest landing distance.

Chapter 6

Table 6-3. Aircraft communications capabilities

	MC-130E	MC-130H	AC-130H	AC-130U	MC-130P	EC-130E
UHF Voice/Data	Yes/Yes	Yes/Yes	Yes/No	Yes/No	Yes/No	Yes/No
Secure Voice/Data	Yes/Yes	Yes/Yes	Yes/No	Yes/No	Yes/No	Yes/No
VHF-AM Voice/Data	Yes/Yes	Yes/Yes	Yes/No	Yes/No	Yes/No	Yes/No
Secure Voice/Data	Yes/Yes	Yes/Yes	Yes/No	Yes/No	Yes/No	Yes/No
VHF-FM Voice/Data	Yes/Yes	Yes/Yes	Yes/No	Yes/No	Yes/No	Yes/No
Secure Voice/Data	Yes/Yes	Yes/Yes	Yes/No	Yes/No	Yes/No	Yes/No
HF-AM Voice/Data	Yes/Yes	Yes/Yes	Yes/No	Yes/No	Yes/No	Yes/No
Secure Voice/Data	Yes/No	Yes/No	Yes/No	Yes/No	Yes/No	Yes/No
SATCOM Voice/Data	Yes/Yes	Yes/Yes	Yes/No	Secure Voice Only	Yes/No	Yes/No
Secure Voice/Data	Yes/Yes	Yes/Yes	Yes/No		Yes/No	Yes/No
Emergency Locator Transmitter	Yes	Yes	Yes	Yes	Yes	Yes
Havequick	Yes	Yes	No	Yes	Yes	Yes
Saber	No	No	Yes	No	No	No

Table 6-4. Aircraft navigation capabilities

	MC-130E	MC-130H	AC-130H	AC-130U	MC-130P	EC-130E
GPS	Yes	Yes	Yes	Yes	Yes	No
LORAN	No	No	No	No	No	No
VOR/Distance Measuring Equipment	Yes (2)	Yes (2)	Yes	Yes	Yes	Yes
TACAN	Yes (2)	Yes (2)	Yes	Yes	Yes	Yes
ILS	Yes (2)	Yes (2)	Yes	Yes	Yes	Yes
Internal Navigational Units	Yes (2)	Yes (2)	Yes	Yes	No	NA
Mission Computers	Yes (2)	Yes (2)	No	No	No	NA
Self-Contained Navigation System (SCNS) and ADF	No	No	No	No	No	Yes

Table 6-5. Additional capabilities

	MC-130E	MC-130H	AC-130H	AC-130U	MC-130P	EC-130E
AN/APQ-112(V) 8 Multimode Radar System	Yes	Yes	No	No	No	No
In-Flight Refueling	No	Yes	Yes	Yes	Yes	Yes

AC-130H GUNSHIPS

6-18. The AC-130H provides precision fire and other support for SOF and general purpose forces, including CAS, armed reconnaissance, interdiction, escort convoy or helicopter, surveillance, and search and rescue. The AC-130H is equipped with the following:

- Side-firing 105-mm howitzer.
- 40-mm cannon.
- Twin 20-mm Gattling guns.
- Fire control computers.
- ECM.
- All-weather targeting.
- Extensive navigation and sensor suites.

AC-130U GUNSHIPS

6-19. The AC-130U provides precision fire and other support for SOF and general purpose forces, including CAS, armed reconnaissance, interdiction, escort convoy or helicopter, surveillance, and search and rescue. The AC-130U is equipped with the following:

- Side-firing 105-mm howitzer.
- 40-mm cannon.
- 25-mm Gattling gun.
- Fire control computers.
- ECM.
- All-weather targeting.
- Extensive navigation and sensor suites.
- 30-mm will replace the 40-mm and 25-mm in the future.

MC-130E COMBAT TALON

6-20. The MC-130E provides low-level, long-range, night, single-ship, or formation refueling of SOF rotary-wing aircraft and infiltration, exfiltration, or resupply of SOF via airland or airdrop. The MC-130E is equipped with the following:

- Aerial refueling system, which extends its range.
- Terrain-following or terrain avoidance radar.
- Precision navigation and defensive suite.

Note. The MC-130E will be upgraded to the MC-130W in the future.

MC-130H COMBAT TALON II

6-21. The MC-130H provides low-level, long-range, night, single-ship or formation infiltration, exfiltration, or resupply of SOF via airland or airdrop. The MC-130 H is equipped with the following:

- Aerial refueling system, which extends its range.
- Terrain-following and terrain-avoidance radar.
- Precision navigation and defensive suite.

MC-130P COMBAT SHADOW

6-22. The MC-130P provides low-level, long-range, night, single-ship or formation refueling of SOF rotary-wing aircraft and limited infiltration, exfiltration, or resupply of SOF via airland or airdrop. The MC-130P is equipped with the following:
- An aerial refueling system, which extends its range.
- Enhanced navigation.
- Defensive avionics and threat warning systems.

EC-130J COMMANDO SOLO

6-23. The EC-130J (CS) provides broadcasting capabilities primarily for PSYOP missions, supports disaster relief operations, and performs communications jamming in military spectrum and intelligence gathering. The capabilities of the EC-130J include reception, analysis, and transmission of various electronic signals to exploit electronic spectrum for maximum battlefield advantage. Secondary capabilities include jamming, deception, and manipulation techniques. Broadcasts in frequency spectrums include AM/FM radio, shortwave, television, and military command, control and communications channels.

CV-22 OSPREY

6-24. The CV-22 is a multi-engine, dual-piloted, self-deployable, medium-lift, advanced vertical and/or short takeoff and landing aircraft. The CV-22 conducts SOF missions worldwide. The capabilities of the CV-22 include low-visibility, clandestine penetration of medium- to high-threat environments employing terrain-following and terrain-avoidance radar, robust self-defense avionics, and secure, anti-jam, redundant communications used by C2 and ground forces. The CV-22 will be fully capable of operations in adverse weather (day or night), in all climates worldwide, and in a variety of conventional and unconventional contingency or combat situations including CBRN warfare conditions. The CV-22 will operate from main operating bases, austere forward operating locations, and aircraft-capable ships. It is air-refuelable from C-130s, KC-135s, and KC-10s.

HH-60G PAVE HAWK

6-25. The HH-60G provides medium-range, low-level, day or night infiltration, exfiltration, or resupply of SOF. Unique capabilities permit selected rescue and recovery missions. The capabilities of the MH-60G include the following:
- Single-ship or formation infiltration, exfiltration, or resupply of SOF via airland, airdrop, or alternate insertion or extraction procedures.
- Extended range due to in-flight refueling.
- Precision navigation.
- Armor protective systems.
- Defensive avionics and threat warning systems.
- Complete shipboard compatibility.

COMBAT AVIATION ADVISORY TEAMS

6-26. Combat aviation advisory teams provide the unique capabilities for training and advising host nation forces in every theater under foreign internal defense (FID), CS, and UW mission areas. Capabilities include light- to medium-airlift operations of low-level, day and night, single-ship or formation infiltration, exfiltration, or resupply of forces via airland or airdrop. Fighter capabilities include air intercept and CAS operations. Helicopter capabilities include low-level, day or night single-ship or formation infiltration, exfiltration, or resupply of forces via airland, airdrop, or alternate insertion and extraction procedures and land and water CSAR. In support of these air operations, the teams are capable of working the following:
- All maintenance and logistics for the different airframe types.
- Air base defense, intelligence, pararescue and combat control unique capabilities. Mission oversight capabilities include all joint air operations center activities.

SPECIAL TACTICS SQUADRON

6-27. Special tactics squadrons are ground combat forces assigned to AFSOC. Special tactics squadrons consist of combat control, pararescue, and combat weather personnel who are uniquely organized, trained, and equipped to establish and control the air-ground interface and provide airman skills in the objective area. Functions of the special tactics squadron include the following:

- Assault zone assessment, establishment, and control.
- Trauma medical treatment and personnel recovery.
- SO terminal attack control.
- Tactical weather observation and forecasting.
- Deploy with air and joint ground forces in the execution of DA, CT, FID, HA, SR, austere airfield operations missions, and CSAR.
- Conduct tactical assault zone survey, and position and monitor terminal NAVAIDs.
- Provide long-range, secure C2 communications.
- Remove obstacles with demolitions.
- Gather and report ground intelligence.
- Conduct sensitive recovery operations and casualty transload and evacuation operations.
- Generate mission-tailored forecasts.
- Determine the impacts of meteorological and oceanographic conditions on current and planned operations.
- Deploy into SO areas to collect weather intelligence, to train and equip host nation military and guerrilla forces to take weather observations, and to set up remote weather networks.

MC-130E COMBAT TALON I AND MC-130H COMBAT TALON II

6-28. The MC-130E Combat Talon (Figure 6-1, page 6-12) aircraft is required to support the range of activities from crisis response to wartime commitment in the SO mission. The mission of the Combat Talon is to conduct day and night infiltration, exfiltration, resupply, PSYOP, and aerial reconnaissance into hostile territory using airland or airdrop. Combat Talons are capable of in-flight refueling, giving it an extended range. The Combat Talon I is capable of in-flight refueling specially modified helicopters for extended helicopter operations. The MC-130 missions may be accomplished either single-ship or in concert with other SO assets in varying multiaircraft scenarios. The MC-130s can airland and airdrop personnel and equipment on austere, marked, and unmarked LZs or DZs, day or night. MC-130 missions may require overt, clandestine, or low-visibility operations. The MC-13CE and MC-130H are Type I aircraft.

EQUIPMENT

6-29. The standard MC-130E has the equipment listed below. Aircraft configurations may vary.
- Terrain-following and terrain-avoidance radar.
- Precision ground mapping radar.
- Precision navigation system (INS, Doppler, and GPS).
- Automatic CARP.
- ECM, CRS, FLIR, and HSLLADS.
- Infrared countermeasures (IRCM).
- Ground-to-air responder interrogator.
- In-flight refueling, receiver operations.
- Secure voice HF, UHF, VHF-FM, and SATCOM radios.
- Helicopter refueling, tanker operations.
- Internal fuel tanks (Benson tanks).

Chapter 6

Figure 6-1. MC-130E

GENERAL PLANNING FACTORS

6-30. Mission planners should consider the following factors:
- *Infiltration and Exfiltration:*
    - For static-line low-altitude airdrops, pilot flies aircraft at 130 knots and 500 feet (minimum) AGL. Combination drops are static-line drops along with CRSs. For combination drops, troops exit from the ramp immediately after ejection of equipment from the ramp.
    - When a free-fall is planned before parachute opening, HALO airdrops are made above an altitude of 5,000 AGL. The navigator /MFFJM will determine the HARP. HAHO airdrops normally occur at an altitude above 10,000 feet AGL, but with no free-fall, to travel large distances. Airspeed for HALO and HAHO is 130 KIAS. For drops above 18,000 feet, a physiological training technician must accompany the flight.
    - Minimum runway length is takeoff and landing roll, plus 152 meters (about 914 meters). Minimum runway width is 18 meters. NVG operations require a runway at least 1,158 meters long and 18 meters wide.
    - A CRRC can be airdropped using low-level procedures (800 feet AGL minimum). The CRRC weighs about 2,500 pounds and is 180 inches by 75 inches. Nineteen parachutists can be airdropped along with one CRRC or eighteen parachutists with two CRRCs. This combination is due to the number of static lines per pass.
- *Resupply:*
    - HSLLADS is the primary method of low-level (250 feet AGL minimum) resupply, since it minimizes risk to the aircraft and aircrew and avoids compromise of the DZ. The HSLLADS airdrop occurs at a maximum airspeed of 250 knots. Automatic CARP or special sight procedures (marked point of impact) are used to determine the RP. High-speed containers must weigh between 250 to 600 pounds each. Four containers may be dropped on one pass, providing the total weight does not exceed 2,200 pounds.
    - CRS is used for low-level (minimum 500 feet AGL), low-speed (130 knots) gravity drops. The CRS is employed to airdrop A-series containers. A total CRS weight not to exceed 6,667 pounds may be dropped on a single pass. A CRS combination drop allows parachutists and bundles to be dropped. Parachutists exit the aircraft from the ramp after the load exits.
    - A door bundle is a container weighing less than 500 pounds, and personnel release it from the jump door or off the ramp. Load dimensions must be 48 inches by 30 inches by 66 inches or less. When dropping a door bundle with parachutists, the door bundle will be the first object to exit the aircraft. When dropping a door bundle off the ramp, it must have a skid board on the bottom of the load and personnel use the intermediate roller conveyers. When dropping a bundle from the aircraft ramp, personnel rig a breakaway parachute to the bundle or rig a pilot chute in a T-10 bag. When dropping a bundle from the jump door, personnel rig a nonbreakaway parachute with a drogue parachute.

- *PSYOP Conducted Through Leaflet Drops.* Psychological warfare units will supply necessary information, leaflets, and other materials, as required.
- *Aerial Reconnaissance Systems and Training.* Some aircrews have training to conduct visual reconnaissance (for example, sea surveillance). A portable camera system may be mounted on the Combat Talon aircraft. This system provides a limited daylight photo capability for use in a low-threat environment. A portable video tape recorder can also be used to tape video from the FLIR. This provides a limited night photo capability.

PERFORMANCE CONSIDERATIONS

6-31. Planners consider the following factors when using SO aircraft to support a mission:
- Aircraft range depends upon several factors, including configuration, payload, length of time spent low level, en route winds, and weather. For planning purposes, range of aircraft (without refueling, 2 hours low level) is 2,300 NMs. Only crew duty day limitations and availability of tanker support limit the range of aircraft capable of in-flight refueling.
- Mission duration will depend on aircraft basing location, aircraft configuration, crew composition, target location, availability of tanker support, and routing required for successful mission accomplishment.
- For a basic crew, their duty day is 16 hours, providing no tactical events occur after 12 hours and no refueling occurs after 14 hours.
- For an augmented crew, their duty day is 20 hours, providing no tactical events occur after 16 hours and no air refueling occurs after 18 hours.
- Normal fuel load for takeoff will not exceed 55,000 pounds for JP8 or its equivalent.
- Normal takeoff GW will not exceed 155,000 pounds. However, GWs up to 175,000 pounds can be employed with a waiver. Empty aircraft weight varies from 95,000 to 100,000 pounds, depending on aircraft configuration.

MC-130W COMBAT SPEAR

6-32. The MC-130W Combat Spear conducts infiltration, exfiltration, and resupply of U.S. and allied SOF in direct support of unified and theater SO commands and USSOCOM contingencies. Detailed missions include—
- Refueling of special operations vertical lift assets.
- Forward arming and refueling.
- Specialized ordnance delivery.
- Airdrops in support of PSYOP.
- Limited C2 capabilities.

Its worldwide mission is performed primarily at night to reduce operational risk.

6-33. The MC-130W is a modified C-130H(2) featuring improved navigation, threat detection and countermeasures, and communication suites. The navigation suite is a fully integrated GPS/INS that interfaces with the AN/APN-241 low power color radar and AN/AAQ-17 infrared detection system. The improved threat detection and countermeasures systems include advanced radar and missile warning receivers, chaff and flare dispensers, and active infrared countermeasures that protect the aircraft from both radar and infrared-guided threats. The communication systems upgrades include dual SATCOM suite with data burst capability. The aircraft has both interior and exterior night vision goggle compatible lighting.

6-34. Structural improvements to the basic C-130H include the addition of the universal aerial refueling receptacle slipway installation (UARRSI), and a strengthened tail empennage. The UARRSI allows the aircraft to conduct in-flight refueling as a receiver, and strengthening of the tail will allow high speed, low level aerial delivery system (HSLLADS) airdrop operations. The MC-130W has Mk32B-902E refueling pods. These pods are part of the most technologically advanced refueling system available, and provide the ability to refuel SO helicopters and the CV-22.

AC-130H/U GUNSHIP

6-35. The AC-130 (Figures 6-2 and 6-3) is a C-130 modified with gun systems, electronic and electro-optical sensors, fire control systems, enhanced navigation systems, sophisticated communications, defensive systems, and in-flight refueling capability. These systems give the gunship crew the capability to acquire and identify targets day or night, coordinate with ground forces and C2 agencies, and deliver surgical firepower in support of conventional and SO missions. The AC-130 is a Type III aircraft. Further information, in a classified format, is contained in the AC-130H and AC-130U tactics manuals.

Figure 6-2. AC-130H

Figure 6-3. AC-130U

SPECIFIC EMPLOYMENT

6-36. The gunship has many capabilities and uses but CAS is the primary mission of the AC-130. It is best suited for this mission and has the unique capability to deliver ordnance extremely close to friendly forces in a troops-in-contact situation. Other missions that the gunship can be used for are as follows:
- *Interdiction.* The gunship is best suited to strike small targets in a permissive environment. The accuracy of the gunship, low-yield munitions, and target identification capability reduce the risk of collateral damage. Because many high-value interdiction targets are well defended, the gunship may

be unsuitable or unable to engage them. Also, the gunship lacks great hitting power and area coverage capability, which limits the potential for damage to hard targets or large area targets.
- *Armed Reconnaissance.* The gunship is well suited to search lines of communication. Capabilities are similar to those for interdiction. The narrow field of view of the sensors limits the ability of the gunship to search large areas. The time required to perform armed reconnaissance must be considered with respect to the threat and terrain. The limitations associated with interdiction apply.
- *Specialized Missions.* Point defense, convoy, naval, and helicopter escort and vectoring are versions of CAS. The AC-130 can provide surveillance and limited protection of friendly forces from enemy ambush, but must maintain communications with escort commander. Mission planning should include a prebriefed route, en route checkpoints, and communications nets. Ground parties using electronic beacons greatly aid in force vectoring.
- *Forward Air Controller (FAC).* The AC-130 can control fighter aircraft. The FAC and mission commander conduct a joint prebriefing of each mission. Communication between aircraft is essential. The AC-130 can mark a target with the laser target designator (code 1688 only with the AC-130H) or by "sparkling" the target with gunfire. Also, the gunship can provide offsets from visible terrain features, sound markers, or easily distinguished features in the area.

CREW QUALIFICATIONS

6-37. Mission crew size is 14 personnel (13 on the AC-130U). The augmented crew size is 18 personnel (17 on the AC-130U). The basic crew consists of the pilot, copilot, navigator, fire control officer (FCO), electronic warfare officer, engineer, loadmaster, television (TV) sensor operator, infrared sensor operator, and five gunners (four on the AC-130U). Selected crews are qualified in NVG low-level missions.

EQUIPMENT

6-38. To perform the various gunship missions, the MC-130 has the equipment listed below. Aircraft configurations may vary.
- Dual fire control computers.
- Enhanced navigation system (dual INS, GPS, and Doppler).
- Secure communications: two HF, two VHF AY-1FM, two UHF, and SATCOM data burst.
- In-flight refueling.
- Low-light-level television (LLLTV) with laser (AC-130H).
- Illuminator.
- All-light-level television (ALLTV) with laser illuminator (AC-130U).
- Infrared detection set (IDS).
- ASD-5 Black Crow (BC) direction-finding set (AC-130H).
- APQ-150 beacon tracking radar (BTR) (AC-130H).
- APQ-180 strike radar (AC-130U).
- Two fixed 20-mm cannons (AC-130H).
- 25-mm cannon, trainable (AC-130U).
- 40-mm Bofors automatic cannon, trainable.
- 105-mm cannon, trainable.
- Battle damage assessment record.

Fire Control Computers

6-39. Fire control computers resolve all variables to complete fire control solutions. They can store preselected targets and sensor sight line targets. Both aircraft have two computers to provide redundancy. Additionally, the AC-130U has two separate fire control channels. This allows independent targeting for dual target attack (DTA). The aircrew can use a DTA against two separate targets or deliver maximum ordnance on a single target by allocating an individual sensor and gun to each fire control channel. The

Chapter 6

aircrew can use DTA with one sensor and two guns on the same target. For dual target engagements where targets are more than about 800 meters apart, the aircrew may not be able to engage targets during all portions of the orbit. A live-fire bore sight area is required before firing near friendly forces. Employment altitude depends on terrain, threat environment, and weather.

Navigation

6-40. Navigation aids include two TACANs and VORs, the GPS, dual INS, and Doppler feed into a Kalman filter that produces excellent navigational accuracy. Additionally, the navigation computers can store targets and navigation points.

Communications Radios

6-41. Two HF, three (two for the AC-130U) VHF AM/FM, two UHF, and one SATCOM. Secure voice is available for HF (KY-75), VHF, UHF, and SATCOM (KY-58). Both UHF radios have Havequick II.

Sensors

6-42. Electro-optical and visual sensors require VMC. The TV and IR systems degrade during thermal crossover. All images in the aircraft are black and white. The ability to acquire and identify a given target depends on the size of the target, slant range, target contrast with the environment, weather conditions, and sensor or FCO proficiency. Normally, a crew can detect a vehicle-size target from 18,000 feet AGL. Sensors include—

- *LLLTV (AC-130H)*. The LLLTV can be used in extremely low light, below about 30 percent illumination, with the gated laser illuminator (GLINT). Normally, the crew would use this laser illuminator only in a permissive environment. The LLLTV has a wide and narrow field of view (FOV). The narrow FOV magnification is about 25 to 1. This gives an FOV of about 70 meters at 6,000 feet AGL, 8,000 feet slant range.
- *ALLTV (AC-130U)*. The ALLTV also uses a laser illuminator. The ALLTV has a wide, medium, and narrow FOV. The narrow FOV magnification is about 30 to 1.
- *IDS*. Both aircraft use the AAQ-17 IDS, which requires no visible light. The AC-130H has an improved version designated the AAQ-17E. This system has a wide and narrow FOV. The narrow FOV of the AAQ-17E magnification is about 30 to 1.

6-43. Gunships have a limited capability to deliver firepower under conditions of low ceilings and/or poor visibility. The APQ-150 BTR and ASD-5 BC give the AC-130H a limited all-weather capability. The APQ-180 strike radar gives the AC-130U a good capability against radar-significant targets, including those marked by beacon. This enables the AC-130U to have the capability to shoot through clouds. For beacon offsets with either aircraft, a ground controller must be present to correct the ordnance of the gunship for target, range, and magnetic bearing from the location of the beacon. The systems tend to be more accurate with shorter offset distances.

Defensive Equipment

6-44. The AC-130 is equipped with a variety of defensive equipment. Though some basic differences exist between the AC-130H and AC-130U, both aircraft are equipped with essentially the same capabilities. Table 6-6, page 6-17, shows the current equipment. All defensive equipment on the gunship is designed to defeat a single engagement, but not allow prolonged exposure to a threat. The primary means of survival is mission planning to avoid a threat envelope—by circumnavigation or by altitude separation. The AC-130H and AC-130U tactics manuals list the threat environments against which the gunship can be employed. The specific capability of defensive systems and tactics against a given threat is classified. Further, risk assessment decisions are a complex balance of mission priority, available options, threat system, proficiency of the threat system operators, and weather conditions. Therefore, a list of the probability of survival against a specific threat is inappropriate. The crew tasked to fly a mission, additional planners, and intelligence support provide the best vehicle to determine the probability of survival and mission success.

Table 6-6. AC-130 defensive equipment

Radar Warning	ALR-69/A1R-56M (AC-130U), QRC 84-05
ECM	ALQ-131 Pods/ALQ-172 (H Model After Fiscal Year 01)
Chaff Dispensers (10)	ALE-40
Flare Dispensers (8)	ALE-40
IRCM	ALQ-84-02A, Engine Heat Shield
Missile Warning	AAR-44

6-45. Gun selection depends on target type and damage desired. There are four basic target categories—soft area, soft point, hard area, and hard point (Table 6-7). Table 6-8 depicts the basic ammunition load for the gunships. Following are weapons and the target category for which they are used:

- *20-mm.* Used for soft area targets from altitudes of 3,000 to 7,500 feet AGL. The standard load is 3,000 rounds high-explosives incendiary (HEI).
- *25-mm.* Used for soft area and point targets from 3,000 to 15,000 feet AGL. The standard load is 3,000 rounds HEI.
- *40-mm.* Used against soft area and point targets from altitudes of 4,500 to 15,000 feet AGL. The standard load is 256 rounds HEI.
- *105-mm.* Used for soft and hard area targets and soft point targets from altitudes of 4,500 to 18,000 feet AGL. The standard load is 100 rounds HE (fuzes include the FMU-153 hardened improved penetrator, M-732 proximity, and a standard M-557 point-detonating) and white phosphorus mix.

Table 6-7. AC-130 target categories

Category	Area Target	Point Target
Soft Targets	Troops in the Open, Area Suppression	Unarmored Vehicles, Antiaircraft Artillery Pieces
Hard Targets	Concrete Buildings	Armored Personnel Carrier, Light Armor

Table 6-8. AC-130 ammunition loads

Type of Load	AC-130H	AC-130U
Training	2,000 Rounds 20-mm 96 Rounds 40-mm 20 Rounds 105-mm	3,000 Rounds 25-mm 256 Rounds 40-mm 100 Rounds 105-mm
Maximum Ammunition	3,000 Rounds 20-mm 256 Rounds 40-mm 100 Rounds 105-mm	3,000 Rounds 25-mm 256 Rounds 40-mm 100 Rounds 105-mm

Beacons

6-46. One type of ground marking equipment used with the AC-130 is the beacon. The two electronic sensors on the AC-130H, the APQ-150 BTR, the ASD-5 BC, and the APQ-180 strike radar on the AC-130U work with several types of beacons. Gunship crews are proficient in beacon use. Actual offset firing with friendly forces at the beacon location is not recommended in training situations. Types of beacons are as follows:

- Radar beacons used with the BTR (AC-130H) and strike radar (AC-130U):
  - PPN-I9 (I-band, codes B and G only).
  - SST-181X (codes 1 through 10).

- PRD-7880 selectable strike beacon single sideband/tactical electromagnetic ignition generator used with the AC-130H BC *only*.

Visual Marking Methods

6-47. Another type of ground marking includes the visual methods. The types listed below mark a position for either the IR or TV sensors:
- *Lights*. A standard survival strobe light with an IR filter can be used with the LLLTV or ALLTV. Also, flashlights, vehicle lights, chemical lights, tracer fire, and so on work with the TV sensors. Positive identification can be made by turning one or more lights off and on in response to radio instructions from the aircrew.
- *Heat Sources*. Heat sources such as meals ready to eat (MREs), heaters, stoves, and so on can normally be identified by the IDS.

PERFORMANCE CONSIDERATIONS

6-48. Mission planners must consider the following factors when using AC-130 gunships to support operations:
- AC-130 takeoff and high-altitude performance are marginal at high-pressure altitudes and high GWs.
- Ferry range is about 1,000 NMs, at 245 KIAS with ammunition and no in-flight refueling. Fuel burn is about 6,000 pounds per hour; 7,000 pounds per hour at low level. Fuel load is contingent on number of crew, ammunition load, and mission configuration.
- Normal maximum GW is 155,000 pounds. Operations with GW of 155,000 to 175,000 pounds require a waiver.
- Required crew rest is 12 hours. Basic crew day is 16 hours, 12 hours tactical. Augmented crew day is 18 hours, 16 hours tactical. Waivers are required to exceed these limits.

6-49. Normal mission profiles are as follows:
- Takeoff or sensor alignment: 0.5 hr.
- En route or TOT, including return to base: 4 hr.
- Descent and landing: 0.5 hr.

MC-130P COMBAT SHADOW

6-50. The mission of the MC-130P, formerly known as the HC-130P/N, is clandestine formation or single-ship intrusion of hostile territory to provide aerial refueling of SO helicopters and the infiltration, exfiltration, and resupply of SOF by airdrop or airland operations. To perform these missions, the primary emphasis is on NVG operations. The MC-130P is a Type III aircraft (Figure 6-4, page 6-19). The MC-130P capabilities listed below are not meant to be all-inclusive:
- The aircraft was originally modified for CSAR. It maintains most of its rescue capability. High-intensity parachute flares, various smoke-producing pyrotechnics, and sea dye are still carried aboard this aircraft for helicopter overwater escort and rescue.
- Some aircraft have been modified with the universal air refueling receptacle slipway installation system for in-flight refueling as a receiver and all have the SCNS. These modifications greatly increase the range and navigational accuracy of the MC-130P. The aircraft normally carries eight crew members. Depending on mission profile and duration, additional crew members are carried. The crew members are NVG-qualified. Although no continuation training is performed, all crew members are considered qualified in CSAR. The special qualifications are HARP airdrop, NVG landings, and in-flight refueling.
- The aircraft has a radar-warning receiver, flares and chaff for diverting enemy missiles from the aircraft. No active jamming capability exists.

- Depending upon the enemy threat, navigation is accomplished through visual and electronic means or visual means only.
- Aircraft tactical missions require VMC. Weather minimums for low-level missions are a ceiling of 1,500 feet and a visibility of 3 miles.

6-51. Comprehensive mission employment information can be obtained in a classified format in the MC-130P tactics manuals.

Figure 6-4. MC-130P refueling two MH-53Js

EQUIPMENT

6-52. The following equipment is installed on the MC-130P:
- In-flight refueling system for helicopters.
- Internal fuel tanks (Benson tanks).
- Airborne radar (APN-59E).
- SCNS.
- RWR (ALR-69).
- Chaff and flare dispensers (ALE-40).
- Missile warning system (AAR-44).
- FLIR (AAQ-17).
- Secure speech (KY 58/75) UHF, VHF, VHF-FM, HF, and SATCOM.
- Havequick II radios.
- NVG (ANVIS-VI and 4949).
- Digital data burst.
- NVG heads-up display.

SPECIFIC EMPLOYMENT

6-53. MC-130P is employed according to the considerations discussed in the following paragraphs.

Chapter 6

Low Level

6-54. The MC-130P employs night tactical operation (NTO) procedures. The aircrew flies NTO missions in VMC by using NVG. The aircrew flies the mission profile at 500 feet AGL by using terrain masking. If necessary, the aircrew can fly the mission with minimal visual and electronic emissions. The range of the mission depends on several factors—length of time on the low-level route, en route weather, winds, and the air refueling offload requirements. The aircrew may fly portions of the profile at high altitude to minimize fuel consumption. To avoid enemy detection in a nonpermissive environment and get the aircraft to the objective area, the aircrew will use NTO procedures.

Formation

6-55. The MC-130P normally flies a formation of aircraft to provide multiple simultaneous refueling of large helicopter formations. An airborne spare tanker is also a part of the formation. Aircraft are flown with 200 to 500 feet separation.

Air Refueling

6-56. Air refueling is the primary mission of the MC-130P. The Combat Shadow can simultaneously refuel two helicopters from a single aircraft, which significantly decreases the time required to refuel helicopters. The aircrew uses NVG for night refueling.

Airdrops

6-57. The MC-130P can airdrop personnel, bundles, CRRC, and CDS. The DZ PI must be marked. The marking can be overt or covert. The location, size, and marking of DZs must follow the guidelines in AFI 11-202, Volume 3, *General Flight Rules*.

Personnel Drops

6-58. The MC-130P can be used for static-line and free-fall jumps as follows:
- For static-line low-altitude airdrops, the aircraft flies at 130 KIAS at a minimum of 800 feet AGL. The number of jumpers dropped per pass depends on the static-line configuration. When using the Combat Shadow as a drop platform, the user must coordinate the number of jumpers so that the aircrew can properly configure aircraft.
- HALO airdrops occur above 5,000 feet AGL where a free-fall is planned before parachute opening. The navigator/MFFJM will determine the HARP. HAHO airdrops normally occur above 10,000 feet AGL and without a free-fall jump, which allows aircraft to travel large distances. Aircrews fly HALO and HAHO airdrops at 130 KIAS.

Equipment Drops

6-59. At very low altitudes, the aircrew directs parabundle and free-fall door bundle drops. The aircrew drops parabundles at 300 feet AGL with parachutes or 150 feet AGL without parachutes. Whether or not bundles have parachutes attached, the aircrew flies at 130 KIAS during the drops. Most MC-130Ps are configured for dropping CDS bundles. Good coordination is necessary so that the right crew qualifications and aircraft configurations are available.

Airland

6-60. Infiltration and exfiltration may be conducted on overtly or covertly (IR) marked LZs. The SOF-modified aircraft are capable of blacked out (unmarked) NVG landings. LZs and lighting must conform to AFI 11-201, *Flight Information*.

PLANNING CONSIDERATIONS

6-61. Three hours are required before takeoff for briefings, final planning, aircraft preflight checks, engine start, taxi, and takeoff. Most missions are 5 to 6 hours long, to include 3 to 4 hours of low-level flight.

Airland infiltration and exfiltration may be conducted on overt, covert, or unmarked (SOF aircraft) LZs. For airland operations, the LZ should be a hard surface. Runway length less than 1,158 meters for NVG landings will not be used. Minimum runway width is 18 meters. Minimum taxiway width is 9 meters.

PLANNING FACTS

6-62. Basic crew duty day is 16 hours, providing no tactical events occur after 12 hours and no air refueling occurs after 14 hours. Crew duty day for an augmented crew is 20 hours, providing no tactical events occur after 16 hours and no air refueling occurs after 18 hours.

EC-130J COMMANDO SOLO

6-63. The EC-130J Commando Solo (Figure 6-5), a specially-modified four-engine Hercules transport, conducts information operations, PSYOP, and civil affairs broadcasts in AM, FM, HF, TV, and military communications bands. A typical mission consists of a single-ship orbit offset from the desired target audience—either military or civilian personnel. The 193rd SOW in Middletown, Pennsylvania, has total responsibility for the Commando Solo missions. Many modifications have been made to the EC-130J. These include enhanced navigation systems, self-protection equipment, air refueling, and the capability of broadcasting radio and color TV on all worldwide standards. This aircraft may also be used to—

- Support disaster assistance efforts by broadcasting public information and instruction for evacuation operations.
- Provide temporary replacement for existing transmitters or expand their areas of coverage.
- Support other requirements that involve radio and television broadcasting in its frequency range.

Figure 6-5. EC-130J

PEACETIME OPERATIONS

6-64. Commanders use the EC-130J in peacetime operations. The following paragraphs explain how this aircraft can operate in various types of missions.

Civil Action

6-65. EC-130J capabilities can support civil actions. The EC-130J broadcasts educational programs and telecasts, messages and speeches by government officials of friendly countries, and entertainment and cultural programs.

Chapter 6

Civil Disturbances

6-66. If rioters disrupt operations of local radio, or TV stations, either through destructive acts or by capture, the EC-130J can be employed to communicate to the affected community and appropriate authorities.

Disaster Control

6-67. During disasters resulting from natural phenomena such as storms, floods, and earthquakes, Commando Solo can transmit on assigned civil defense frequencies and can provide vital communications links.

C-5/C-17 SOLL II

6-68. The C-5/C-17 SOLL II forces from the AMC are required to conduct clandestine formation or single-ship intrusion of hostile territory to provide highly reliable, self-contained, precision airdrop or airland drop of personnel and equipment. The assumed mission concept will be day or night, low-level, adverse weather, and without the use of external aids. Minimum lighting, minimum communications, deceptive course changes, and preplanned avoidance of enemy radar/air defenses and populated areas enhance mission success. The aircraft are well suited for many special operations applications because of their load-carrying capability, ability to operate into short runway operations (1,524 meters), and worldwide signature. These aircraft operate in a low-threat environment as defined by the tactics manual for these aircraft.

SOLL CAPABILITIES

6-69. SOLL II crews consist of three pilots, two navigators, two loadmasters, and a flight engineer. The C-5 has four loadmasters. Aircrews use NVG for navigation, DZ acquisition, and airdrop. Night VMC routes may be flown at 500 feet AHO within 3 NMs of route centerline. LZs will be appropriately marked with standard overt RCL markings. The C-5 can accomplish blacked-out landings. Weather minimums are a ceiling of 1,500 feet and visibility of 3 miles for SOLL missions.

EMPLOYMENT OPERATIONS

6-70. Because of OPSEC considerations, rapid-response requirements, and/or lack of suitable forward operating bases, many C-5, C-17 SOLL II missions will require long-range employment flights. C2 communications are necessary and will be accomplished by secure SATCOM and line-of-sight radios. Mission planners select landfall points to minimize detection by hostile forces. Precise navigation positioning after extended over water flights is required. On these long missions, it is imperative that aircrew and user fatigue are minimized. By minimizing aircrew and user fatigue, human errors are reduced during critical phases of the mission, such as the low-level portion and the objective area operations.

Low-Level Infiltration or Exfiltration Operations

6-71. C-5 and C-17 SOLL II forces are required to penetrate hostile or sensitive airspace under blacked-out or adverse weather conditions without EW or ground control intercept radar detection, either single-ship or in formation with other aircraft. The aircrew constructs low-level routes to minimize detection. Strict navigational tolerances and terrain avoidance capabilities are critical factors for mission success. C-5 and C-17 SOLL II forces will fly the low-level route at as high an altitude as the threat and other detection factors will allow.

Airdrop Operations

6-72. SOLL II forces will be used to conduct clandestine airdrops of personnel, supplies, and equipment into very small unmarked water and land DZs at night during blacked out and/or adverse weather conditions. These airdrops may be conducted either single-ship, in formation, or in concert with other aircraft. Aircrews may conduct drops at low altitudes (for example, 300 to 1,500 feet AGL) or from aircraft normal cruise altitude (for example, C-17 flight level of 350 feet). SOLL II airdrops are typically conducted in a hostile ground threat environment and are required to navigate to a precision airborne RP without the assistance of external aids. Because of ground threats and various OPSEC considerations, multiple passes on a DZ are not acceptable and negatively affect mission success. Airdrops may be on unmarked DZs relying solely on visual identification of DZs with covert visual markings. Aircrews may conduct these airdrops with or without NVGs, depending on the ground tactical situation. During visual airdrops and while operating with NVGs, the pilots must be capable of independently determining aircraft attitude, airspeed, heading, and course deviations, and visual formation position. SOLL II crews are also required to conduct clandestine, single-pass, and low-altitude combination airdrops of small boats and personnel.

Airland Operations

6-73. SOLL II forces are required to conduct self-contained precision approaches to minimum clandestinely prepared LZs under marginal weather conditions. While on final approach, SOLL II aircrews must be capable of independently determining if the LZ is clear of obstructions placed by hostile forces. Throughout the approach and especially during the critical final phase while using NVGs, both pilots must be capable independently of determining the following:

- Aircraft barometric and radar altitude.
- Indicated airspeed.
- Ground speed.
- Aircraft descent rate.
- Heading.
- Course deviations.

The pilots must perform the entire approach and landing without displaying any overt external lights.

Ground Operations

6-74. SOLL II forces must be capable of safe, rapid, clandestine off-loading and onloading of personnel, equipment, vehicles, and other cargo without displaying any overt lights. Depending on the mission, SOLL II forces may conduct blacked out hot refueling operations with other aircraft.

6-75. The nature of the missions listed above results in the SOLL II mission being highly dependent upon accurate, complete, all-source, real-time intelligence. The prudent use of complementary assets such as Airborne Warning and Control System, Compass Call, or Wild Weasel may improve the chance of mission success.

CV-22 OSPREY

6-76. The CV-22 Osprey (Figure 6-6, page 6-24) is a tilt-rotor aircraft that combines the vertical takeoff, hover, and vertical landing qualities of a helicopter with the long-range, fuel efficiency, and speed characteristics of a turboprop aircraft. Its mission is to conduct long-range infiltration, exfiltration and resupply missions for SOF.

6-77. This versatile, self-deployable aircraft offers increased speed and range over other rotary-wing aircraft, enabling AFSOC aircrews to execute long-range SO missions. The CV-22 can perform missions that normally would require both fixed-wing and rotary-wing aircraft. The CV-22 takes off vertically and, once airborne, the nacelles (engine and prop-rotor group) on each wing can rotate into a forward position.

6-78. The CV-22 is equipped with integrated threat countermeasures, terrain-following radar, forward-looking infrared sensor, and other advanced avionics systems that allow it to operate at low altitude in adverse weather conditions and medium- to high-threat environments. Table 6-9, page 6-25, list the general and technical characteristics of the Osprey.

Chapter 6

Figure 6-6. CV-22 Osprey

Table 6-9. Osprey characteristics

Capabilities	CV-22 Osprey
Length	57 feet, 4 inches (17.4 meters)
Height	22 feet, 1 inch (6.73 meters)
Wingspan	84 feet, 7 inches (25.8 meters)
Rotary Diameter	38 feet (11.6 meters)
Thrust	More than 6,200 shaft horsepower per engine
Powerplant	Two Rolls Royce AE1107C turboshaft engines (max and intermediate, shp (kw)—6,150 (4,586)
Rotor System	Blades per hub—3; Construction—graphite/fiberglass; Tip speed, fps (mps)—661.90 (201.75); Diameter, ft (m)—38.08 (11.6); Blade folding—automatic, powered
Transmissions	Takeoff VTOL, Normal Power–4580 rhp; Takeoff VTOL, Interim Power–5183 rhp
Performance @ 47,000 lbs	Max speed, SL, kts (km/h)—250 (463); Max rate of climb, SL, fpm (m/m) 3,200 (975); Service ceiling, ft (m)—25,000 (7,620); Service ceiling, one engine inop, ft (m)—10,300 (3,139)
Range	Unrefueled mission radius with 24 troops, nm (km)—390 (722); Self-deployment, nm (km)—2,100 (3,892) with refueling
Crew	Cockpit—crew seats—2/3; Cabine—crew chief and troop seats/litters—1 + 24/12
Dimensions (internal)	Length, max, ft (m)—20.8 (6.34); Width, max, ft (m)—5.7 (1.74); Height, max, ft (m)—5.5 (1.67)
Weights	Takeoff, vertical, max, lbs—52,600 (23,859); Takeoff, short running, max, lbs (kg) 60,500 (27,443); Cargo hook, single, lbs (kg)—10,000 (4,536); Cargo hook, dual, lbs (kg)—15,000 (6,147)
Dimensions (external)	Length, fuselage, ft (m)—57.33 (17.48); Width, rotors turning, ft (m)—84.6 (25.78); Length, stowed, ft (m)—63.1 (19.23); Width, stowed, ft (m)—18.5 (5.64); Height, nacelles fully vertical, ft (m)—22.1 (6.74); Height, vertical stabilizer, ft (m)—17.65 (5.38)
Fuel Capacity	MV-22 gals (liters)—1,721 (6,513); CV-22 gals (liters)—2,037 (7,710); Aux, self-deployment, gals (liters)—up to 3 tanks, each 430 (1,628)

HH-60G PAVE HAWK

6-79. The HH-60G Pave Hawk (Figure 6-7) is a modern, medium-lift, SO helicopter for missions requiring medium- to long-range infiltration, exfiltration, and resupply of SOF on land or sea. In addition, the unique mission equipment of the SOF allows this aircraft to be used for recovery of injured SO personnel. The Pave Hawk can be employed in a low- to medium-threat environment.

6-80. Aircrews maintain qualification in NVG tactical operations, NVG aerial refueling, NVG shipboard operations, and NVG overwater operations, to include rubber boat deployment, low and slow fast-rope infiltration, and hoist or rope ladder exfiltration. Aircrews also maintain qualification in weapons employment, using 7.62-mm minigun and side-firing .50 caliber machine guns.

Figure 6-7. HH-60G Pave Hawk

EQUIPMENT

6-81. The HH-60G is a highly modified variant of the UH-60A Blackhawk. It offers increased capability in range (endurance), navigation, communications, and defensive systems.

Refueling

6-82. An air-fueling probe for in-flight aerial refueling increases mission endurance. In addition, pressure or gravity feed systems at FARPs or onboard ships allows refueling on the ground. The HH-60G has a choice of internal auxiliary fuel tanks for extended range operations. Either the aircraft can be equipped with the single 117-gallon tank or the dual 185-gallon tanks. The single tank provides 3.3 hours of aircraft operations, and the double tank 4.5 hours of operations without refueling.

Insertion and Extraction Systems

6-83. The HH-60G has standard insertion and extraction devices for hoist, fast rope, rappelling, and SPIES operations. The standard devices include the following:
- Internal cargo tie-down rings.
- Externally mounted rescue hoist.
- FRIES.

Navigation

6-84. HH-60G navigation equipment includes the following:
- A Doppler.
- An inertial navigation unit with a laser gyro/GPS.
- TACAN.
- KG-IO map display unit.
- Weather avoidance radar.
- FLIR.

Communications

6-85. Communications systems include secure HF, UHF, and FM radios with SATCOM and digital data burst capabilities. The Pave Hawk is equipped with PLS for rescue operations and secure communications with ground force C2.

Altitude Hold and Hover Stabilization

6-86. The aircraft is equipped with the altitude hold and hover stabilization (AHHS). The AHHS allows the HH-60 to hover without outside visual references.

DEFENSIVE SYSTEMS

6-87. Defensive equipment includes the following:
- ALQ-144 IRCM system.
- Hover infrared suppressor subsystem.
- Improved flare and chaff dispensing systems.

WEAPONS EMPLOYMENT

6-88. Defensive armaments include a forward cabin-mounted 7.62-mm minigun firing either 2,000 or 4,000 rounds per minute and cabin-mounted .50-caliber machine guns. With the addition of the ESSS, the aircraft can carry fixed forward-firing armaments for use as a defensive and escort aircraft. Each ESSS wing carries two 7- or 19-shot 2.75-inch folding fin aerial rocket pods or dual 20-mm cannons and .50-caliber machine guns.

PLANNING CONSIDERATIONS

6-89. The HH-60G can be successfully employed in the low- to medium-threat environment. As the level of threat increases above this, the chance of detection will increase, decreasing the probability of success. The probability of success will also decrease as the total number of aircraft in the mission increases, because of an increased chance of detection. The requirement to operate from a FARP will also decrease the probability of success, because of the extended exposure time.

6-90. The HH-60G is capable of operating in a variety of weather conditions. Because of the aircrew's use of night optical devices (NVGs and FLIR) and color weather radar, the aircraft can operate in very low visibility with low cloud ceilings. However, the HH-60G is a VMC platform with weather avoidance capability.

6-91. The time required to adequately plan for a mission varies with the complexity and length of the mission (that is, flight time, number of other aircraft, types of aircraft involved in the formation, threat, and location of the objective). Generally, comprehensive mission planning requires a minimum of 6 hours. Ideally, a tasking arrives while the crews are in crew rest, and mission planners accomplish primary mission planning. The crews arrive about 3 hours before their mission departure time and fine-tune the planning.

SPECIFIC EMPLOYMENT

6-92. The HH-60G can support a full range of special air warfare activities, to include SO, PSYOP, and Civil Affairs operations. Mission effectiveness is highly dependent upon accurate, complete, all-source, real-time intelligence. The HH-60G has weather avoidance radar, but this equipment does not replace the use of detailed, highly accurate, timely weather forecasts for premission planning.

DEPLOYMENT OPERATIONS

6-93. The HH-60G can be deployed by airlift and sea lift or can be self-deployed, as discussed below. The preferred deployment option is airlift, using C-5, and is essential if rapid deployment is required. A C-5 can transport a maximum of five HH-60Gs. The aircraft can be loaded for shipment in less than 1 hour and downloaded, rebuilt, and in the air in less than 1 hour.

- The optimum deployment package is four HH-60Gs via C-5. The rapid tear down and buildup times make air transportation a faster means of deployment. When distances exceed 1,500 NMs using aerial refueling or 1,000 NMs using ground refueling, transporting the aircraft by air is faster than self-deploying.
- Deployments can be worldwide by using a main base or a limited or standby base with host support.
- Deployments can be conducted in a deceptive or low-visibility mode.
- Self-deployment aerial refueling requirements are as follows:
  - One tanker aircraft and one spare per four HH-60Gs.
  - Two tanker aircraft plus one spare per six HH-60Gs.

EMPLOYMENT OPTIONS

6-94. The HH-60G will operate at low altitudes over land and water. The aircraft will normally be employed as part of a larger vertical lift package, which may require dissimilar multiship formations. The HH-60G will operate into unprepared, unlighted, uncontrolled LZs 50 meters or larger in diameter. The aircraft is capable of transporting 12 combat-equipped troops in an alternate-loading configuration without internal auxiliary fuel tanks. With internal fuel tanks installed, maximum troop capacity is ten, with an optimum load of six.

CHARACTERISTICS

6-95. The HH-60G has the following characteristics:
- *Range:*
  - Unrefueled 117-gallon tanks with 400 pounds reserve fuel: 150-NM radius or 2.8 hours.
  - Unrefueled dual 185-gallon tanks with 400 pounds reserve fuel: 225-NM radius or 4.1 hours.
  - Aerial refueling: Unlimited.
- *Speed:*
  - Cruise: 110 to 130 knots.
  - Dash: 140 knots.
- *Gross Weight:* a maximum of 20,250 pounds.
- *Ammunition Load:*
  - Maximum: 4,500 rounds 7.62-mm per canister or 800 rounds.
  - Standard: 3,000 rounds 7.62-mm per canister and 800 .50-caliber rounds.

Chapter 6

- *Duration of Mission:* lasts for a maximum of 14 hours (limited by crew duty day regulation).
- *Typical Profile:* would include flying from the point of departure to a point at which the low-level environment would be at altitudes from 50 to 300 feet AGL. Once the low-level environment is entered, flight operations would be conducted 50 to 150 feet AGL (all altitudes are dictated by threat), arriving at the objective at a prearranged TOT (+/–30 seconds). After meeting the objective, the HH-60G would return to a preplanned recovery base at altitudes commensurate with the threat. The departure and recovery base can be on land or on a ship. Throughout the mission, the HH-60G flies in formation with the other aircraft required to complete the mission. The option to rendezvous with other aircraft at various points along the profile is available as required. The typical mission profile would exploit the ability of the HH-60G to operate at night, in blackout, over land or sea, in marginal weather conditions, and with minimum or no communications.
- *Fuel Loads:*
  - Single 117-gallon tank installed: 3,200 pounds maximum.
  - Dual 185-gallon tanks installed: 4,500 pounds maximum.
- *Standard Crew:* two pilots, one flight engineer, and one aerial gunner.

6-96. Tables 6-10 through 6-13, pages 6-28 and 6-29, list the aircraft, communications, and navigation capabilities and the aircraft's transportability factors.

Table 6-10. USAF rotary-wing aircraft capabilities

Capabilities	HH-60G Pave Hawk[1]
Maximum Airspeed Velocity Never Exceed (in Knots)	163 137
Planning Cruise Speed (in Knots)	110
Maximum GW In-Ground Effect/ Out-of-Ground Effect	22,000
Aircraft Weight Empty, Average	16,100
Total Payload in lb (Fuel, Cargo, and/or Passengers)	5,900
Total Fuel Capacity, Main Only, in lb	2,200
Total Fuel Capacity for Main Plus Two Auxiliary Tanks (If Applicable), in lb	4,500
20-Min Reserve	400
Endurance (hr)/Range (NM) +20-Min Reserve (Main Tanks Only)	2.0/220 NMs
Endurance (hr)/Range (NM) +20-Min Reserve (Main and Two Auxiliary Tanks, Maximum GW)	4.5/500
Maximum Passengers at 290 lb Each, No Auxiliary Tanks (Seats/No Seats)	12/14, With Guns Installed 10/12
Maximum Passengers at 290 lb Each, With Two Auxiliary Tanks (Seats/No Seats)	8/10, With Guns Installed 5/9
Standard Ammunition Load	7.62 mm x 6,000 .50 Caliber x 800

[1] Average CONUS day (2,000-ft pressure altitude, 20°C).

Air Force Special Operations Organization and Aircraft

Table 6-11. USAF rotary-wing aircraft communication capabilities

Capabilities	HH-60G
UHF Voice/Data Secure Voice/Data	Yes/No Yes/No
VHF-AM Voice/Data Secure Voice/Data	Yes/No Yes/No
VHF-AM Voice/Data Secure Voice/Data	Yes/No Yes/No
HF Voice/Data Secure Voice/Data	Yes/No Yes/No
SATCOM Voice/Data Secure Voice/Data	Yes/No Yes/No
PLS	Yes/No No/No
Emergency Locator Transmitter	No
Havequick/DF	Yes
MX Data Burst	KY-879 Digital Message Device

Table 6-12. USAF rotary-wing aircraft navigation capabilities

Capabilities	HH-60G
GPS	AN/ASN-151
LORAN	No
VOR/DME	AN/ARN-123V
TACAN	AN/ARN-118V
ILS	AN/ARN-123V
INS	SNU-34-1
Doppler	AN/ASN-137

Table 6-13. USAF rotary-wing aircraft air transportability

Capabilities	HH-60G
C-130	0
C-17	2
C-5	5

This page intentionally left blank.

Chapter 7

Nonstandard Aircraft Used During Airborne Operations

This chapter contains aircraft descriptions, JM procedures, and aircraft preparation techniques for nonstandard rotary-wing and fixed-wing aircraft. Usually, the loadmaster or crew chief and JM jointly prepare the aircraft. The JM should install a field-expedient anchor line cable. These nonstandard aircraft are service-tested and approved for personnel airdrop operations

C-23B/B+ SHERPA

7-1. The C-23B/B+ Sherpa is a twin-engine, nonpressurized turboprop aircraft (Figure 7-1). The aircraft has a cruise speed of 180 knots with a range of 800 NMs. The aircraft can drop 12 combat-equipped parachutists or 16 parachutists without combat equipment in the airdrop configuration. This is a base planning figure. Actual troop capacity may vary because of aircraft limitations based on weight, density, altitude, and fuel loads. Troops may be loaded over the ramp or through the portside door.

Figure 7-1. C-23B/B+ Sherpa

DROP PROCEDURES

7-2. The primary jump door for personnel and cargo is the rear ramp. Because of the inability to recover towed parachutists through the Sherpa troop doors, the only authorized static-line use of the troop doors is for in-flight emergencies. Additionally, there exists the possibility of damage to the aircraft when a static-line deployment bag makes contact with the tail section of the plane after a troop-door exit. Only MFF parachutists are authorized to use the troop doors for a non-emergency exit.

7-3. For static-line operations, the primary method for determining the exit point for parachutists is using the wind drift indicator. GMRS and VIRS can also be used based on the situation and the mission. The aircraft is also capable of a GPS release if the pilot has the RP coordinates. A thorough briefing between the aircrew and all key personnel is mandatory before any operations involving the C-23B/B+. Standard drop altitude and speed is 1,500 feet AGL at 105 knots. MFF operations can be conducted up to 17,500 feet MSL.

Chapter 7

SEATING CONFIGURATION

7-4. Parachutists sit in two sticks along the port side and starboard side of the aircraft. Parachutists 1 through 8 sit on the port side, and Parachutists 9 through 16 sit on the starboard side.

ANCHOR LINE CABLE ASSEMBLIES

7-5. There are two anchor line assemblies located overhead and running the length of the cabin down the center. The cables run from the reinforced anchor line attachment plate on the forward bulkhead to the anchor line connector at the center of the ramp hinge. For personnel parachute drop operations, parachutists use only the starboard side anchor line cable. Either cable can be used for cargo parachute drop operations.

STATIC-LINE RETRIEVAL SYSTEM

7-6. The static-line retrieval system is a 5,000-pound winch located forward in the cabin on the floor and against the bulkhead (Figure 7-2). The retrieval cable runs up from the winch and along the portside anchor line cable and is attached to the starboard side anchor line cable forward of the anchor line stop. The flight engineer operates the retrieval system from the rear of the cabin. The flight engineer uses the retrieval system only in case of a towed parachutist.

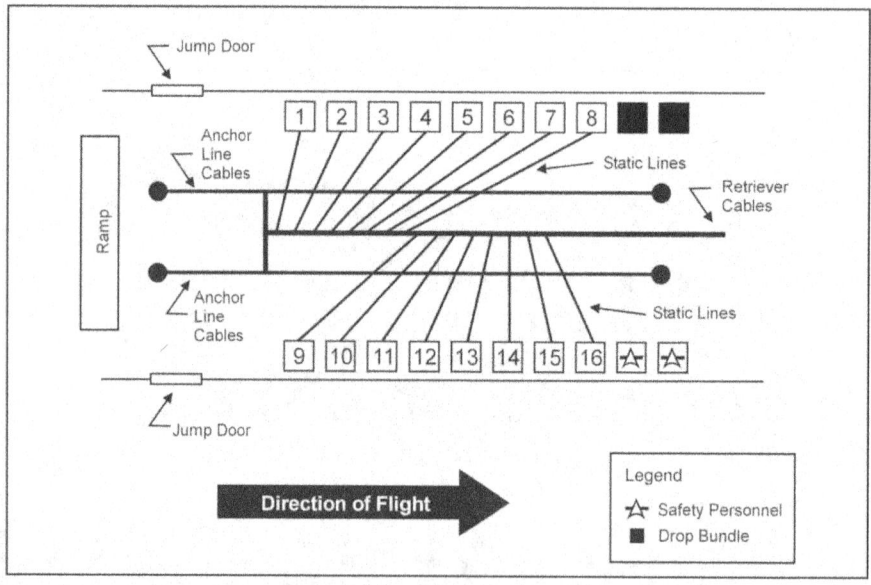

Figure 7-2. C-23B/B+ static-line retrieval system

SUPERVISORY PERSONNEL REQUIRED

7-7. The C-23B/B+ aircraft require the following supervisory personnel:
- The *JM* will lead the stick out when jumping. The JM may be nonjumping or static.
- The *safety* will be nonjumping. Either one or two safeties may be used.
- The *flight engineer* oversees all operations in the cabin.

Nonstandard Aircraft Used During Airborne Operations

PREPARATION AND INSPECTION

7-8. The JM and flight engineer jointly inspect the aircraft. JMs should follow basic aircraft inspection criteria as outlined in their JM handbooks. At a minimum, the inspection should include the following:
- Seat configuration and seat belts (correct number and location).
- Static-line retrieval cable and winch (attached correctly and secured).
- Jump lights (may have to wait until aircraft is powered up).

LOADING PARACHUTISTS

7-9. Parachutists require a step or ladder when loading through the portside door. The flight engineer directs the parachutists as to when they may load the aircraft. Designated personnel escort the parachutists to the aircraft. When loading the aircraft through the portside door or the ramp door, parachutists may wear rucksacks.

Cold Loading

7-10. Cold loading occurs when the aircraft is shut down and engines are not turning. Parachutists may load the aircraft over the rear ramp or the portside door. The C-23B or B+ has a double-hinged ramp that operates differently from conventional ramps on other aircraft with which JMs may be familiar. During ground operations, the aircrew can lower the ramp door to its lowest position (resting on the ground). It is in this configuration that equipment and parachutists can easily load. The ramp can also be opened to the half-lowered position. Parachutists may also use the portside door with steps during cold loading.

Hot Loading

7-11. Hot loading occurs when the engines are turning. During multilift operations, the aircraft may be hot-loaded to expedite the airborne operation. Parachutists may load either by the ramp door or through the portside door. The ramp cannot be lowered to the ground during hot-load operations. After loading the aircraft, the parachutists must immediately take their seats and fasten their seat belts. The safety ensures the parachutists are secured. The safety signals the JM once the parachutists are secured. The JM then signals the flight engineer that he is ready for takeoff.

> **CAUTION**
> JM ensures parachutists immediately move forward in the aircraft cabin when loading to prevent aircraft tail from striking ground.

JUMP COMMANDS AND TIME WARNINGS

7-12. The JM issues the following jump commands:
- GET READY.
- PORTSIDE PERSONNEL, STAND UP.
- STARBOARD SIDE PERSONNEL, STAND UP (if required).
- HOOK UP (gates toward starboard skin).
- CHECK STATIC LINES.
- CHECK EQUIPMENT.
- SOUND OFF FOR EQUIPMENT CHECK.
- STAND BY (30 seconds).
- FOLLOW ME (jumping JM).
- GO (static JM).

Chapter 7

7-13. The time warnings begin at 20 minutes from the DZ. The jump procedures (time warnings and jump commands) are given in Appendix F, Table F-1, pages F-1 and F-2 and Table F-3, pages F-3 and F-4.

CARGO OPERATIONS

7-14. The C-23B/B+ is capable of low-level and high-altitude cargo delivery operations. Bundle weight on the ramp should not exceed 500 pounds.

Cargo Airdrops Without Personnel

7-15. The pilot sets up approach, airspeed, and altitude. He commands the flight engineer to STAND BY. The flight engineer ensures bundle static line is hooked up and moves cargo to the edge of the ramp. The pilot gives the countdown to the flight engineer FIVE, FOUR, THREE, TWO, ONE, NOW. The flight engineer releases cargo over the ramp and retrieves the static line and clevis (Figure 7-3).

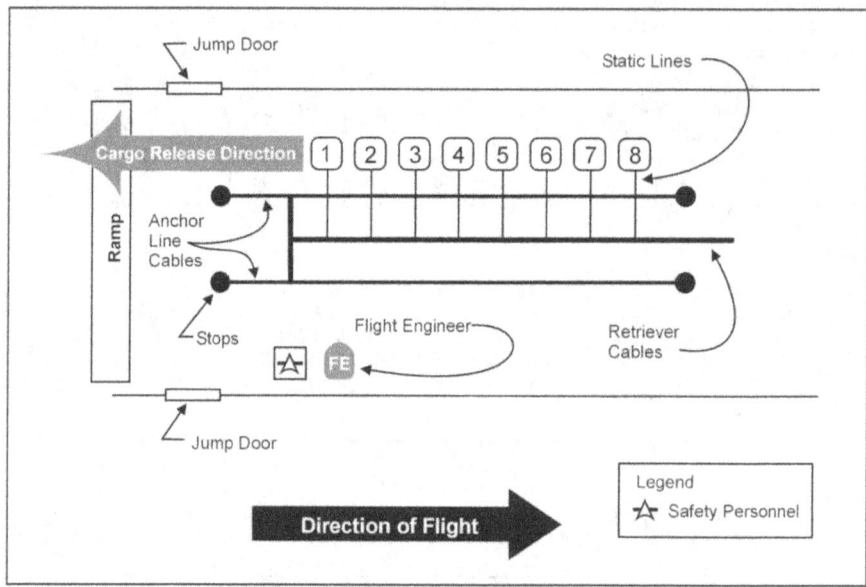

Figure 7-3. C-23B/B+ static-line exit procedures

Cargo Airdrops With Personnel

7-16. The JM and safety coordinate and rehearse cargo release procedures with the flight engineer before the mission. Cargo may be released before or after parachutists. The safety retrieves static line and clevis after the drop.

> *Note.* Either breakaway or nonbreakaway 15-foot static lines may be used. Based on the mission, the JM will select the appropriate 15-foot static line to be used (breakaway verses nonbreakaway).

MFF OPERATIONS

7-17. All MFF operations are conducted IAW FM 3-05.211.

> **CAUTION**
>
> No more than four parachutists are authorized aft of the portside jump door before exit. Excessive weight load and cargo shift in the ramp door area before exit should be avoided.

C-27A SPARTAN

7-18. The C-27A (Figure 7-4) is a pressurized, medium-range transport aircraft developed from the Aeritalia G-222. The C-27A is a twin-engine, high-wing-mount, tailgate-equipped aircraft that is similar to a smaller C-130. The C-27A can carry 34 fully equipped combat troops, 28 static-line parachutists, 34 MFF parachutists, or 16 MFF parachutists on oxygen. It can airdrop up to six CDS bundles. Typical internal loads are one high mobility multipurpose wheeled vehicle or three full-sized 463L pallets that are turned sideways. Static-line parachutists may be dropped using either of the two jump doors, but may not use the ramp. MFF parachutists may use both jump doors or the ramp. Unless specified otherwise in this section, standard jump procedures for SF are used.

Figure 7-4. C-27A Spartan

SAFETY PRECAUTIONS

7-19. Safety precautions for the C-27A are as follows:
- *Equipment.* At 36 by 75 inches, the jump doors are wide enough for standard door bundles (that is, A-7A and A-21). The 15-foot static line with drogue is used. Troops may follow.
- *Door Bundles.* When personnel follow door bundles, the door bundle static line will be outfitted with a drogue.
- *Platforms.* The JMs must ensure jump platforms and windscreens are available. This equipment is mandatory for each aft personnel door that is to be used. The safeties help the loadmaster

Chapter 7

install the door platforms if they are to be installed in flight. They ensure personnel hook up consecutively and that Parachutists 13 and 26 are in the correct position. The safeties control the static lines as the parachutists approach the door to exit. They help the loadmaster retrieve the deployment bags.

- *Jumpmasters.* The JMs inspect the door platforms after the loadmaster opens the doors. The JMs hook up to the cables on their side of the aircraft. They control and observe the personnel as they exit. JMs exit last.
- *Aircraft.* The drop speed of the aircraft is 125 knots. Parachutists cannot exit both doors at the same time. Static-line parachutists cannot use the ramp for jumping.
- *Movement Into the Door.* Parachutists cautiously move to the door, avoiding the static lines of preceding parachutists. This precautionary action may slow movement into and out of the door.

SUPERVISORY PERSONNEL REQUIRED

7-20. For jumps out one jump door, one JM, one nonjumping safety, and an airdrop-certified USAF loadmaster are required to ensure C2. These personnel requirements double when using both jump doors.

SEATING CONFIGURATION

7-21. Parachutists form two sticks. Parachutist 28 (primary jumpmaster [PJM]) is seated on the port side of the aircraft forward of the jump door. Forward of him are Parachutists 1 through 13. Parachutist 27 (AJM) is seated on the starboard side of the aircraft just forward of the jump door. Forward of him are Parachutists 14 through 26. The safeties sit on each side to the rear of the jump doors.

ANCHOR LINE CABLES

7-22. There are two anchor line cable assemblies in the C-27A. The anchor line cables run from the attachment point on the forward bulkhead, through the anchor line support bracket just behind both doors, and to the side of the aircraft over the tailgate. Each parachutist is issued main and reserve parachutes. Each parachutist must inspect his parachute for safety wires and for fitting of the parachute harness.

LOADMASTER BRIEFING

7-23. The loadmaster briefs the parachutists once they are seated. The loadmaster briefs them on aircraft safety, emergency procedures, and comfort facilities.

STATIC-LINE OVER-THE-RAMP OPERATIONS

7-24. The C-27A cannot be used for static-line, over-the-ramp operations. The erratic behavior of the deployment bags poses a serious safety hazard.

EQUIPMENT DROP

7-25. Personnel can push door bundles off the ramp. Personnel can install rollers on the ramp to aid in handling larger bundles.

MILITARY FREE-FALL

7-26. When both doors are closed, MFF parachutists can exit over the ramp. The using unit must provide two console positions for the loadmasters to use during MFF jumps above 10,000 feet. During ramp exits, the JM has a very difficult time spotting the RP from the aircraft. Therefore, the JM should not wear an all-purpose, lightweight individual carrying equipment (ALICE) pack for this type operation, and the unit should use a nonjumping JM.

JUMP PROCEDURES

7-27. The JM issues jump commands. Appendix F, Table F-3, pages F-3 and F-4, provides a list of the commands.

JUMPMASTER CHECKLIST (C-27A)

7-28. The JM inspects the airplane. The JM follows the checklist in Appendix F, Table F-7, pages F-9 and F-10.

CASA-212

7-29. The Aviocar CASA-212 (Figure 7-5) is a twin-engine, high-wing, multipurpose light transport designed for operations involving short, rough airfields. The aircraft lift capability is dependant upon the environmental conditions and fuel requirements for the mission. The number of personnel and parachutists will not exceed—

- 22 personnel in the troop lift mode (200 pounds per passenger).
- 15 in the combat lift mode (300 pounds per passenger).
- 14 static-line, nontactical parachutists (260 pounds per parachutist) with one static jumpmaster and one safety on board (total 16 passengers).
- 12 static-line, combat-equipped parachutists (350 pounds per parachutist) with one static jumpmaster and one safety on board (total 14 passengers).
- 14 combat-equipped MFF parachutists (350 pounds per parachutist).

Figure 7-5. CASA-212

SEATING CONFIGURATION

7-30. Parachutists are seated in two sticks of parachutists (Figure 7-6, page 7-8). The odd-numbered personnel are seated on the starboard side and the even-numbered personnel are seated on the port side.

ANCHOR LINE CABLE ASSEMBLY

7-31. There is one anchor line cable assembly in the CASA-212. It runs from the reinforced anchor line attachment plate on the forward bulkhead to the anchor line connector near the right side of the aft starboard emergency door.

Chapter 7

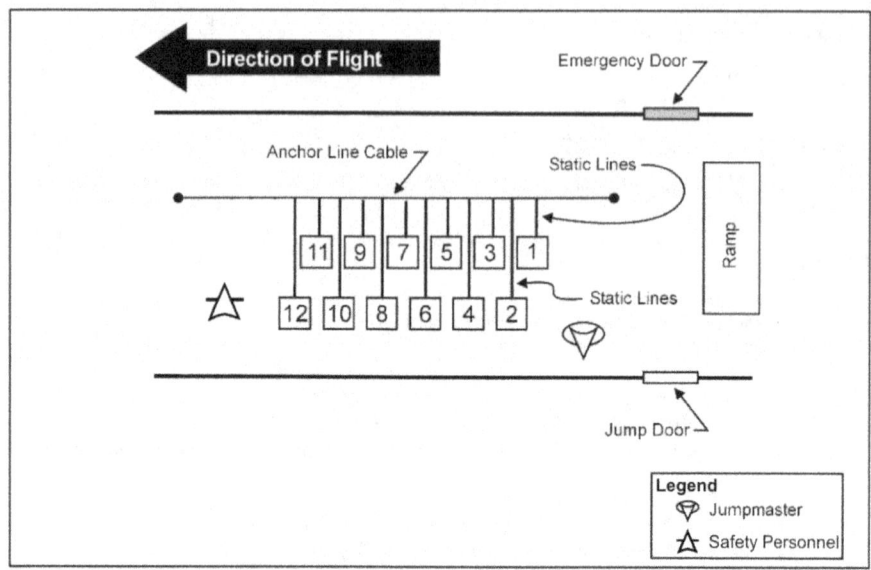

Figure 7-6. CASA-212 seating configuration for combat-equipped parachutists

SUPERVISORY PERSONNEL REQUIRED

7-32. The CASA-212 requires one JM and one safety for airdrop operations.

Note. If the JM is not static, two safeties must be aboard the CASA 212.

JUMP COMMANDS

7-33. The following jump commands are used with the CASA-212 aircraft:
- GET READY.
- STARBOARD SIDE PERSONNEL, STAND UP.
- PORTSIDE PERSONNEL, STAND UP.
- HOOK UP. On this command, the odd-numbered personnel hook up between the even-numbered personnel to form a continuous stick of parachutists, hooking the open portion of the snap hook facing inboard over the left shoulder. All parachutists take up a reverse bight.
- CHECK STATIC LINES, CHECK EQUIPMENT, and SOUND OFF FOR EQUIPMENT CHECK. These commands are executed in the same manner as with other fixed-wing aircraft.
- STAND BY (Ramp). A proper exit position is taken by the parachutist.
- GO. Personnel exit the aircraft at 1-second intervals at a 45-degree angle toward the port side.

SAFETY PRECAUTIONS

7-34. Safety precautions for the CASA 212 are as follows:
- *Parachutists—*
  - Secure all seats in the up position when they stand to hook up. During extreme air turbulence, parachutists take a short bight on the static line to steady themselves.
  - Remain off the ramp while it is being lowered.

Nonstandard Aircraft Used During Airborne Operations

Note. To assist the JM in looking for the DZ, the troop door may be removed before the airborne operation begins. The safety restraint harness is attached to the 2,500/5,000-pound tie-down positions on the floor of the aircraft, out of the way of the parachutists.

- *JM,* or safety, ensures all personnel properly hook up.

Note. On aircraft that do not have a positive communication system, the following safety measure is recommended: one ring on the alarm bell signals the JM to look at the jump light or communicate with the cockpit.

- *Equipment:*
  - When accompanying supplies and equipment are dropped from the door, the bundles must be standard air delivery containers no larger than 40 by 24 by 36 inches.
  - When ramp bundles are dropped, either the 15-foot static line with drogue or the breakaway static line may be used.
  - When door bundles are dropped, the 15-foot static line with drogue is used with cargo parachutes.
- *Aircraft:*
  - Aircraft speed during the jump is 90 to 110 knots.
  - When parachutists are jumping over the ramp, the aft port troop door may be removed for spotting.

Note. When aft port troop door is removed, a safety net must be installed.

TOWED PARACHUTIST PROCEDURES

7-35. Towed parachutist procedures are the same as those for high-performance aircraft. (FM 3-21.220 provides more information.) If the JM or safety cannot retrieve or cut away the parachutist, the preferred method is to try a landing on a grass or foamed runway. There is a retrieval system for the CASA-212, and it must be present and operational for static-line personnel airborne operations.

7-36. When notified of a towed parachutist, the aircrew will climb to 500 feet above the drop altitude (weather permitting), maintain drop airspeed (90 to 110 KIAS), make only shallow bank angle turns (to prevent parachutist spin), and avoid flying over or upwind of water or built-up areas (should the parachutist fall free).

7-37. The towed parachutist will indicate consciousness and readiness to be cut free by maintaining a tight body position, and by placing both hands on the reserve parachute. If the towed parachutist is unconscious and requires retrieval, the JM and safety's safety harness anchor strap must be preset to the proper length or they may fall out of the aircraft during the retrieval process. The duties of the safety may require him to release his safety harness anchor strap during the retrieval process. The safety must ensure he is secured before approaching the open ramp.

7-38. The towed parachutist procedures for the CASA-212 are as follows:
- The JM stops the remaining parachutists, and the safety notifies the aircrew of the towed parachutist condition.
- The aircrew turns the green light off and the red light on.
- The JM and safety retrieve the deployment bag(s) and secure the remaining parachutists on the port side of the aircraft (opposite the anchor line cable).
- The JM attaches the retrieval strap around the towed parachutist's static line and then attaches the retrieval cable hook to the metal buckle of the retrieval strap.
- The JM signals the safety to inform the aircrew to close the ramp (the upper rear cargo door remains open).

- The JM removes the working end of the 1-inch tubular nylon strap, routs it around the static line, and attaches it to the nonworking end attaching point. (This procedure may require manhandling of the static line toward the starboard side of the aircraft.)
- The JM places the retrieval bar under the static line and 1-inch tubular nylon strap and installs it into the starboard side attaching point.
- The JM (with the assistance of the safety, if required) installs the retrieval bar into the portside attaching point.
- The JM installs a security pin into each retrieval bar attaching point.
- The JM removes the 1-inch tubular nylon strap from around the static line and reattaches it to the working end attaching point.
- The JM signals the safety to inform the aircrew to open the ramp to jump attitude.
- The JM signals the safety to winch-in the towed parachutist.
- As the parachutist reaches the retrieval bar, the JM (with the assistance of the safety, if required) pulls the parachutist under the bar and onto the ramp.
- As more of the parachutist's weight is supported by the ramp, the JM signals the aircrew to raise the ramp in small increments to allow slack to form in the winch cable. During this process, the JM continues to pull the parachutist under the retrieval bar.

WARNING

The ramp must be moved in small increments to prevent the parachutist's body from being pinned between the ramp and the retrieval bar.

- The JM signals the safety to change the direction of the winch and let out slack.
- The aircrew continues to incrementally raise the ramp as the parachutist is pulled in. This helps the JM transfer the parachutist's weight to the ramp.
- When the parachutist is safely inside the aircraft, the safety removes the parachutist's static line from the anchor line cable.
- The JM and safety render first aid, if required, and secure the parachutist in a seat with a seatbelt.
- The JM removes and secures the retrieval bar, and reinstalls the port and starboard retrieval bar security pins.
- The JM signals the aircrew to close the ramp and upper rear cargo door. If the retrieval bar cannot be removed the ramp and upper rear cargo door should be closed as much as possible.

CAUTION

Severe damage to the ramp, aircraft frame and hydraulic system will occur if the ramp and rear cargo door are fully closed with the retrieval bar installed.

- The JM monitors the movement of the ramp and cargo door to ensure they are clear.
- The aircraft will land IAW the prebriefed and preaccident plan.

Nonstandard Aircraft Used During Airborne Operations

AIRCRAFT CONFIGURATION FOR RAMP STATIC-LINE PERSONNEL AIRDROP

7-39. The aircrew configures the aircraft. The JM verifies the configuration of the aircraft. Static-line ramp parachute operations are authorized only when the retrieval system is operational. The aircraft is configured for a static-line personnel airdrop. Figure 7-7 shows the items of equipment (1 each) that are needed.

• Hand winch.	• Static-line deflector block.
• Retrieval strap.	• Retrieval bar.
• Hook knife.	• Extended interphone cord.
• One 3-foot length of 1-inch tubular nylon.	• One roll cloth-backed adhesive tape.
• Anchor cable.	• Two restraint harnesses.

Figure 7-7. Equipment for static-line airdrop

7-40. The loadmaster configures the aircraft. However, the JM inspects the aircraft before loading parachutists. The JM ensures the proper aircraft configuration for the operation. The loadmaster—

- Attaches a hand winch to the right tie-down row in Zone 1 and checks that it is secure.
- Inspects cable for broken wires or kinks and checks its operation.
- Ensures static-line deflector block is attached to the right side of the ramp and covers the bolt head with tape.
- Inspects retrieval base onboard and attaching brackets.
- Installs and checks extended interphone cord for operation.
- Fits and adjusts restraint harnesses.
- Ensures that the 3-foot length of 1-inch tubular nylon and the retrieval strap are secured and available for immediate use.

IN-FLIGHT PROCEDURES

7-41. The CASA-212 does not have standard in-flight procedures. Jump commands and time warnings for the CASA-212 are listed in Appendix F, Table F-5, pages F-6 and F-7.

CASA-212 JM CHECKLIST

7-42. The JM performs a check of the CASA-212 to ensure proper configuration and readiness. A JM checklist is provided at Appendix F, Table F-8, page F-11.

EMERGENCY PROCEDURES

7-43. The parachutists, pilot, and JM are involved in the CASA-212 emergency procedures. The aircraft pilot explains the crash or emergency bailout procedures to the JM. The JM relays the procedures to the parachutists. The normal alarm bell warnings are as follows:

- Three short rings: aircraft is having trouble.
- Three short rings, followed by one long ring: prepare for impact.
- Six short rings: aircraft is having trouble but has altitude enough for a safe bailout. Exit on JM command.

Reserve Parachute Activation

7-44. If a reserve parachute accidentally activates inside the aircraft and the pilot parachute and canopy catch air and exit the aircraft, the parachutist must immediately exit the aircraft, regardless of his position in the stick. If a reserve parachute accidentally activates inside the aircraft and the pilot parachute and canopy do not catch air and exit the aircraft, all parachutists near the pilot parachute and canopy yell "Parachute!" and immediately try to collapse and control the parachute. The JM or safety notifies the pilot to close the ramp (if open), and the JM will control the spotting door (if open) while the safety and the parachutist with the deployed reserve parachute bundle up and secure the exposed canopy. The JM or safety moves the parachutist to the front of the aircraft. The JM or safety issues the parachutist a new reserve parachute, reinspects the parachutist's equipment, and places the parachutist back into the stick.

Towed Parachutist

7-45. Towed parachutist procedures are the same as those for high-performance aircraft (FM 3-21.220). If the JM or safety cannot retrieve or cut away the parachutist, the preferred method is to attempt a landing on a grass or foamed runway. There is a retrieval system for the CASA-212, and it must be present and operational for static-line personnel airborne operations.

Jump Refusal

7-46. Jump refusal procedures are the same as for high-performance aircraft (FM 3-21.220).

SUPERVISORY PERSONNEL REQUIRED

7-47. Four personnel supervise safety procedures. They are one JM who performs standard aircraft check procedures, one AJM, and two safeties. Duties of supervisory personnel are the same as for a C-130.

ANCHOR LINE CABLE ASSEMBLIES

7-48. There are four anchor line cables, two on each side. There are also two static-line retrievers, one on each side of the aircraft. Parachutists use only one anchor line cable of the anchor line assembly when conducting aft end or ramp jumps.

JUMP PROCEDURES

7-49. The CASA-212 jump commands and procedures are the same as for the C-130. The commands are—
- GET READY.
- OUTBOARD PERSONNEL, STAND UP.
- INBOARD PERSONNEL, STAND UP.
- HOOK UP.
- CHECK STATIC LINES.
- CHECK EQUIPMENT.
- SOUND OFF FOR EQUIPMENT CHECK.
- STAND BY (door or ramp).
- GO.

A ramp jump is conducted in the same manner as for a C-130 with the changes listed in Appendix F, Table F-4, page F-5.

EQUIPMENT PREPARATION

7-50. At the 2-minute warning, or as directed by the pilot, the aircrew removes the jump doors. The aircrew stows the jump doors on the cargo ramp.

7-51. Personnel use the 15-foot static line with drogue to drop equipment containers from the jump doors when parachutists follow. Personnel use the 15-foot static line with drogue or the breakaway static line to drop container loads from the ramp when the parachutists follow.

Note. Personnel lash loose equipment and the removed jump doors to the cargo ramp or to the rear of the forward bulkhead (if flight time is less than 20 minutes). They ensure the equipment and jump doors do not obstruct parachutists.

Aircraft Inspection

7-52. The JM checks the CASA-212 to ensure proper configuration and readiness. A JM checklist is in Appendix F, Table F-8, page F-11.

UV-18B TWIN OTTER AND DE HAVILLAND DHC-6 TWIN OTTER

7-53. The UV-18B, also known as the Twin Otter (Figure 7-8), is a light STOL twin-turboprop, multipurpose aircraft. It can fly various missions to include personnel, bundle, and MFF deliveries (without oxygen), and message or materiel pickup operations. The civilian version of this aircraft is the De Havilland DHC-6 Twin Otter, series 300. These aircraft are very similar and most procedures are the same except the UV-18B and some old versions of the DHC-6 have floor-mounted anchor cables. Most versions of the DHC-6 Twin Otter are equipped with an overhead anchor line cable. Physical characteristics of the UV-18B and DHC-6 include the following:

- Drop speed—75 knots at 1,500 feet AGL.
- Flaps—set at 15 to 20 degrees for parachute operations.
- Aisle length—18 feet 5 inches.
- Aisle width—6 feet 1 inch.
- Door width—56 inches.
- Door height—50 inches.

Figure 7-8. UV-18B Twin Otter

Supervisory Personnel Required

7-54. Regardless of the number of parachutists, two personnel supervise safety procedures—one JM who performs standard aircraft check procedures and one safety. The safety is located front and center of the aircraft behind the pilot and copilot seats. The JM sits by the jump seat at the starboard rear bulkhead near the jump door. The JM and safety wear a safety harness.

Seating Configuration

7-55. The configuration of both Twin Otters is the same. The number of parachutists that can jump in one pass depends on whether or not the seats are in the plane, whether or not the parachutists are jumping with

Chapter 7

combat equipment, and the weight of the parachutists and equipment. The anchor line cable is limited to 3,500 pounds. The various seating configurations are as follows:

- When there are sidewall folding seats, 16 static-line parachutists without combat equipment can be carried. The side-facing troop seats are set up with a 20-inch pitch for normal operations. Each parachutist has a seat belt attached to the wall-mounted Douglas track. The multiple attachment points of this track permits variation in the number and spacing of parachutists (Figure 7-9).

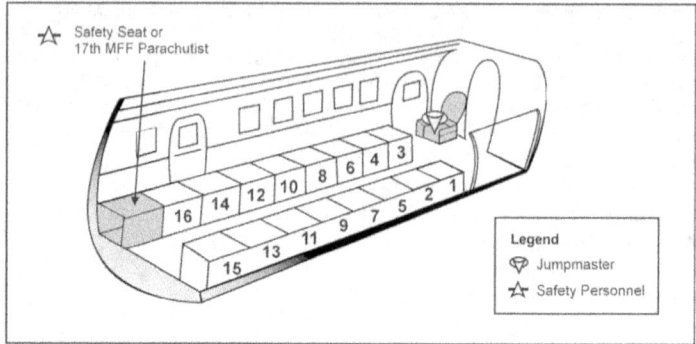

Figure 7-9. Seating configuration with seats and without combat equipment

- With sidewall folding troop seats, 17 MFF parachutists and a jumping JM can be carried.
- When there are no side-facing seats, 12 parachutists without combat equipment can be carried. All parachutists sit on the floor facing aft, legs outstretched (Figure 7-10). Seven parachutists sit on the starboard side, and five parachutists sit on the port side of the air-craft. Each parachutist sits between the legs of the previous parachutist.

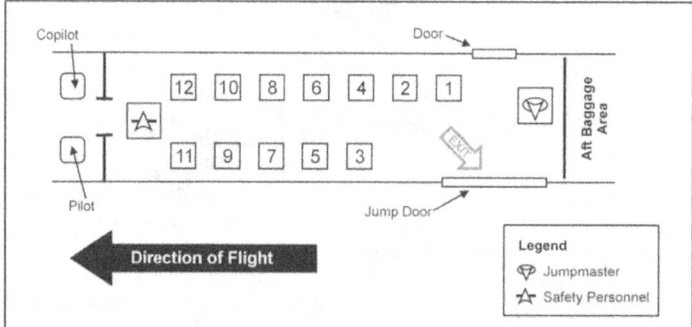

Figure 7-10. Seating configuration without seats or combat equipment

- When there are no seats, 12 MFF parachutists with combat equipment can be carried.
- When parachutists jump with combat equipment, eight static-line parachutists can be carried. All parachutists sit on the floor of the aircraft, facing aft (Figure 7-11, page 7-15). Five parachutists sit on the starboard side and three sit on the port side of the aircraft. Each parachutist sits between the previous parachutist's legs.
- For cargo missions, the side-facing troop seats can be quickly folded and stored against the sidewalls, permitting the full use of cabin volume.

Nonstandard Aircraft Used During Airborne Operations

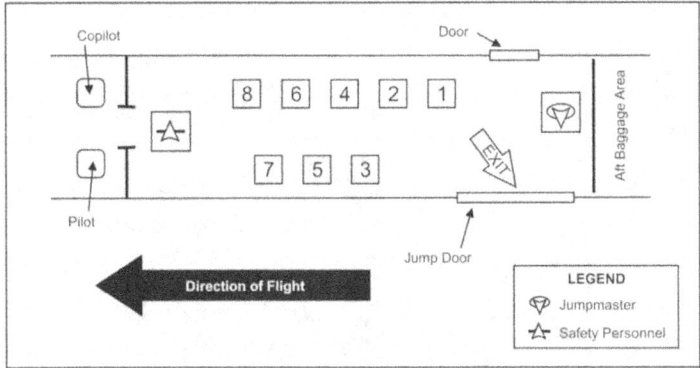

Figure 7-11. Seating configuration without seats and with combat equipment

WARNING

This aircraft may not be equipped with an anchor line retriever cable. The number of parachutists on any one pass is limited to the number of static lines and deployment bags the JM or safety can safely retrieve. The number of parachutists is also limited by the maximum static-line load of 3,500 pounds.

ANCHOR LINE CABLE ASSEMBLIES

7-56. Depending on the type or version of aircraft, the aircrew will mount the anchor line cable on the ceiling or on the floor. Usually the ceiling-mounted anchor line will be the one that is used. The aircrew mounts the anchor line cable according to the following:

- The overhead static-line anchor cable runs the length of the cabin along the starboard top side, about 6 inches from the headboard. The anchor line cable runs from behind the copilot seat to the rear bulkhead.
- The floor-mounted anchor line cable extends the length of the cabin, which permits each parachutist to connect his static line regardless of his position in the stick.
- Static-line load and assemblies handle a maximum of 3,500 pounds.
- The recommended length for the static line is 12 feet. These aircraft do not require static-line extensions.

INWARD FOLDING JUMP DOOR

7-57. The jump door consists of two hinged segments designed to fold inward and upward against the ceiling (Figure 7-12, page 7-16). The aircrew straps the door to the ceiling and secures it for paradrop operations. The lower half of the door folds to allow jettison of small objects. The aircrew removes the bubble window used for search operations and stores it for maximum headroom before the drop. The Federal Aviation Administration (FAA) authorized this aircraft to land with the aircraft door open. However, the FAA did not authorize this aircraft to take off with the aircraft door open. The crew must coordinate beforehand with the appropriate authority.

27 February 2009   FM 3-05.210   7-15

Chapter 7

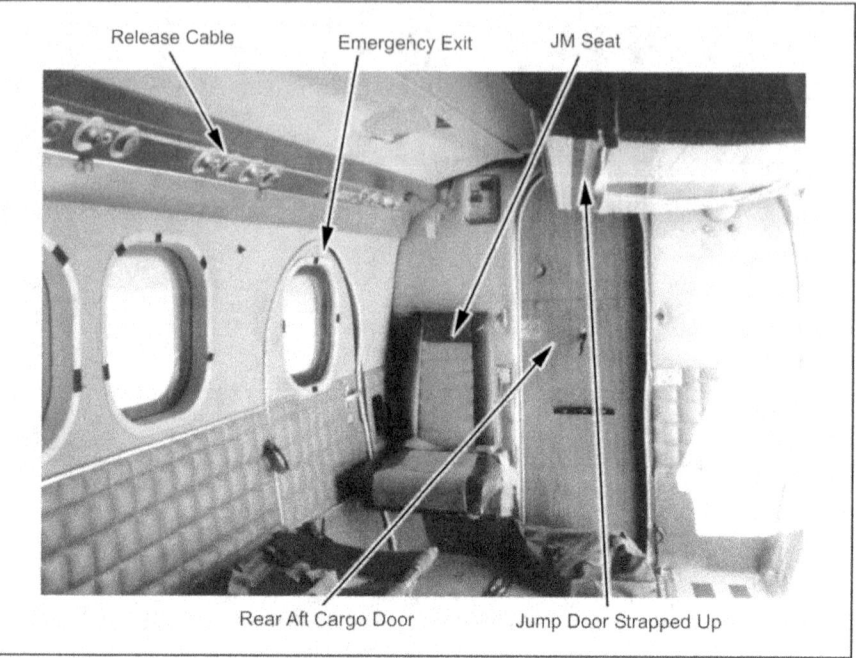

Figure 7-12. Twin Otter rigged for static-line jump

JUMP LIGHTS

7-58. The UV-18B uses the standard red and green lights as jump signals and the emergency bailout bell. The lights and bells are located on the port side of the aft bulkhead near the jump door. The JM will have interplane communications with the pilot, if available. Pilot-controlled lights can be provided on the rear bulkhead adjacent to the jump door. When using some DHC-6s, warning lights may not be available. The pilot may modify the time warnings and signals. The JM must coordinate with the pilot and then brief the parachutists on any changes from the standard time warnings and signals.

BAGGAGE COMPARTMENT

7-59. The JM has ready access to the aft baggage compartment. When time permits, the JM can use this compartment for the stowage of static lines retrieved between successive sticks of parachutists.

LOADING PROCEDURES

7-60. The loading procedures for the UV-18B vary. Whether or not the parachutists use the seats and if the anchor line cable is floor- or ceiling-mounted determine the loading procedures.

7-61. Loading procedures for aircraft with a ceiling-mounted anchor line and seats installed are as follows:
- The safety should load first and ensure that his safety harness reaches all the way to the rear of the aircraft. The JM and the safety are static and secure themselves at an anchor point or to the anchor line cable before takeoff. They ensure the safety lines do not extend past the jump or cargo door.

Nonstandard Aircraft Used During Airborne Operations

- The parachutists use the aircraft ladder or step platform to climb into the aircraft. The aircraft is loaded one parachutist at a time in reverse order. The aircraft is loaded from front to rear. The safety saves the first seat on the port side for himself.
- Each parachutist hands his static-line snap hook to the JM as he enters the aircraft. The JM hooks up each parachutist to the anchor line cable with the opening gate of the snap hook facing the port side of the aircraft.
- Each parachutist takes a normal bite on his static line under the supervision of the safety. Each parachutist covers his reserve parachute rip cord with his right hand to protect it from accidental activation.
- As each parachutist hooks up, he moves toward the front of the aircraft and sits. The safety assists and positions each parachutist.
- If two sticks are jumping, they sit in a staggered fashion with the second pass toward the front and the first pass toward the rear.
- The JM makes a final safety check of the parachutists and his safety harness. The JM notifies the safety of the number of parachutists and that all is ready for takeoff. The safety will pass this on to the pilot.

7-62. For aircraft with a ceiling-mounted anchor line and without seats installed, the procedures for loading and hooking up are the same except for the following:

- After hooking up and getting a normal bite on the static line, the parachutists sit down facing the rear of the aircraft. The parachutists slide toward the front of the aircraft and sit between the legs of the previous parachutist. The safety assists and positions each parachutist.
- Each parachutist covers his reserve parachute rip cord with his right hand to protect it from accidental activation.
- If the aircraft makes more than one pass, the parachutists in the rear will make up the first pass and the parachutists toward the front will make up the second pass.

7-63. For aircraft with a floor-mounted anchor line cable, the same procedures will be followed except that the parachutists will hook up their static lines to the cable on the floor and maintain a reverse bite. When sitting on the floor, parachutists avoid tangling their static lines.

JUMP PROCEDURES

7-64. The jump command and procedures for the UV-18B and DHC-6 vary depending on the configuration of the aircraft. The time warnings and communications between the pilot and JM can also differ. When using different versions of the UV18-B, the intercom and warning lights may not be available. The pilot may modify the time warnings and signals. The JM must coordinate with the pilot, then brief the parachutists on any changes to the time warnings and signals.

ABORT PROCEDURES

7-65. The pilot announces, NO DROP over the intercom system and turns on the red light. Using his right hand, the copilot signals the JM with a throat-cut motion.

AIRCRAFT INSPECTION

7-66. The JM checks the UV-18B or DHC-6 to ensure proper configuration and readiness.

EQUIPMENT

7-67. Personnel use a drogue or breakaway static line on cargo parachutes. Personnel may deliver equipment and supplies in standard air delivery containers rigged with light cargo parachutes from the door of the aircraft. Personnel attach the static-line snap hook of the cargo parachute to the anchor line cable.

Chapter 7

EMERGENCY PROCEDURES

7-68. Personnel must observe emergency procedures. The emergency procedures are discussed in the following paragraphs.

Towed Parachutist

7-69. Towed parachutist procedures are the same as those for high-performance aircraft (FM 3-21.220). There is no built-in retriever system; however, some DHC-6 aircraft have a standard hand winch and a 10,000-pound nylon strap on hand to assist in the retrieval. If the JM or safety cannot retrieve the parachutist inside the cabin or the parachutist is unconscious, the preferred method is to land on a grass runway or a foamed runway.

Accidental Reserve Activation

7-70. If a reserve parachute deploys in the aircraft, any Soldier who sees the deployed parachute yells "PARACHUTE!" Soldiers closest to the parachutist with the deployed parachute make every effort to contain the reserve parachute. If the pilot parachute gets out of the aircraft, the parachutist will exit as quickly as possible. When the parachute can be contained, the JM will secure the aft door and inform the pilot to land ASAP.

Emergency Signals

7-71. The emergency signals are as follows:
- Three short rings: Aircraft is having trouble.
- Three short rings followed by one long ring: Prepare for impact.
- Six short rings: Aircraft is having trouble but has enough altitude for bailout. Exit on JM command only.

U-21A UTE

7-72. The U-21A (Figure 7-13) is a twin-engine, low-wing, utility aircraft capable of transporting nine personnel in a troop configuration. This aircraft can carry six parachutists in the airdrop configuration, or six personnel in the staff transport configuration. In the air ambulance configuration, this aircraft can carry three ambulatory and three litter patients.

Figure 7-13. U-21A Ute

SEATING CONFIGURATION

7-73. Parachutists enter the aircraft in reverse order and with their static lines over their left shoulders. The parachutists sit facing the rear of the aircraft. The JM supervises the loading and hookup of the parachutists. The JM then hooks up and secures the safety belt across the rear of the cargo compartment (Figure 7-14).

WARNING

As each parachutist hooks up, the JM ensures there is no entanglement of the parachutists' equipment with the static line. After being hooked up and seated, the parachutist is handed his static line and then takes a reverse bite to control the static line.

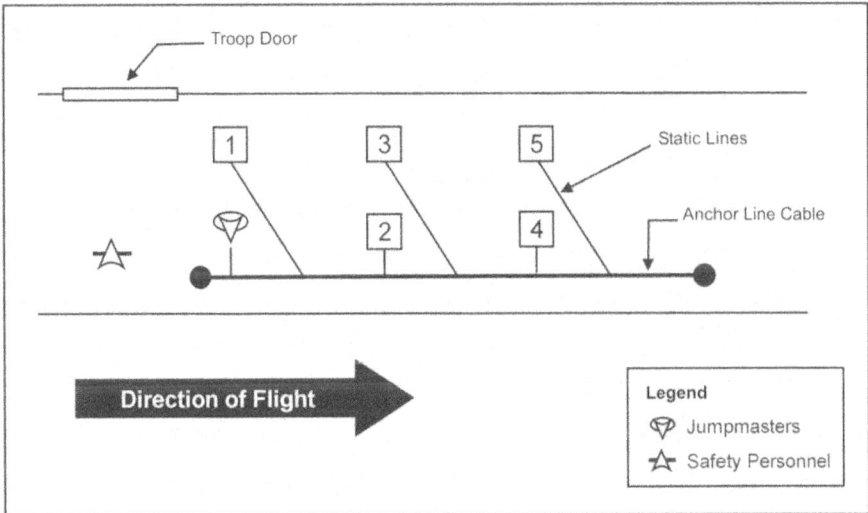

Figure 7-14. In-flight seating configuration

SUPERVISORY PERSONNEL REQUIRED

7-74. Two personnel supervise safety procedures. They are one JM and one safety.

ANCHOR CABLE ASSEMBLIES

7-75. The ground crew assembles the anchor line by using an A-7A strap (Figure 7-15, page 7-20) attached through the first four cargo tie-down rings on the right side of the cargo compartment floor. Six (10,000-pound test) D rings are located between the second and third cargo tie-downs. The ground crew secures the A-7A strap to itself by using the friction adapter and tapes any excess strap.

Chapter 7

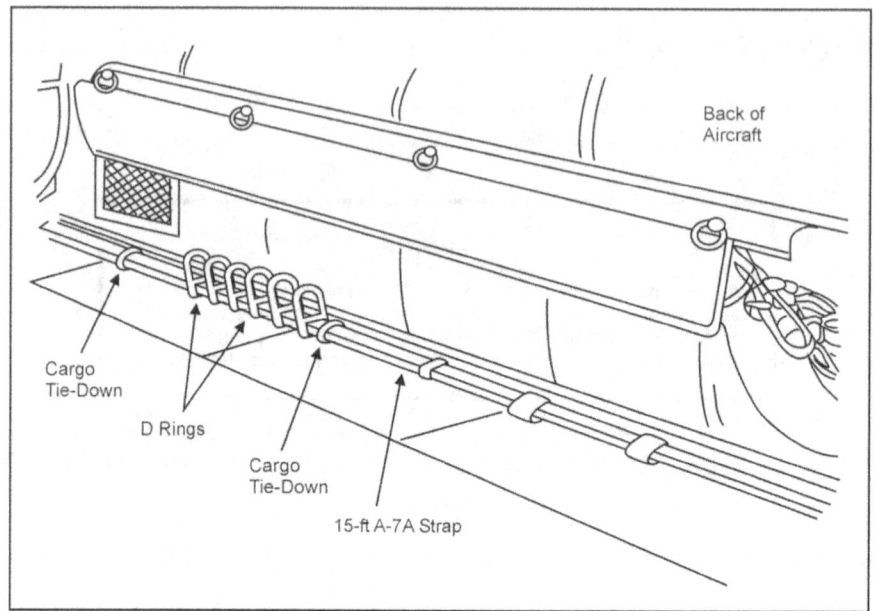

Figure 7-15. Example of anchor line installation

AIRCRAFT PREPARATION AND INSPECTION

7-76. Before boarding the aircraft, the JM and pilot, or pilot's representative, inspect the aircraft. A checklist is provided in Appendix F.

JUMP COMMANDS AND TIME WARNINGS

7-77. The JM issues the jump commands. The JM jump commands and time warnings are listed in Appendix F.

SAFETY CONSIDERATIONS

7-78. A maximum of six parachutists with combat equipment may safely jump from the aircraft and only one may carry a container, weapon, individual equipment (CWIE). In addition to six parachutists, one 500-pound or two 250-pound door bundles may be dropped on each sortie (within aircraft load constraints as determined by the aircraft commander). A safety is on board for static-line jumping to control and retrieve static lines.

7-79. The JM ensures these procedures are followed:
- The aircraft is correctly configured.
- Parachutists properly hook up static-line snap hooks to the anchor line cable. Parachutists do not use the overhead anchor line cable, as a component of the aircraft, because it allows deployment bags to interfere with the aircraft stabilizer.
- Parachutists' weapons do not exceed 39 inches.
- Speed of the aircraft is not less than 100 knots or more than 115 knots, aircraft flap setting is 60 percent, and the position is straight and level, never a tail-low attitude.

C-208B CARAVAN

7-80. The C-208B is a light STOL, single-turboprop, multipurpose aircraft capable of various missions (Figure 7-16). Examples of these missions are personnel, bundle, and MFF deliveries (without oxygen or combat equipment) and message or materiel pickup operations. It can carry eight static-line parachutists without equipment, a static JM aft and a static safety forward or six combat-equipped static-line parachutists, a static JM aft, and a static safety forward. It can also carry 10 MFF parachutists, including a parachutist JM. Drop speed is 90 to 100 knots.

7-81. The C-208B uses the standard red and green lights as jump signals and the emergency bailout bell. The lights and the bell are located on the port side of the aft bulkhead near the jump door. The JM has interplane communications with the pilot (if available) or uses the directional button located on the leading edge of the jump door to let the pilot know to move left or right for proper lineup to the RP.

7-82. Physical characteristics of the C-208B are as follows:
- Aisle length—13 feet 8 inches.
- Aisle width—5 feet 2 inches.
- Door width—48 inches.
- Door height—50 inches.

Figure 7-16. C-208B Caravan

SEATING CONFIGURATION

7-83. All parachutists sit on the floor of the aircraft, facing aft. Four parachutists sit on the port side and four sit on the starboard side. Parachutists will sit between the legs of other parachutists.

7-84. The safety uses a safety harness and is located center front of the aircraft behind the pilot and copilot seats. The JM also wears a safety harness and is located center rear of the aircraft by the jump doors.

7-85. Safety personnel attach a single safety strap across the center of the jump door and will remain attached to the trail edge of the door at all times. They use the leading edge attaching point to make the door secure or unsecure.

Anchor Line Cable Assemblies

7-86. The static-line anchor line cable is located on the starboard side of the aircraft about 4 inches from the floor and runs behind the copilot's seat to the leading edge of the jump door. This aircraft does not require static-line extensions.

JM Inspection

7-87. The JM inspects the aircraft before loading parachutists to ensure the aircraft is properly configured for the operation. A JM checklist is in Appendix F.

Loading Procedures

7-88. The aircraft is loaded one parachutist at a time in reverse order. The parachutists use the aircraft ladder or step platform to climb into the aircraft.

7-89. The safety should load first and ensure that his safety harness reaches all the way to the rear of the aircraft. The JM and the safety are static and will secure themselves at an anchor point or to the anchor line cable before takeoff. The JM and safety ensure the safety lines do not extend past the troop and cargo doors.

7-90. Each parachutist hands his static-line snap hook to the JM as he enters the aircraft. The parachutists then sit down, facing the rear of the aircraft. The JM hooks up each parachutist to the anchor line cable with the opening gate of the snap hook facing up. Each parachutist covers his reserve parachute rip cord with his right hand to prevent it from accidental activation.

7-91. The portside parachutists load first, followed by the starboard side. As each parachutist hooks up and sits, he slides toward the front of the aircraft, assisted and positioned by the safety. The two sticks are offset.

7-92. Each parachutist takes a reverse bite on his static line under the supervision of the safety. The JM makes a final safety check of the parachutists and his safety harness. The JM then notifies the safety of the number of parachutists and that all is ready for takeoff. The safety passes on this information to the pilot.

Jump Commands and Time Warnings

7-93. The JM issues the jump commands, which are slightly modified. The C-208B jump commands and time warnings are listed in Appendix F, Table F-6, pages F-7 and F-8.

Equipment

7-94. Personnel may deliver equipment and supplies in standard air delivery containers rigged with light cargo parachutes from the door of the aircraft. Personnel attach the static-line snap hook of the cargo parachute to the anchor line cable. Personnel use cargo parachutes with breakaway static lines. Personnel place rifles in adjustable, individual weapon cases. The weapon case must not exceed 36 inches in length. Only Parachutist 1 can carry a CWIE when an air delivery container is not used as part of the interior load. The bag must not exceed 36 inches in length. Personnel use a drogue or breakaway static line on cargo parachutes.

Emergency Procedures

7-95. Personnel must observe emergency procedures. The emergency procedures are discussed in the following paragraphs.

Crash or Emergency Bailout Procedures

7-96. The aircraft pilot will explain the crash or emergency bailout procedures to the JM. The JM explains the procedures to the parachutists. The normal alarm bell warnings are as follows:
- Three short rings: Aircraft is having trouble.
- Three short rings followed by one long ring: Prepare for impact.
- Six short rings: Aircraft is having trouble but has enough altitude for a safe bailout. Exit on the JM's command.

Nonstandard Aircraft Used During Airborne Operations

Reserve Parachute Activation

7-97. If a reserve parachute accidentally activates inside the aircraft and the pilot chute and canopy catch air and exit the aircraft, the parachutist must immediately exit the aircraft, regardless of his position in the stick. If a reserve parachute accidentally activates inside the aircraft and the pilot chute and canopy do not catch air and exit the aircraft, all parachutists near the pilot chute and canopy immediately try to collapse and control the chute from catching air and exiting the aircraft. The JM controls the jump door while the safety and the parachutist with the deployed reserve bundle up and secure the exposed canopy. Any Soldier who sees the deployed reserve parachute will tell the pilot to land the aircraft immediately. Because of the confined space aboard the aircraft, the JM does not move the parachutist or issue him a new reserve in flight.

Towed Parachutist

7-98. Towed parachutist procedures are the same as for high-performance aircraft, as detailed in FM 3-21.220 and USASOC Reg 350-2. If the JM cannot retrieve or cut away the parachutist, the preferred method is to try landing on a grass or foamed runway. There is no built-in retrieval system for the C 208B, although a standard hand winch and a 10,000-pound nylon strap should be present to assist in towed parachutist retrieval.

Jump Refusal

7-99. Jump refusal procedures are the same as for high-performance aircraft, as detailed in FM 3-21.220 and USASOC Reg 350-2. If the JM cannot remove the parachutist refusing to jump from the door, the JM will instruct the pilot to land immediately. The individual refusing to jump will remain hooked up. The individual will not be attached to the aircraft with any kind of restraining strap.

This page intentionally left blank.

Chapter 8

Cargo Slings, Airdrop Containers, and Poncho-Expedient Parachute

This chapter discusses the use of the A-7A cargo sling, the A-series airdrop container, CTU-2/A high-speed aerial delivery container, the poncho-expedient parachute, steel strapping, and rigging knots. The containers may be packed with supplies, disassembled equipment, or small items of ready-to-use equipment prepared for airdrop. The container load may require cushioning material such as honeycomb, felt, or cellulose wadding, depending on the load requirements and the method of airdrop. The number and types of parachutes required to stabilize and slow the descent of the load depends on the type of container used, the weight of the load, and the method of airdrop. Chapter 3 contains additional information about delivery systems.

A-SERIES CONTAINERS

8-1. Personnel use the A-7A cargo sling and A-21 cargo bag when rigging door bundles. Personnel can use the A-22 cargo bag for rigging a bundle for a ramp drop. They can rig containers with a drogue or breakaway static line. For Army aircraft, personnel rig a container load to be airdropped from a shackle (wing load), helicopter door, or utility aircraft with a breakaway static line. For high-performance fixed-wing aircraft, personnel normally rig loads with parachutes having nonbreakaway static lines. Jump door loads that are to be followed immediately by parachutists must be rigged with parachutes having nonbreakaway static lines. Each static line must have a drogue attached to it as outlined in appropriate technical manuals (TMs). Personnel place loads in jump doors so the largest dimension is upright or vertical. Personnel position the parachute on top of the load or toward the inside of the aircraft. Personnel rig a ramp load to be followed immediately by parachutists with a T-10 parachute (converted for cargo) or a parachute having a breakaway static line.

RIGGING PROCEDURES

8-2. Personnel rig door bundles so that when they are placed on the balance point of the jump platform, the parachute is on top or facing the center of the aircraft, based on the largest dimension, and not on the side. The maximum weight of the bundle is 500 pounds (not including parachute weight). Exceptions to this rigging technique are allowed for the 90-mm recoilless rifle and the Stinger missile. In both cases, personnel place the bundle upright with the parachute facing the center of the aircraft. Personnel rig the 90-mm recoilless rifle and the Stinger missile by using the A-21 container. The personnel place the skid board on the Stinger inside the canvas cover.

8-3. When rigging an item, personnel must pack all components needed for its assembly in the same airdrop bundle. (For example, a radio and battery are packed together.) When personnel rig items such as radio equipment, they individually wrap each item. Personnel place padding or honeycomb under the item being prepared and between the items comprising the load to prevent contact. Personnel must use cellulose wadding, felt, or other energy absorbing material to avoid metal-to-metal or metal-to-wood contact.

8-4. Personnel roll all excess lengths of webbing and tie them with 1/4-inch cotton webbing in a surgeon's locking knot. Doing so reduces the danger of bundles becoming snagged when ejected or released from the aircraft.

Chapter 8

8-5. If personnel place hazardous materials inside bundles, they must have a shipper's certificate. Personnel complete the shipper's certificate IAW Technical Manual (TM) 38-250, *Preparing Hazardous Materials for Military Air Shipments*. Personnel attach the shipper's certificate to the manifest, not the bundle.

A-7A CARGO SLING

8-6. The A-7A cargo sling is a 188-inch long strap that weighs 1.5 pounds. Each sling strap has a stationary parachute quick-fit adapter (commonly called a friction adapter) and a floating D ring. Personnel use the A-7A cargo slings to drop non-fragile supplies. They use two, three, or four A-7A cargo slings to make bundles for airdropping. Maximum weight capacity of the bundle is 500 pounds (not including the parachute). The minimum weight depends on the parachute used. The maximum dimensions of the bundle are 30 inches wide by 48 inches deep by 66 inches high (to include cargo parachute) or 69 inches high to accommodate the 90-mm recoilless rifle or Stinger missile.

8-7. Personnel use a combination of two, three, or four sling straps for rigging a load, depending upon the size, weight, and shape. Two A-7A sling straps have a maximum weight limit of 300 pounds; three straps, 400 pounds; and four straps, 500 pounds (Figure 8-1).

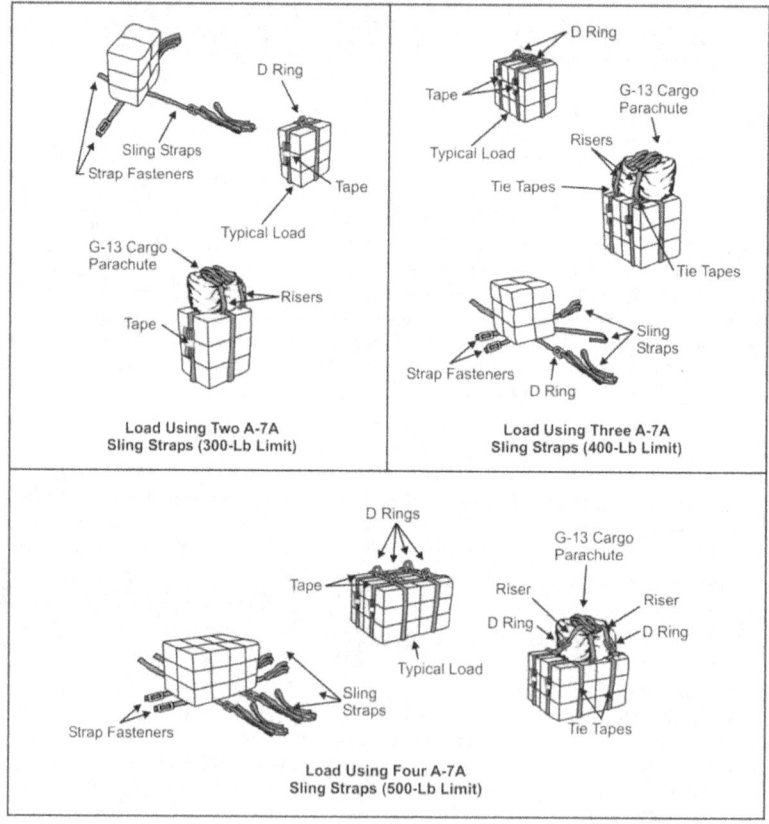

Figure 8-1. Rigging of the A-7A cargo slings

TWO-STRAP BUNDLE

8-8. The JM or safety personnel lay out one strap perpendicular (lengthwise) to the bundle with the thick-lip portion of the friction bar on the strap fastener facing down. The JM or safety personnel lay out one strap parallel to the bundle with the thick-lip portion of the friction bar on the strap fastener facing down and over the top of the perpendicular strap. When the straps are in place, they are ready to receive the bundle. The JM or safety personnel—

- Center the bundle on the perpendicular strap.
- Route the perpendicular strap over the top of the bundle and through the single D ring (through the rectangular portion of the D ring), fold, and secure.
- Route the parallel strap through the D ring (through the rectangular portion of the D ring), roll, and secure.
- Tighten all straps.
- Tie off the excess webbing above the strap fastener by using one turn of 1/4-inch cotton webbing secured with a surgeon's locking knot.
- Ensure the excess webbing is not above the top of the bundle. The bundle has one smooth side for ease in ejecting from the aircraft.

THREE-STRAP BUNDLE

8-9. The JM or safety personnel lay out one strap parallel (lengthwise) to the bundle. The JM or safety personnel lay out two straps parallel to each other on top of the parallel strap, ensuring strap fasteners are on the same side, at least 16 inches from each other, and centered. The JM or safety personnel then—

- Center the bundle on the parallel strap.
- Route the parallel strap over the top of the bundle and through the two D rings, fold, and secure.
- Route the parallel strap through the D rings from the inside toward the outside so that the D rings are pointing to each other, fold, and secure.
- Tighten all straps.
- Tie off the excess webbing above the strap fastener by using one turn of 1/4-inch cotton webbing tied in a surgeon's locking knot.
- Ensure the excess webbing is not above the top of the bundle. The bundle has one smooth side for ease in ejecting from the aircraft.

FOUR-STRAP BUNDLE

8-10. The JM lays out two straps parallel to the bundle and centered. He lays out two straps parallel to each other on top of the parallel straps and centered. He centers the bundle on the parallel straps. The JM—

- Routes the parallel straps through the two D rings (one D ring per strap), folds, and secures.
- Routes the parallel straps through the D rings, ensuring both D rings point in the same direction. He folds and secures the parallel straps (one D ring for each strap), ensuring the strap fasteners are on the same side.
- Tightens all straps.
- Ties off the excess webbing above the strap fastener by using one turn of 1/4-inch cotton webbing tied in a surgeon's locking knot.
- Ensures the excess webbing is not above the top of the bundle. The bundle has one smooth side for ease in ejecting from the aircraft.

A-21 CARGO BAG

8-11. The A-21 cargo bag consists of the following components:

- Canvas cover: cotton duck material, 97 inches by 115 inches, with eight strap keepers.
- Sling assembly with scuff pad: one 188-inch main strap, two 144-inch side straps, scuff pad (30 inches by 48 inches), and four lifting handles.

Chapter 8

- Quick-release assembly: quick-release device with safety clip, three quick-release straps, and one fixed quick-release strap.
- Two ring straps: one 9-inch strap that has a friction adapter and one 7-inch strap with a D ring.

CHARACTERISTICS

8-12. Container components weigh 18 pounds with a maximum weight of 500 pounds, not including the parachute. The minimum load capacity is dependent on the type parachute used and the method of airdrop. Personnel should use the A-21 cargo bag to drop fragile and nonfragile supplies. Dimensions are a maximum 30 inches wide by 48 inches deep by 66 inches high or 69 inches high for the Stinger missile. (FM 10-550, *Airdrop of Supplies and Equipment: Rigging Stinger Weapon Systems and Missiles*, provides additional information.) Dimensions include the cargo parachute.

Note. See FM 10-500-3 for further information on rigging containers.

METHOD OF RIGGING

8-13. Figure 8-2, page 8-5, shows the rigging of the A-21 cargo bag. The JM or safety personnel—
- Spread the canvas cover on a level surface with all strap keepers facing up.
- Position the sling assembly webbing straps down on the canvas cover and threads the straps through the keepers.
- Turn the sling and canvas cover over as a unit so the sling is beneath the cover.
- Center the load on the canvas cover, using cushioning material, as needed.
- Wrap the load in the canvas cover, side flap first, and fold all excess material under.
- Attach the two ring straps to the 188-inch main strap, keeping both D rings touching and centered.
- Attach the four quick-release straps to the 144-inch side straps.
- Ensure the rotating disk is facing up when the quick-release assembly is placed on top of the load (thick-lip portion of the friction bar facing out).
- Thread the fixed, quick-release strap with the quick-release assembly attached through the nearest steel rod ring.
- Thread the remaining quick-release straps through the nearest steel rod rings.
- Insert the lugs into the quick-release assembly.
- Tighten the quick-release straps and the two-ring straps; roll all excess webbing. Ensure the webbing is tied off below the friction adapter with a surgeon's locking knot and the quick-release device is centered on the bundle.

A-22 CARGO BAG

8-14. The A-22 cargo bag (Figure 8-3, page 8-5) is an adjustable cotton duck cloth and webbing container. The A-22 cargo bag consists of a cotton or nylon webbing sling assembly, a cover, and four cotton or nylon suspension webs. The A-22 cargo bag has a maximum load capacity of 2,200 pounds.

8-15. The maximum allowable dimensions for a rigged load are 48 inches wide and 53.5 inches long. Maximum height is normally 83 inches, but with USAF approval may extend up to 100 inches. For a low-velocity airdrop, a standard cargo bag skid (48 by 53.5 inches) serves as a base for the container load. For a high-velocity airdrop, the standard cargo bag skid or an appropriate size piece of plywood serves as the base of the container load. The weight of the A-22 cargo bag and skid is about 58 pounds.

Note: The low-velocity CDS with A-22 cargo bag is not organic to SF.

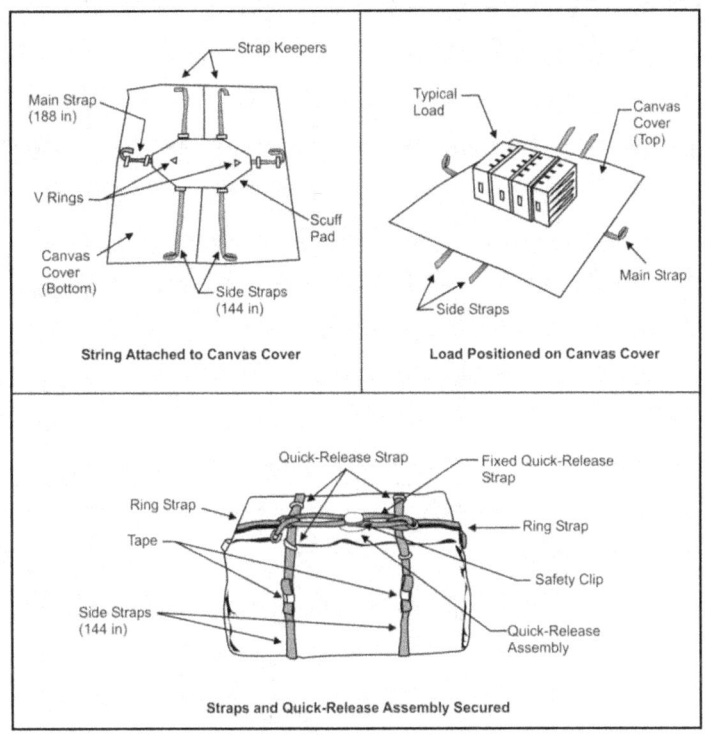

Figure 8-2. Rigging of the A-21 cargo bag

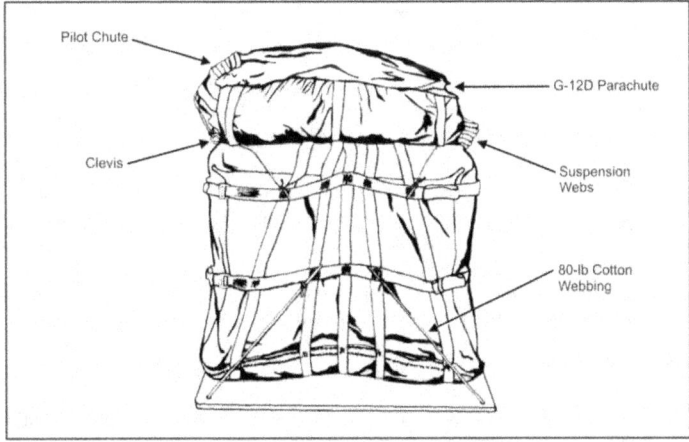

Figure 8-3. Rigging of the A-22 cargo bag

Chapter 8

CARGO PARACHUTE RIGGING ON A-SERIES CONTAINERS

8-16. After the JM or safety personnel rig the A-series containers. Then the JM inspects the cargo parachutes and attaches them to the load.

INSPECTION

8-17. The parachutist places the cargo parachute on the center of the bundle. The JM inspects the bundle for the following:

- Four tie-down straps.
- Two risers complete (clevis, clevis pin, and safety wire).
- Static line, complete with drogue device (clevis, clevis pin, and safety wire).
- Drogue device is attached to the break cord attaching loop, unless a breakaway static line is used.

ATTACHMENT

8-18. The JM ensures—

- Risers go directly to the attaching point (D ring).
- Tie-downs are attached (tied to side straps).
- Static line is free to deploy.
- Risers are not routed around or under any part of the bundle.

Note. The cargo parachute should be attached with the side of the pack where the risers come out, collocated to the rough side of the bundle.

8-19. The container measures 21 inches in diameter by 106 inches long, weighs 213 pounds empty (with parachute), and is made of glass-wound resin acrylic that allows easy destruction by burning. The CTU-2/A container can be used to deliver—

- Critical supplies, such as food, water, ammunition, and medicine.
- Surface-to-air recovery (STAR) kits.
- DZ marking equipment or RTs for airdrop or air strike direction.

PONCHO-EXPEDIENT PARACHUTE

8-20. The poncho-expedient parachute (Figure 8-4) can be used to drop up to 65 pounds of equipment. It lessens the need for expensive parachutes. The following illustrates and describes how the JM or safety personnel rig the poncho:

- Pull the hood drawstring loop to close the hood opening, then wrap the excess drawstring tightly around the base of the hood and tie it off so no air will escape.
- Fold the poncho in half (bottoms together) with the snaps down.
- Cut eight 6-feet long suspension lines of 550 cord.
- Tie one suspension line to each of the grommets on the poncho with a bowline knot.
- Make sure there are no tangles in the suspension lines and that they are the same length.
- Tie the free ends of the suspension lines to a snap link with one large overhand knot that is further secured by one or two half-hitch knots.
- Fold the poncho. Lay the half-folded poncho flat. On both long sides of the poncho, make S folds 6 to 8 inches wide to meet in the center (there should be the same number of folds on both sides). Next, fold the narrow-folded poncho into an M fold.
- Cut a piece of 15-foot 25-pound test cord (or a lightweight string that will break when the bundle is deployed from the aircraft) for use as a static line.
- Tie the loop end of the static line to the drawstring (wrapped around the hole of the poncho), and tie with a square knot.
- Attach the load to the snap link attached to the suspension lines.

- Fold the suspension lines on top of the load.
- Place the poncho-expedient parachute folded into an M on top of the folded suspension lines.
- Affix the poncho-expedient parachute to the top of the load with one wrap of 25-pound test cord in the same manner as tying a package, ensuring the cord goes through the loop in the static line. Tie with a square knot. This will deploy the suspension lines before breaking loose from the aircraft.

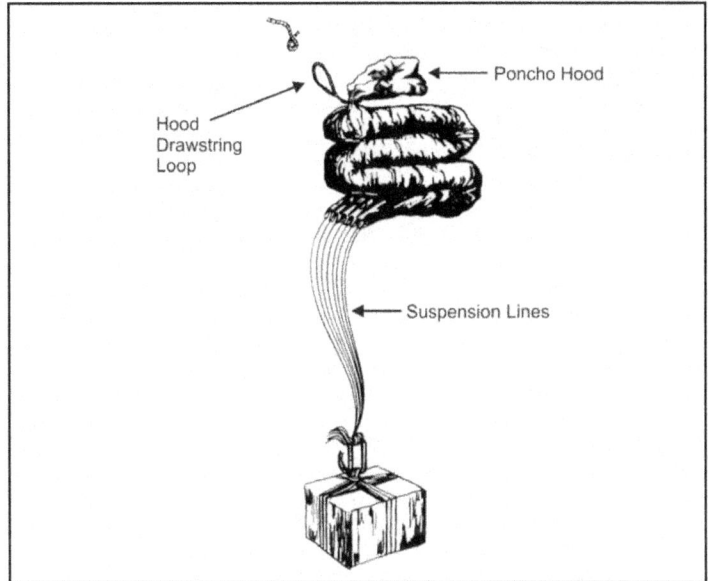

Figure 8-4. Poncho-expedient parachute

STEEL STRAPPING

8-21. The steel strapping commonly used for rigging airdrop loads is made of flat steel, 0.020 inch thick by 5/8 inch wide, with a breaking strength of 1,000 pounds. Personnel may use steel strapping to make a container or to bind equipment items together for packing in container loads. When using steel strapping to make a container, personnel use double thickness. The load limit for steel strapping is 250 pounds.

RIGGING KNOTS

8-22. A good knot must be easy to tie or untie and must hold without slipping. Personnel must use the proper knot during the rigging of loads for airdrop. Figure 8-5, page 8-8, shows the knots used most frequently.

Chapter 8

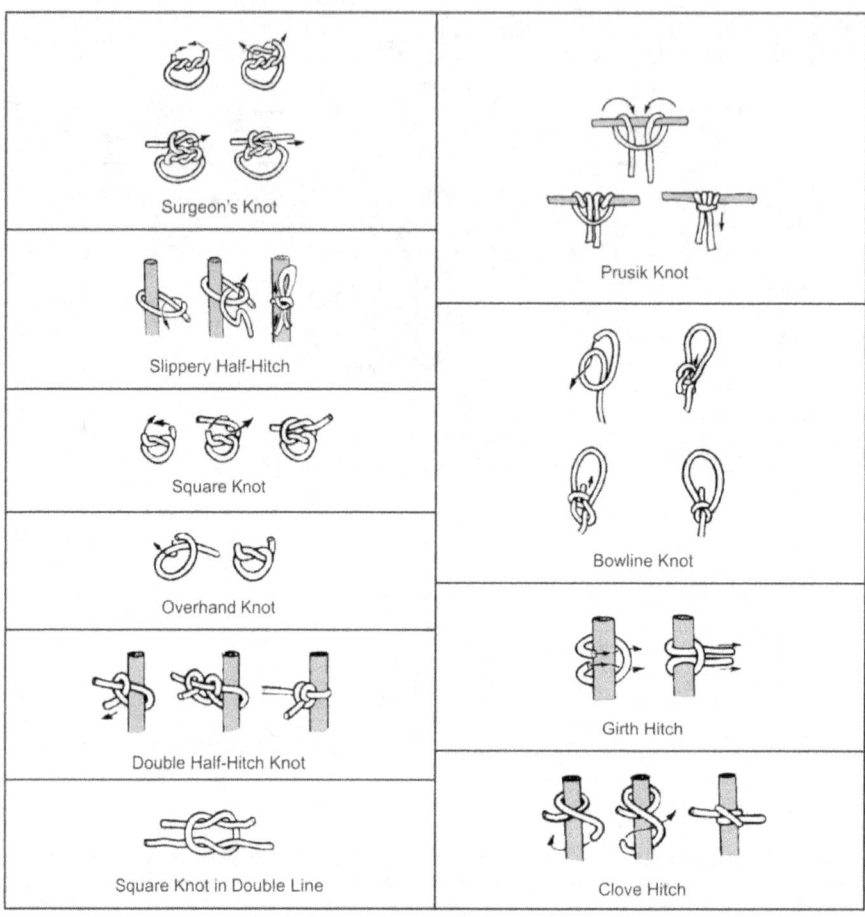

Figure 8-5. Rigging knots

Chapter 9
Special Patrol Infiltration and Exfiltration System

The USMC designed SPIES for use in inserting and extracting patrols where a helicopter landing was impracticable. The system provides a means of exfiltrating up to 14 personnel over short distances. It is not recommended for infiltration, because team members are exposed the entire time. Because of the nature of SPIES operations, a thorough briefing is required for all participants before the operation. This briefing is crucial when additional assets (gunships, aerial observers, or artillery support) are used with the extraction helicopter. SO units under the command of USSOCOM may conduct SPIES operations.

TRAINING OBJECTIVES

9-1. Personnel being extracted must receive training in the SPIES extraction procedures before infiltration. The objectives of SPIES training are to safely conduct and maintain maximum proficiency in the execution of SPIES operations.

PREOPERATIONS BRIEFINGS AND PROCEDURES

9-2. Before conducting a SPIES training mission, the participants must have a basic understanding of the requirements. The SPIES master conducts briefings to ensure the personal extracted and the pilot know the procedures. As SOPs are developed and units train together, the SPIES master simply refers to the SOP. He always gives a safety briefing and conducts an equipment inspection.

SAFETY BRIEFING

9-3. As in all training, a safety briefing must precede operations using SPIES. The briefing should consist of, but not be limited to, a review of the following:
- Area hazards.
- General aircraft safety.
- Characteristics of equipment associated with the SPIES.
- Equipment inspection and proper donning of the harness.
- Method of extraction and insertion to be used.
- Hand-and-arm signals and emergency signals.
- Medical coverage.
- Communications requirements.
- Night operation requirements.

TRAINING REQUIREMENTS

9-4. All personnel must complete initial, sustainment, and/or refresher training. Personnel must also complete the appropriate training to become SPIES-qualified.

Chapter 9

Initial Training

9-5. Soldiers will be SPIES-qualified when they—
- Have been thoroughly briefed on SPIES, its purpose, capabilities, limitations, and emergency procedures.
- Have been thoroughly briefed on the duties and responsibilities of the pilot in command, SPIES master, and GSO or NCO.
- Have completed a minimum of three satisfactory SPIES extractions, to include one with combat equipment and weapon.
- Know the procedures, techniques, and equipment necessary to conduct SPIES extractions by demonstrating confidence and proficiency.

Sustainment Training

9-6. Units routinely conduct sustainment training to maintain acquired skills. Units will receive training on SPIES procedures within 24 hours before the operation. At a minimum, this training will include the following:
- Rigging and inspecting individual equipment.
- Rigging and inspecting aircraft and accompanying equipment.
- Hand-and-arm signals.
- Safety requirements and emergency procedures.
- Rehearsals, as needed.

Refresher Training

9-7. Refresher training is mandatory for Soldiers who have not participated in SPIES operations during the previous 12 months. Soldiers must receive sustainment training and conduct at least one SPIES operation under the observation of a current SPIES master.

KEY PERSONNEL QUALIFICATIONS

9-8. Before conducting SPIES operations, the SPIES master and GSO or NCO must possess certain qualifications. They must meet the criteria discussed in the following paragraphs to be SPIES-qualified.

SPIES MASTER

9-9. Selection of personnel to be qualified as SPIES master should be based on the individual's demonstrated leadership capabilities, maturity, and knowledge of SPIES operations. Individuals selected must participate in at least three SPIES operations (observe twice and execute SPIES master duties once under supervision by a qualified SPIES master). For example, personnel configure the hookups in the helicopter, help prepare an operation, and conduct a successful operation under the supervision of a qualified SPIES master. Personnel know all aspects of a SPIES operation. They can give an effective pilot's brief, use the aircraft communications equipment, and understand aviation terminology. Personnel will be qualified to perform the duties of SPIES master when they have—
- Completed the initial SPIES training.
- Received instructions on and demonstrated proficiency in rigging the helicopter, inspecting and preparing SPIES, and donning the SPIES harness.
- Received instructions and demonstrated proficiency in the performance of the following SPIES master duties:
    - Coordination responsibilities.
    - Troop or aircrew briefings.
    - Organization of the personnel to be extracted.
    - Instruction to pilots in maintaining the aircraft in position over the target.

- Throwing and retrieving SPIES.
- Hand-and-arm signals.
- Emergency procedures.

GROUND SAFETY OFFICER (GSO) OR NCO

9-10. A GSO or an NCO must be a current SPIES master. A GSO or NCO is required during a SPIES operation. The GSO or NCO may also be a member of the extracted team.

PERSONNEL DUTIES AND RESPONSIBILITIES

9-11. SPIES training and operations require the designation of key personnel to perform assigned tasks. The positions are unit commander, SPIES master, GSO or NCO, air mission commander, and pilot in command.

UNIT COMMANDER

9-12. The unit commander or designated representative will ensure the Soldiers' supervisors have screened all Soldiers before participating in SPIES training. Sscreening ensures the Soldiers are physically and professionally able to participate in SPIES operations. Minimum standards include the following:
- Be assigned or attached to a USSOCOM unit.
- Have a current medical examination.
- Have passed the Service fitness test.
- Have no injury or physical condition that would cause a potential safety hazard during SPIES operations.

SPIES MASTER

9-13. The SPIES master ensures the conduct of safe and efficient extraction missions. Each extraction helicopter will have at least one SPIES master at a minimum. The SPIES master—
- Ensures all personnel and equipment are at the proper place for rehearsals and operations.
- Rigs the helicopter under the supervision of the aircraft crew chief. Conducts a thorough inspection of SPIES, anchor system, and aircraft tie-downs.
- Conducts a briefing with the aircrew members.
- Communicates with the pilot or crew through the aircraft intercom system and monitors communications between the pilot and the ground.
- Directs pilot in command to maneuver the aircraft into the proper position for deployment of SPIES.
- Prepares and deploys SPIES manually to ensure the system lands in the proper location. Recovers and redeploys the system if the desired area is missed or if the mission is aborted.
- Receives hand-and-arm signals or radio instructions from personnel being extracted and relays them to the pilot in command.
- Directs the pilot in command out of the extraction area until extracted personnel have cleared obstacles.
- Observes extracted personnel from the extraction site to a safe letdown area, and monitors aircraft speed.
- Directs emergency procedures in releasing Soldiers at ground level on command from the pilot.
- Directs the landing of Soldiers.
- Collects equipment after the aircraft lands and repacks the equipment after completing the maintenance checks.
- Aborts any portion of the operation because of the following conditions:
  - Aircraft behind or below the extraction aircraft.
  - Unsafe conditions within the aircraft, precluding a safe extraction.

Chapter 9

- Defective equipment.
- Any other unsafe condition.

Note. Currently, the crew chief handles the duties of controlling the rope and communicating with the pilot. However, during a tactical operation, the crew chiefs will be serving as observers and gunners and the SPIES master will control the rope and serve as the link to the pilot. During training, the SPIES master should be performing as many of his duties as possible to remain proficient should a tactical situation arise.

GROUND SAFETY OFFICER OR NCO

9-14. During training, a GSO or NCO will be located on the ground at the extraction point and the letdown area. Depending on the mission requirements, the GSO or NCO may be a member of the extraction team. This officer—

- Ensures each Soldier has properly donned the harness.
- Ensures radio or visual signals communication with the SPIES master or aircrew.
- Ensures each Soldier has properly hooked up to the extraction ropes, and verifies hookup of the personnel safety sling.
- Ensures Soldiers and ropes are clear from all obstacles.
- Signals the SPIES master that personnel are ready for extraction.
- Assists personnel as they land at the letdown area.

AIR MISSION COMMANDER

9-15. When more than one helicopter is involved in the operation, the employing aviation unit designates the air mission commander. He—

- Ensures all aircraft and aircrews are at the appropriate locations for training, rehearsals, and the operation.
- Ensures all aircraft are properly configured for SPIES operations.
- Ensures the aircrews and all personnel not a member of the aircrew are briefed and understand their responsibilities during SPIES operations, including aircraft safety and actions in case of an emergency.
- Emphasizes procedural techniques for clearing, recovering, and jettisoning SPIES personnel at ground level, or for the aircraft prematurely departing the PZ or extraction zone (EZ).
- Ensures all aircraft deploy SPIES on the designated target.

PILOT IN COMMAND

9-16. The pilot in command assumes the duties of the air mission commander on single-ship missions. He also—

- Ensures the crew chief has inspected the anchoring device assembly for completeness and functionality and installed the SPIES properly.
- Keeps the aircraft centered over the PZ or EZ with corrections from the SPIES master or crew chief, as required.

AIRCRAFT CREW CHIEF

9-17. The aircraft crew chief ensures the SPIES master has rigged the helicopter for the operation and that he is performing his duties correctly. The crew chief—

- Conducts a thorough inspection of SPIES, anchor system, and aircraft tie-downs.
- Rigs or supervises the rigging of the helicopter.
- Communicates with the pilot or SPIES master through the aircraft intercom system and monitors communications between the pilot, SPIES master, and the ground.

Special Patrol Infiltration and Exfiltration System

- Ensures the SPIES master knows and understands the aircraft SOPs.
- Is capable of performing the duties of the SPIES master.
- Supervises the SPIES master in the performance of his duties.

SPIES EQUIPMENT

9-18. SPIES consists of a specially modified rope, a harness worn by the Soldiers, snap links, and cargo slings. Figure 9-1 shows all the SPIES equipment.

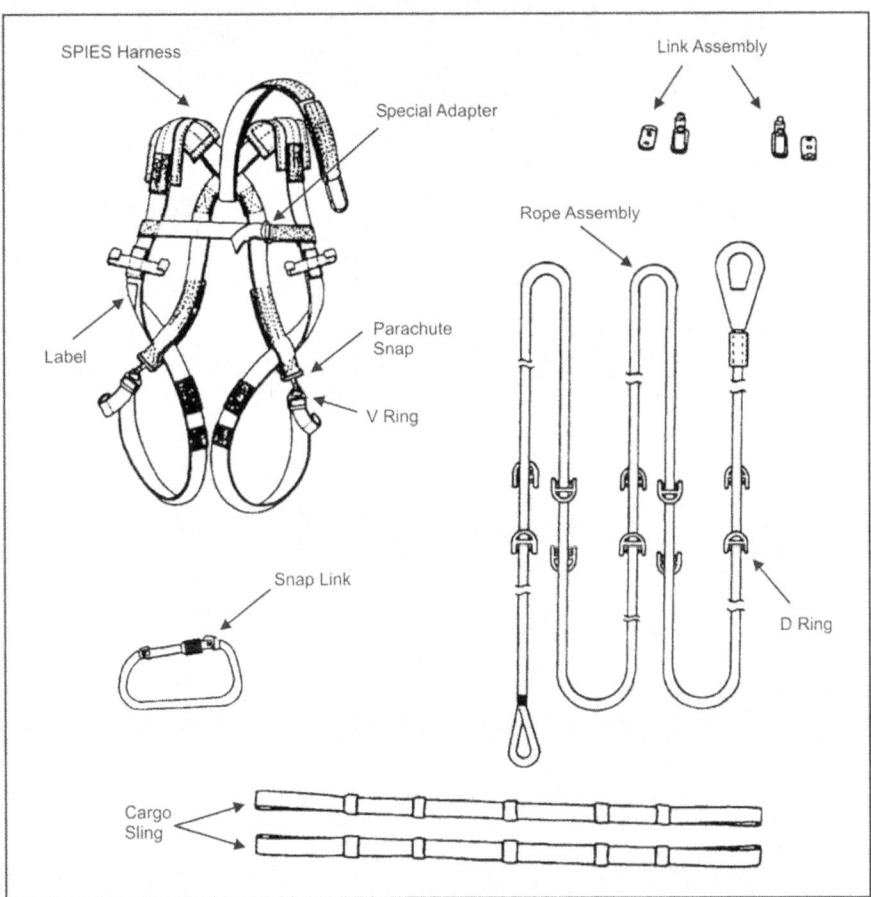

Figure 9-1. SPIES equipment

SPIES ROPE

9-19. The SPIES rope is a two-in-one type, braided, nylon extraction rope 1 inch in diameter and 120 feet long. There are 10 separable D rings inserted at the lower end of the rope to serve as individual attachment points. The SPIES master inserts in pairs through the core of the rope the 10 separable D rings. The SPIES

Chapter 9

master arranges the D rings 1 foot apart from each other on opposite sides of the rope (Figure 9-2). He spaces each pair of D rings 7 feet apart from the succeeding pair; the first pair is located 7 feet from the lowered end of the rope. If needed, the SPIES master can add four additional D rings to the rope.

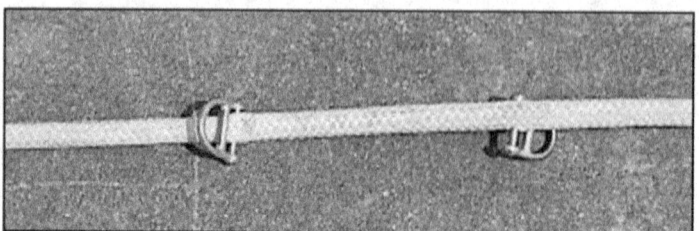

Figure 9-2. D rings attached to the SPIES rope

SPIES HARNESS

9-20. The SPIES individual harness is constructed of Type VII nylon webbing. Each harness is equipped with leg straps connected with parachute harness ejector snaps and parachute harness V rings. The chest strap requires a reversible quick-fit adapter. A 20-inch strap is connected to the crossover portion of the backstrap, which in turn, is attached to a D ring on the rope, using a mountain piton snap link.

INSPECTION OF SPIES

9-21. A certified SPIES master or rigger inspects SPIES at 6-month intervals or whenever the serviceability of the equipment is in doubt. The items for which the SPIES master or rigger inspects are discussed below.

SERVICE LIFE

9-22. The SPIES master and riggers check ropes, harnesses, and suspension slings for expiration of service or total life. Expiration of service is 7 years (opening manufacturer's package), and total life is 15 years from date of manufacture.

SPIES HARNESS

9-23. The SPIES master inspects the harness and suspension sling webbing for the following signs:
- Contamination from oil, grease, acid, or rust.
- Cuts.
- Twists.
- Fading.
- Excessive wear.
- Fusing (indicated by unusual hardening or softening of webbing fibers).
- Fraying.
- Burns.
- Abrasions.
- Loose or broken stitching.

9-24. Riggers must repair loose or broken stitching when more than three stitches are loose or broken. The SPIES master removes the damaged harness or suspension sling from service and returns it to the riggers for repair or appropriate disposition.

9-25. The SPIES master inspects all hardware for signs of corrosion, pitting, ease of operation, security of attachment, bends, dents, nicks, burrs, and sharp edges. Riggers must replace hardware (except chest strap adapter) that requires them to remove the stitching from the webbing.

9-26. Riggers who are familiar with SPIES repair the SPIES harness. The riggers replace the V ring by cutting the strap above the stitching and then folding and stitching a new end section of the chest strap.

SPIES ROPE

9-27. The SPIES master or riggers inspect the rope surface for splices, cuts, excessive abrasions, and snags. Cuts on the rope are excessive when there are four or more cut strands in any 5-inch length. The two-to-one braided rope has 12 pairs of strands (24 individual strands) around the circumference. Abrasion is extensive when torn yarns are equivalent to that of four strands of any 5-inch length. Rope subjected to heavy loads may display glazed areas where it has worked against hard surfaces. This condition may be caused by paint or fused fibers. After long use, the rope may become fuzzy on the surface (although this should be minimized with the surface coating). However, the effect on the strength of the rope is negligible. Inspect the eye loop at the end of the SPIES rope to ensure it is not broken, frayed, or loose. The SPIES master will determine when outdated, spliced, abraded, or cut ropes are removed from service.

9-28. The SPIES master or riggers check the rope for signs of contamination by acid, alkaline compounds, salt water, fire-extinguishing solutions, and petroleum-based solvents. Although used ropes gradually change color, such changes do not indicate a decrease in strength unless contact with strong chemicals has caused the change. Changes in color caused by chemicals are usually spotty; changes caused by use are uniform throughout the length of the rope.

OPERATIONAL REQUIREMENTS

9-29. Units must follow operational requirements as closely as possible during training under usual conditions and during unusual conditions (adverse weather or terrain conditions and night operations). Personnel must use sound judgment to determine what action to take depending on the nature and severity of the condition.

> **CAUTION**
>
> Acid contamination, cuts, or fraying of harness or sling webbing constitute damage that is not reparable.

SITE SELECTION

9-30. The limitations and capabilities of the mission aircraft are the primary factors in site selection. Mission planners consider site altitude and temperatures as they determine air density that affects the helicopter payload. There are no particular selection criteria for SPIES extraction sites, as any small clearing is ideal. Forested areas may be used but can be dangerous when extracting more than one person because the safety rope can become entangled in the foliage and branches.

MEDICAL COVERAGE

9-31. A qualified and equipped 18D, 91W, or medic or Service-qualified emergency medical technician (EMT) will be at all training sites. All medics will be equipped with the following:
- An M-5 aid bag, or equivalent, packed IAW unit standards.
- A medical transportation vehicle that is covered and large enough to carry a stretcher, a litter or backboard, and any other items deemed necessary.

9-32. Medics must know casualty evacuation (CASEVAC) procedures and have coordinated requirements necessary to expedite evacuation and treatment of personnel on and off military installations. Units cannot conduct training without a medic, medical equipment, and transportation. If the situation warrants and the

Chapter 9

installation cannot support a MEDEVAC mission, the SPIES aircraft may be used as a last-resort CASEVAC vehicle. The medic must coordinate this contingency with the aircrew before the start of training. The medic will develop an evacuation plan that includes, but is not limited to, the following:
- Medical facilities (location and capabilities).
- Emergency telephone numbers.
- Routes to medical facilities.

COMMUNICATIONS REQUIREMENTS

9-33. Radio communications are the primary means of communication between the helicopter and the Soldiers on the ground, unless the situation dictates otherwise. If radio communications are hampered or contraindicated, personnel use special procedures along with hand or light signals. Personnel use the same hand-and-arm signals discussed in Chapter 10 in case of radio failure or poor communications caused by static or noise overriding the audio output of the radio. The flight crew can pass on all communications to the SPIES master through the aircraft intercom system from the GSO. Additionally, the GSO will inform the aircraft to stop operations if an unsafe condition develops. During night operations, if radio communications are hampered, personnel will use special procedures along with hand or light signals.

ADVERSE WEATHER AND TERRAIN CONDITIONS

9-34. Units will **not** conduct SPIES training and operations under the following conditions:
- Wind chill factors caused by the rotor wash of the helicopter or extraction cruise air speeds that may cause cold weather injuries.
- Water or ice on the SPIES.
- SPIES is exposed to the elements long enough to freeze, thereby reducing its tensile strength.
- Blowing particles produced by rotor downwash cause the aircrew or the SPIES master to lose visual contact with the ground.

NIGHT OPERATION REQUIREMENTS

9-35. Night operations require additional safety equipment and precautions. They are as follows:
- The SPIES master attaches one chemical light to the SPIES deployment bag and one chemical light to the harness of each person being extracted.
- The personnel being extracted will not wear night vision devices during the extraction.
- The SPIES master and the aircrew members will wear NVGs as required during night operations. NVG lighting criteria will be IAW Army regulations, specific aircraft aircrew TMs, unit SOPs, or the tactical environment.

HELICOPTER RIGGING

9-36. The SPIES master and crew chief rig the helicopter. Each one conducts an inspection to ensure the helicopter is properly rigged. Because each helicopter requires different material and rigging procedures, the SPIES master and crew chief should refer to the specific section for rigging each aircraft.

RIGGING OF UH-60

9-37. The following equipment is required to rig a UH-60:
- One 120-foot rope with deployment bag.
- Two 11-foot, three- or four-loop cargo slings or two 9-foot, three- or four-loop cargo slings. If the aircraft does not have cargo hooks, use four cargo slings.
- Two Type IV connector links. (If the aircraft does not have cargo hooks, use four cargo slings.)
- Heavy-duty tape (100-mph tape).
- A 12-foot length of tubular nylon or one 12-foot sling rope.

Special Patrol Infiltration and Exfiltration System

- Nine oval snap links.
- A 4-inch by 4-inch block of wood that is 18 inches long.
- A fire ax (for use during emergency cutaway procedures).

9-38. The SPIES master or crew chief rigs the UH-60 for SPIES operations as follows:
- Joins the two 9- or 11-foot cargo suspension slings to form one continuous sling using a Type IV link (Figure 9-3). Tapes the straps for the three- or four-loop cargo sling with 100-mph tape at about 12-inch intervals.

Figure 9-3. Suspension slings connected by Type IV link

- Lowers the cargo hook of the helicopter and leaves the hatch cover off.
- Uses the cargo hook as the primary attachment point for the SPIES rope. Attaches the end of the SPIES rope, which has a polyurethane encapsulated eye, to the cargo hook.
- Uses padding around the edge of the cargo hatch to protect the sling from damage.
- Runs the sling across the helicopter deck. Take one end under the helicopter and through the eye of the SPIES rope (Figure 9-4). Connects the other end of the sling using a Type IV link assembly.

Figure 9-4. SPIES rope connected to cargo hook and sling

Chapter 9

- Tapes down the 4-inch by 4-inch block of wood along the right edge of the doorway (Figure 9-5) so the sling crosses the block perpendicularly at the middle. The wood block serves as a chopping pad in case of an emergency cutaway.
- Positions the Type IV link just inside of but not on the wood block (Figure 9-6).

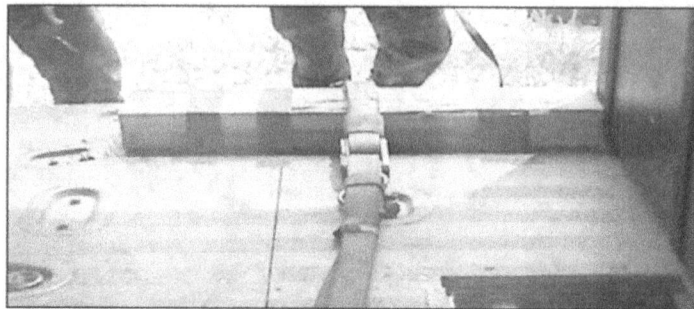

Figure 9-5. Placement of wood clock

Figure 9-6. Placement of Type IV link

- Once the SPIES rope and cargo straps are in place, secures the straps running across the deck of the helicopter in place by four pairs of snap links. Places all snap links to the left of the wood block (or right side if the wood block is on the left door). This placement allows the sling to release if cut during emergency procedures (Figure 9-7, page 9-11).
- Connects the two snap links to the same cargo ring with the swing gates reversed (Figure 9-8, page 9-11).
- Evenly spaces the snap links across the deck. Alternates the snap links from one side of the strap to the other and from top to bottom. Thus, the first snap link can be to the rear of the strap, wrapping around the bottom two straps (Figure 9-9, page 9-12). Places the next snap link in the front of the cargo strap and around the top two sections of the strap. Continues this process until at least four points are established.
- Gathers the excess cargo sling on the side of the aircraft opposite the wood block. Tapes the excess securely to the floor.

Special Patrol Infiltration and Exfiltration System

Figure 9-7. Placement of four pairs of snap links

Figure 9-8. Snap links connected to sling and ring, gates reversed

- Uses a Prusik knot, tie the tubular nylon or sling rope to the SPIES rope about 2 to 3 feet below the cargo hook (Figure 9-10, page 9-12). With the other end, ties a bowline with a half hitch and connect a snap link. Connects the snap link to a cargo ring in the middle of the aircraft floor (Figure 9-11, page 9-12). This line serves as a recovery line for the rope so that the aircrew can retrieve the rope into the aircraft. Ensures the line is long enough so that the weight on the SPIES rope is hanging from the hook and not from the recovery rope.
- Coils the SPIES rope (Figure 9-12, page 9-13) and places it on the opposite side of the aircraft from the wood block. Ensures the SPIES rope is not tangled or in the way.

9-39. If the cargo hook is unavailable or it is not working properly, the SPIES can be used safely by doubling the cargo slings and Type IV links. Two cargo straps are side by side with a total of four slings and four Type IV links.

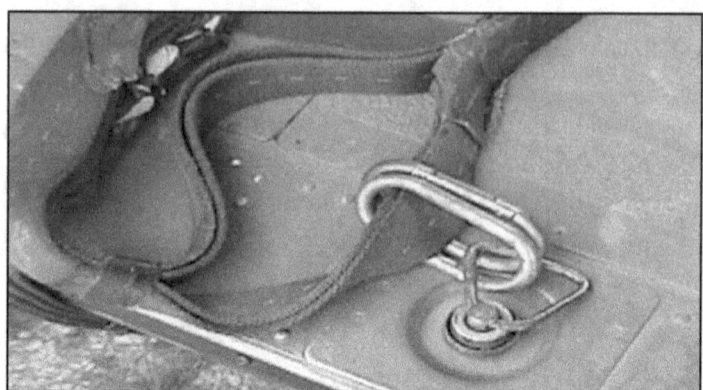

Figure 9-9. Snap links alternating on strap

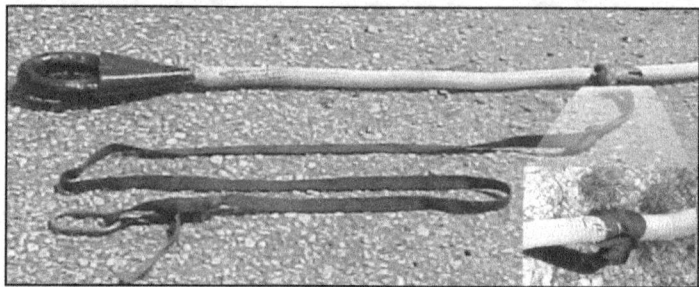

Figure 9-10. Recovery rope tied to SPIES rope

Figure 9-11. Recovery rope connected to SPIES rope and helicopter

Special Patrol Infiltration and Exfiltration System

Figure 9-12. Rigging complete

RIGGING OF MH-47 HELICOPTERS

9-40. The SPIES master or crew chief may rig the MH-47 aircraft for SPIES using one of two ways. The SPIES master or crew chief attaches the SPIES rope to the aft fast rope bars in the same way as the fast rope or rigs the MH-47 using the cargo hook. (See CH-46/47 below.) The preferred method is to attach the rope to the fast rope bars. The SPIES master must not exceed the 2,300-pound limit of the bars when using this method.

RIGGING OF CH-46/47 AND CH-53 HELICOPTERS

9-41. The SPIES rope can be attached to the CH-46/47 and CH-53 helicopters. Rigging the SPIES for these helicopters requires two 9- or 11-foot cargo suspension slings, four Type IV links, and eight snap links (Figure 9-13, page 9-14).

9-42. The following equipment is required to rig a CH-46/47:
- One 120-foot SPIES rope with deployment bag.
- Two 11-foot, three- or four-loop cargo slings or two 9-foot, three- or four-loop cargo slings.
- Four Type IV links.
- Eight oval snap links.
- Heavy-duty tape (100-mph tape).
- A 12-foot length of tubular nylon or one 12-foot sling rope.

9-43. The SPIES master or crew chief rigs the CH-46/47 for SPIES operations as follows:
- Passes the cargo slings through the encapsulated eye of the SPIES rope.
- Places the cargo slings in two U shapes. Places one sling forward of the cargo hole in the center of the aircraft floor and the other sling aft or toward the rear of the helicopter.
- Connects each end of the sling to an outboard cargo tie-down ring on the aircraft floor by a Type IV connector. Uses two tie-down rings for each sling.
- Installs the cargo straps so that they hold the SPIES rope in the center of and slightly below the opening of the cargo hatch.

Note. Not all of the tie-down rings will be in the exact same position on all helicopters.

- Uses a pair of snap links for added security at each point of attachment, with the swing gates reversed. The snap links ensure a backup in case of a faulty tie-down ring and reduces the amount of movement in the cargo suspension straps.
- Uses tape or padding around the edge of the cargo hatch to protect slings from damage.

Chapter 9

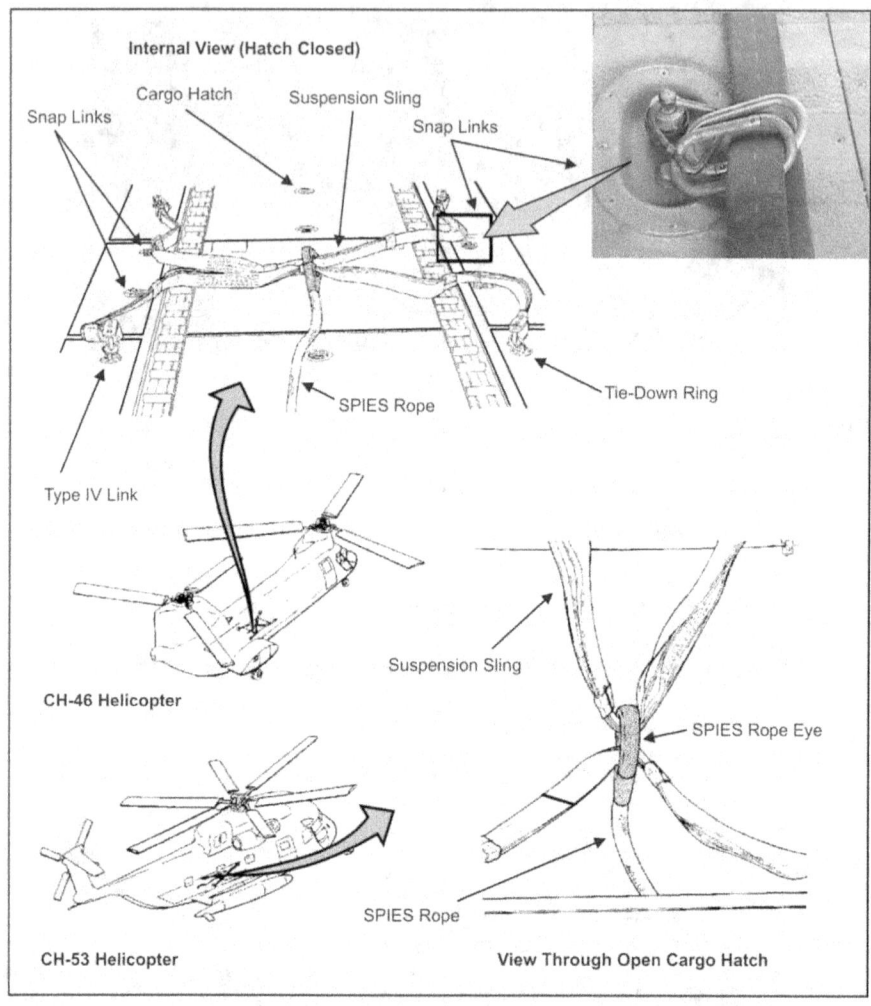

Figure 9-13. SPIES rigged on CH-46/47 and CH-53 helicopters

SPIES OPERATIONS

9-44. The SPIES is used when the team requires immediate extraction or is unable to move to a clear (open) position suitable for helicopter landing. If the situation, mission, or terrain suggests the possibility of a SPIES extraction, the team includes the SPIES harness on its individual equipment list. The team must coordinate with the extraction aircraft and the SF operations base before infiltration to ensure the aircraft and the base can support this extraction method with equipment and trained personnel.

SPIES MASTER OR CREW CHIEF DUTIES

9-45. The SPIES master and the crew chief provide for the safe conduct of the SPIES operation. Because there is considerable overlap between the duties of these two personnel, they must coordinate closely before the operation to determine who is performing which duty. Units must ensure the SPIES master gains as much experience in the performance of his duties as possible, because during a tactical operation, the SPIES master will probably be performing all of the duties.

Preflight Duties

9-46. The SPIES master performs the following:
- Inventories and inspects all SPIES equipment.
- Briefs the pilot and other concerned personnel about details of the operation, especially the extraction and dismounting procedures.
- Ensures an internal communications system (ICS) helmet and a gunner's belt are available for the SPIES master's use. Attaches a sling rope, if available, into the belt. Connects and checks the operation of the ICS to be used. (The SPIES master and the pilot must establish ICS communications between each other on all SPIES operations.)
- Attaches the SPIES rope to the helicopter.
- Ensures all items are secure in the aircraft.
- Checks the location of the emergency ax. Places the ax where readily available, yet secure enough so as not to endanger the men on the SPIES rope. Inspects the ax to ensure it is sharp.

Extraction Duties

9-47. The SPIES master performs the following:
- On arrival at the team's estimated position, helps the pilot determine the exact location of the team members.
- As the aircraft approaches the team's location, aids the pilot (using the clock system) in placing the aircraft directly above the team.
- Requests permission from the pilot to drop the SPIES rope when the aircraft is hovering above the team.
- Drops the rope, taking care to avoid striking team members on the ground.
- Notifies the pilot when the rope is down, and reports all altitude corrections to ensure team members reach all SPIES attachment points.
- Watches for the thumbs-up signal from all team members.
- On receipt of the thumbs-up signal, advises the pilot the team is ready for extraction and requests a vertical liftoff.
- Advises the pilot of the team's position, the location of any potential obstacles, and the avoidance of horizontal movement.
- If a team member becomes entangled with an obstacle during the extraction, notifies the pilot and requests the vertical lift be stopped. If the situation is critical, prepares to cut the SPIES rope (the anchor point or cargo straps) after the team members are secured to the obstacle or on the ground.
- When positive that all obstructions are clear, advises the pilot to obtain a safe altitude (about 350 feet AGL for training purposes or as the situation dictates in combat) or to transition into forward flight.
- At frequent intervals during the flight, advises the pilot on the safety status of all team members. Maintains a constant visual watch on the team and frequently checks security of the SPIES attachments.

Chapter 9

Dismounting Duties

9-48. The SPIES master performs the following:
- On arrival at the dismounting area, gives the pilot the approximate distance of the lower rope end from the ground.
- Once the pilot starts the vertical descent, continually informs him as to the approximate distance of the lower rope end from the ground.
- Informs the pilot of any horizontal drift that occurs and any obstructions near the SPIES rope. Also informs the pilot of any oscillation that may occur.
- Informs the pilot when the rope is about 25 feet above the ground and 10 feet above the ground. Ensures the rate of descent is slow enough to enable the team members to land and safely get out from under team members.
- Reports to the pilot when the first man initially touches down, when the last team member starts to move safely from underneath the helicopter, and when all team members are disconnected.

9-49. On order of the pilot, the SPIES master or crew chief retrieves the SPIES rope into the helicopter or disconnects the SPIES rope and drops it to the ground. When using the UH-60/UH-1H helicopter, the only way to retrieve the SPIES rope while in the air is by having an arranged recovery rope attached with a 16-foot sling rope. In some cases, the SPIES master joins two 12-foot long sling ropes to haul the SPIES rope aboard and attaches the rope about 5 or 6 feet below the cargo hook or cargo strap hookup point. The type of knot used to connect the sling (or recovery) rope to the SPIES rope is self-tightening (for example, the Prusik knot). The SPIES master fastens the standing end of the sling rope to the deck tie-down or uses a snap link. Although it is important to keep the line out of the way, the primary consideration is its length. The rope must be long enough to compensate for any oscillation in the SPIES during flight.

PREPARING FOR EXTRACTION

9-50. If the mission or insertion precludes the wearing of the harness, team members will carry it as per unit SOP. Once the team requests the extraction helicopter, they retrieve the harness and don it. Team members wear the SPIES harness under their load-carrying equipment. However, if the situation does not allow for removal of the load-carrying equipment, team members may wear the harness over their load-carrying equipment.

9-51. The procedures for donning the harness are as follows:
- Using a 12-foot safety line, the Soldier ties one end around his chest with an end-of-the-line bowline. The Soldier ties another bowline in the other end of the rope and connects a snap link to the loop. The Soldier slides the knot around so the knot is between his shoulder blades and the end of the rope with the snap link over one shoulder.
- The Soldier dons the harness by placing his arms through the shoulder straps, by connecting and tightening the chest strap, and by connecting and tightening the leg straps. The Soldier does not route the safety line under the harness.
- The Soldier securely ties or attaches weapons to himself or the harness.

9-52. There are several techniques for extracting rucksacks. Unit SOP, size and weight of the rucksacks, number of personnel and rucksacks, and time to prepare affect the choice of technique used.
- The Soldier may wear small, lightweight rucksacks. When using this technique, the Soldier must loosen the shoulder straps of his rucksack because of the routing of the backstrap and safety line.
- The Soldier may tie rucksacks to a line and hook them by snap links to the bottom loop of the SPIES rope.
- The Soldier can tie a loop in the safety line by using a cat's paw or middle-of-the-line bowline and attach the rucksack to the loop with a snap link. When using this technique, Soldiers must ensure the SPIES rope suspends the rucksack.

Special Patrol Infiltration and Exfiltration System

EXTRACTION PROCEDURES

9-53. The extraction helicopter proceeds to the area and establishes radio or visual contact with the team. If available, a backup helicopter equipped with the SPIES remains aloft and away from the area. The helicopter maintains visual contact with the PZ and monitors radio communications.

9-54. After the team has indicated readiness for pickup, and the tactical situation has stabilized, the extraction helicopter moves to the PZ by the safest route. When the helicopter is above the team's location, the SPIES master drops the SPIES rope on order of the pilot after the aircraft has obtained a stable hover at slightly above treetop height.

9-55. The team's GSO positions himself so he can move and approach the rope as the SPIES master drops it. Once the rope is clear of any obstacles, the GSO signals the team to their assigned positions along the hookup points. Team members sling individual weapons over their shoulders, barrel down and to the front, and secure equipment to withstand the wind. Using the primary or harness snap link, each team member hooks to the D ring on his side of the line. Then the Soldier attaches the safety line to the D ring on the opposite side of the rope to form the secondary hookup point. The GSO physically inspects, if time and situation permit, or hooks himself in on the lowest point, along with the rear security, to ensure the running end is clear of all obstacles. He then faces forward along the line so he can observe his team and see the helicopter.

9-56. With one hand, the team members give thumbs-up signals to allow the GSO and the SPIES master to see they are ready to go and maintain the SPIES rope over the shoulder closest to the rope. With the other hand, the Soldier maintains control of his weapon, with the barrel directed downward and outward at a 45-degree angle. This position allows the team members to fire their individual weapons by using the hip position.

9-57. The GSO gives the thumbs-up signal to the SPIES master. This thumbs-up signal (at night, an arranged light signal) continues until the helicopter lifts the team off the extraction surface. The GSO also signals (arms held straight to the sides) when the team is at safe altitude (about 10 feet above the tallest obstacle at the extraction site).

9-58. During extraction, the team radio operator maintains communications with the extraction helicopter. He gives a verbal backup to the thumbs-up signal and relays all information during the flight. His location is near or at the bottom hookup point. He can then assist in giving accurate information about the extraction, clearing of obstacles, and descent.

9-59. Liftoff of the extraction helicopter is vertical until the SPIES rope has cleared all obstacles. Once the SPIES rope is clear of all obstacles, the extraction helicopter changes to horizontal flight and departs the area by the safest route. With team members attached to the SPIES rope, airspeed is limited to 70 knots in moderate climates and 50 knots in cold climates.

DISMOUNTING PROCEDURES

9-60. When the extraction helicopter has reached a safe dismount area, the pilot transitions into hover flight at an altitude of 250 feet AGL. The pilot starts a vertical descent with the SPIES master continuously providing information to the pilot on the distance from the ground to the lower end of the SPIES rope. The vertical descent rate of the aircraft (at touchdown) is less than 5 feet per second.

9-61. When the team members reach the ground, they immediately move out toward the front of the helicopter. The team members ensure the SPIES rope does not interfere with the aircraft and that the aircraft does not land on the rope. All team members rapidly unhook themselves and their teammates who need assistance. Once unhooked, they move away from the area and set up security, or help clear the rope if the helicopter is going to land.

EMERGENCY PROCEDURES

9-62. During the flight—from extraction until the team is safely and quickly detached from the SPIES rope—each team member should be aware of any problems originating from above or below. The man

Chapter 9

above checks the man below. At the first sign of danger or an emergency, the team leader or a team member places his free hand on his head. Upon observing anyone on the SPIES rope with his hand on his head, the SPIES master tells the pilot to make an emergency landing in the nearest and safest area.

WATER EXTRACTION PROCEDURES

9-63. SPIES supports water extractions. For this procedure, the SPIES master ties on three inflatable life vests or any type of flotation device to the SPIES rope to provide buoyancy for the rope while in the water. One flotation device is tied at each end of the attachment points, and one flotation device is tied in the middle of the attachment point area just above the middle two sets of D rings. Each team member to be extracted wears his SPIES harness under his individual life vest. Each team member can also wear swimming fins, mask, and snorkel (amphibious operations) to ease hooking up to the SPIES rope within the spray area beneath the hovering helicopter.

9-64. After the extraction helicopter has attained a stable hover above the team member's location, the SPIES master drops the SPIES rope (with floatation attached) on order from the pilot. When the team members complete the hookup to the SPIES rope, the team leader signals the SPIES master to start liftoff. Liftoff of the extraction helicopter is vertical until all team members and the bottom end of the rope have cleared the water. During the initial liftoff, the helicopter drags team members through the water. Therefore, team members should prepare to roll on their backs until clear of the water.

9-65. Flight speed and altitude are the same as over land. The dismounting procedures remain the same, except when landing on a ship. Once onboard, the team members take their orders from personnel in charge of the deck.

AFTER-OPERATIONS PROCEDURES

9-66. After-operations procedures include repairing, cleaning, and proper storing of the rappelling equipment. The SPIES master conducts an after action review of the operation. The SPIES master furnishes a copy to the S-3 for reference.

REPAIRING AND CLEANING EQUIPMENT

9-67. After rappelling operations, the Soldiers wash contaminated ropes with mild detergent, such as liquid dish soap, and cold water followed by a rinse in clean, fresh water. They dry the ropes at room temperature not to exceed 140 degrees Fahrenheit. Soldiers can remove stubborn oil, grease, hydraulic fluid, and other petroleum stains with the cleaning agent xylem (Grade A or B, TT-X 916).

STORING EQUIPMENT

9-68. Soldiers should protect the nylon materials from direct sunlight, which can cause ultraviolet deterioration. When not in use, the SPIES rope should be stored in an aviator's kit bag. Soldiers use bins or similar facilities for storage of SPIES equipment. Shelves used for storage should be at least 4 inches from the walls and 12 inches from the floor. Areas used for storage should be well ventilated and free of oil, acid, cleaning compounds, and other contaminants. Equipment must not be stowed above or near hot water pipes or heating apparatus.

Chapter 10
Fast-Rope Insertion and Extraction System

Small units use the FRIES for rapid infiltration and exfiltration using rotary-wing aircraft in confined areas. Using this system, up to a team-sized element can infiltrate directly onto the objective or into an area where a helicopter cannot land. This method is the fastest way of deploying troops from a helicopter unable to land, but the troops have a limited amount of equipment and supplies with which they can deploy. When mission requirements include large amounts of equipment or heavy crew-served weapons, ground force SOPs will determine the technique of employment to ensure personnel safety and equipment required. When using the fast rope for infiltration, Soldiers require very little special equipment. When using the fast rope for exfiltration, the troops require special harnesses, and, therefore, the helicopter can extract only a limited number at one time. HQ, Department of the Army, policy specifies FRIES is not approved for Army-wide use, and names the Commanding General, USASOC, as the executive agent for FRIES doctrine. The use of FRIES is restricted to SOF, pathfinders, long-range surveillance units, and schools approved by HQ, Department of the Army, with a USASOC-approved FRIES program of instruction. Other units wishing to conduct FRIES must send a request to conduct FRIES operations to Commander, USASOC, ATTN: AOOP-TRS, Fort Bragg, NC 28310-5000.

OBJECTIVES

10-1. The objectives of FRIES training are to safely conduct and maintain maximum proficiency in the execution of FRIES operations. The objectives of this chapter are to—
- Prescribe safety and administrative procedures.
- Outline the training requirements.
- Prescribe the qualifications, duties, and responsibilities of commanders and key personnel.
- Outline the operational requirements.
- Cover the inspection and use of the equipment required.

GUIDANCE FOR COMMANDERS

10-2. Units that have approval from HQ, Department of the Army, to perform FRIES operations are authorized to conduct initial FRIES qualification and fast-rope master (FRM) qualification training. This publication, USSOCOM M 350-6, and applicable SOAR policies establish training requirements. The United States Army Technology Applications Program Office specifies aircraft material requirements.

10-3. FRIES training and operations possess inherent risks; therefore, safety is paramount. All training and operations require in-depth attention to detail from planning through preparation and execution.

10-4. The following basic guidance applies to FRIES training and operations:
- Unit commanders must personally approve FRIES training sites.
- The unit S-3 must complete a risk analysis or assessment before FRIES operations.

Chapter 10

- Night FRIES operations are medium risk or higher.
- Units authorized to conduct FRIES operations may incorporate FRIES activities in field training exercises only after personnel have previously become FRIES-qualified.
- Field exercises using FRIES activities are subject to the limitations imposed by applicable directives, including mandatory support requirements.
- Units must submit to HQ, USASOC, a request in writing for waiver of procedures or restrictions prescribed herein. The request must be endorsed by the first general officer in the requesting unit's chain of command.

SAFETY

10-5. Safety is everyone's responsibility. All personnel involved in FRIES operations are responsible for identifying hazardous situations and preventing injuries of personnel. Anyone who observes an unsafe condition or act is authorized to halt the operation and inform the FRM or the pilot in command. Refer to USSOCOM M 350-6 for the most current safety requirements.

BRIEFING

10-6. All units must present a safety briefing before conducting FRIES training or operations. The safety briefing must cover (as a minimum) the following:
- The planned event (day or night, tower or helicopter, high or low hover, insertion or extraction).
- Actions on the aircraft (or tower).
- General tower or aircraft safety.
- Operation (or training) location and any hazards within 500 meters.
- Identification of key personnel, their location, and their duties and responsibilities.
- FRIES equipment and its characteristics.
- FRIES equipment rigging or inspection.
- Personnel preparation (shirttails tucked in, chin strap fastened, loose straps taped, and gloves and goggles on).
- Hand-and-arm signals.
- Emergencies and emergency procedures.
- Medical coverage (medic with aid bags, backboards, litters, and vehicle driver on site).
- Communications requirements (frequencies, call signs, required reports).
- Any other information essential to safety.

PREFLIGHT AND IN-FLIGHT

10-7. Anyone who detects an in-flight emergency will immediately notify the pilot in command, crew chief, or safety and follow aircrew instructions. Below are preflight and in-flight procedures for FRIES:
- Before a FRIES operation, the FRM will coordinate both routine and emergency procedures as well as any unique conditions or potential problems with the pilot in command. He also discusses the procedures used during flight, during roping, and before aircraft departure from the roping site. The FRM then discusses the information with all the assistant fast-rope masters (AFRMs), safeties, and crew chiefs. He reviews the time warnings (including hand-and-arm signals) and appropriate actions—such as who will deploy ropes, verify rope status, and perform rope-clearing procedures.
- The FRM gives a mission prebrief discussing any aircraft equipment that, if disturbed, could cause an aircraft or fast-rope problem. The FRM should make all problem areas known to all personnel concerned and preventive measures taken to neutralize or eliminate them where possible.
- The FRM gives the crew chief the final position adjustment over the target, using hand signals.

- The crew chief securely stows all safety straps and communications cords that are not in use. All safety straps and communication cords in use should be positioned so they do not present a hazard to personnel movement in the aircraft.
- After the pilot in command gives the command DEPLOY ROPES, the FRM deploys the rope and ensures it is in contact with the surface before ropers exit.

Note. The safety is responsible for observing rope and ropers and for making sure the pilot maintains aircraft position until all ropers are on the surface.

CAUTION

IAW AR 95-1, *Flight Regulations*, anytime operations occur over water, all personnel will wear an approved personal flotation device as identified in FM 3-21.220 or USASOC Reg 350-2.

During Roping

10-8. The following safety precautions apply during roping:
- During descent, ropers must watch for obstructions and for lower ropers, and must break as necessary.
- Individual ropers will lock-in during emergencies by wrapping one leg around the rope once or twice and standing on the fast rope with the other foot.
- Ropers must maintain sufficient intervals on the rope to prevent contact with lower ropers. Individual ropers exit when the head of the Soldier in front of them disappears below the ramp.
- Ropers execute descents at speeds commensurate with their experience and proficiency in FRIES infiltration operations. For example, less-experienced ropers should descend slowly until they become more proficient.
- The number of ropers on each rope at one time must not exceed the safe limit of the rope mount point in use, as follows:
  - MH/UH-60: Do not exceed 1,500 pounds per rope.
  - MH-47: Do not exceed 1,500 pounds per rope for extractions, regardless of the mount used. The aft mounts may go up to 2,250 pounds for insertions.
  - MH/HH-53: Do not exceed 1,500 pounds per rope.
  - MH-60G: Do not exceed 1,300 pounds per rope.

Reminders

10-9. Before conducting operations, all key personnel will review the safety guidelines. The safety procedures for key personnel are discussed in the following paragraphs.

10-10. The ropers must follow specific safety procedures. They must—
- Wear a long-sleeved shirt or jacket, full-length pants, laced boots, a helmet, protective goggles, and heavy leather gloves.
- Avoid wearing harnesses for multiple integrated laser engagement system harnesses during fast roping.
- During the actual descent, avoid using NVGs because the goggles limit depth perception and create a tunnel vision effect.
- Avoid using a fast rope longer than 90 feet for FRIES descent training.
- Not carry more than 60 pounds of equipment during FRIES descents. The maximum weight of rucksacks is limited to 35 pounds. When rucksacks exceed 35 pounds (or total load exceeds 60 pounds), gear must be lowered (belayed) on a separate system, preferably from the opposite door.

Chapter 10

- Avoid carrying bulky equipment that could interfere with FRIES operations. During insertion, the FRM or AFRM lowers (belays) equipment that ropers cannot carry during FRIES operations (due to its weight or bulk). During extractions, the FRM or AFRM must attach rucksacks to the extraction loops of the FRIES rope by using snap links.

10-11. The safety procedures for units, unit commanders, and mission planners are as follows:
- Commanders must personally approve the FRIES site selected for night operations. The FRIES site must be large enough to permit all ropers, upon reaching the surface, to move clear of the rope and for units or elements to consolidate or re-form.
- The mission commander, officer in charge, or noncommissioned officer (NCO) in charge (NCOIC) conducts a detailed risk assessment before FRIES operations. All night FRIES operations will be regarded as medium risk, at a minimum.
- Units must not conduct FRIES training in densely wooded areas or areas prone to blowing dust, sand, or snow, which can produce a hazard for ropers or the aircraft.
- Units must not conduct FRIES training and operations during severe weather, such as rain, sleet, ice, or extreme cold.
- Units must not conduct FRIES training if—
  - The aircraft is unstable.
  - The fast rope is not fully deployed (5 feet of the rope must be on the surface).
  - Any obstacle is present that could interfere with the descent of the ropers.
  - The rope is frozen or slippery.
  - Anyone identifies any unsafe condition.

10-12. The safety procedures for safety personnel are as follows:
- Take a position so they can observe active FRIES points without obstructing ropers.
- Conduct final rigging checks of the mount, equipment, and ropes at the 10-minute warning.
- Make sure the aircraft is over the target area and the ropes reach the ground.
- Constantly keep the pilot in command and FRM informed of when—
  - The aircraft is in position for deployment.
  - Anything is being deployed from the helicopter (ropes, bundles, or personnel).
  - Five feet of rope is on the surface.
  - The first roper exits on the fast rope.
  - All ropers on the surface are clear.
  - The ropes are recovered or jettisoned.
  - The aircraft is cleared for flight.
  - Fifteen feet of the extraction rope is on the surface.
  - Extraction personnel are attached to the rope and ready for liftoff.
  - Extracted personnel are safely lowered to the ground.
  - Extracted personnel are out from under the aircraft.
  - Any other unsafe conditions occur.

WARNING

FRMs and safeties use the word "clear" to tell pilots the aircraft is ready to fly. FRMs and safeties must ensure (during insertion) that all personnel are well away from the ropes, and that the ropes have been fully recovered or released. During extraction, FRMs and safeties ensure all personnel are securely attached to the ropes, before any use of the word "clear."

FRIES EQUIPMENT

10-13. FRIES operations require common individual equipment, specialized equipment, and training areas for training. Each Soldier involved in the training provides the individual equipment. The FRM procures the FRIES equipment. Training areas required include a tower equipped for FRIES training and an LZ or a training area suitable for helicopter FRIES operations.

INDIVIDUAL EQUIPMENT REQUIRED

10-14. Each Soldier participating in FRIES training or operations will have the following required minimum personal equipment:

- Heavy leather gloves.
- A helmet with a chin strap.
- Protective goggles.
- A long-sleeved shirt or jacket, long pants, and boots.
- Hearing protection and identification tags for helicopter operations.

UNIT FRIES EQUIPMENT

10-15. The following equipment is required by the unit before conducting FRIES training:

- *Insertion Fast Rope.* The fast rope is a polyester rope, consisting of three 1 3/4-inch strands, olive drab in color, that comes in 20-, 40-, 60-, 90-, and 120-foot lengths. The top of the main rope has an 8-inch eye splice to allow the rope to be attached to specially equipped helicopters.
- *Extraction Fast Rope* (Figure 10-1). The extraction rope is the same as the insertion rope except that at the bottom of the main rope, a 9/16-inch diameter white nylon rope is spliced into the main rope to form three extraction loops. A 9/16-inch diameter black nylon rope is also spliced into the main rope to form three safety loops at the same position as the extraction loops.

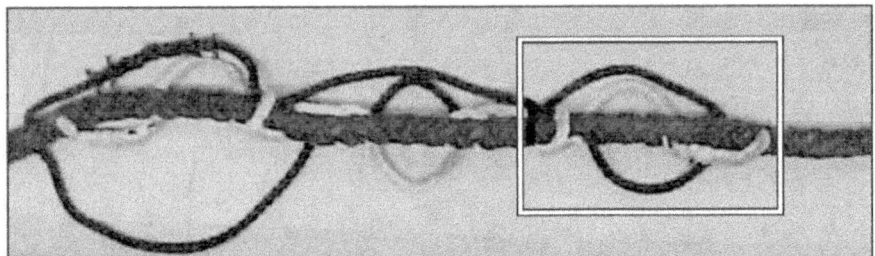

Figure 10-1. Fast-rope extraction loops

- *Extraction Harness* (if conducting exfiltration). Only a service-approved harness and, if required, bridle will be used as the extraction harness for FRIES. A commonly used harness is the SPIES harness (Figure 10-2, page 10-6). This harness has a 12-foot sling rope for use as a safety line. If the STABO harness is used, the bridle is also required.
- *Snap Links.*
- *Approved Belay Devices* (Figure 10-3, page 10-6).
- *Tower Equipped for FRIES Training.*
- *Helicopter Equipped With FRIES Mount Bars.*

Chapter 10

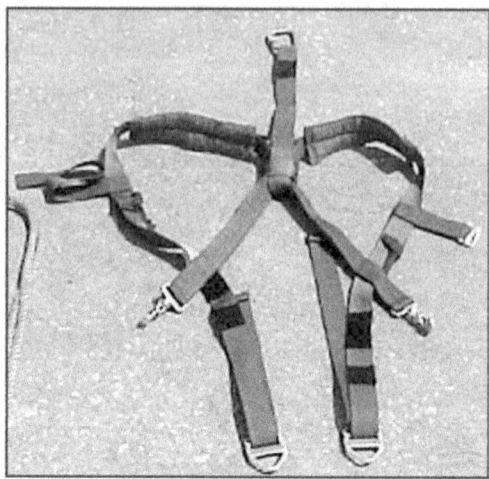

Figure 10-2. SPIES harness

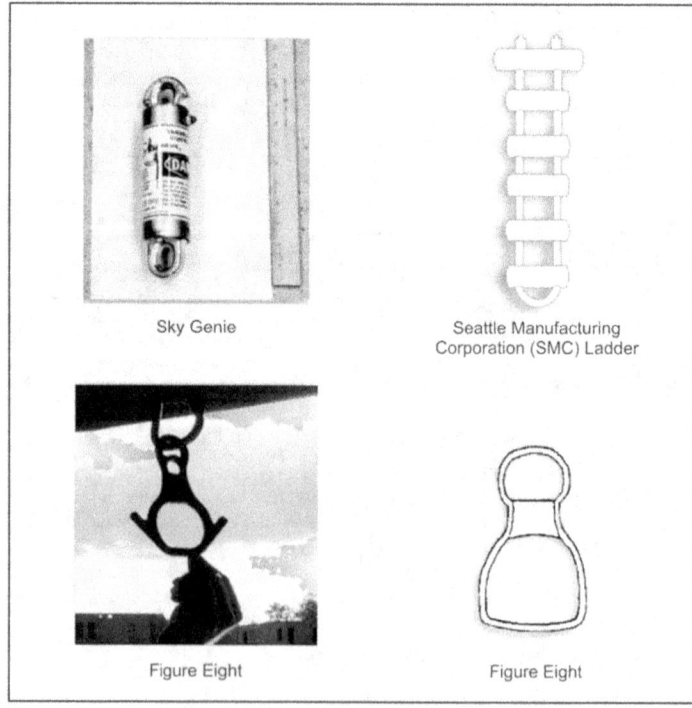

Figure 10-3. Belay devices used to lower equipment
during FRIES operations

Fast-Rope Insertion and Extraction System

MAINTENANCE AND INSPECTION

10-16. Before conducting a FRIES operation, the FRM must inspect all FRIES equipment (ropes, SPIES harnesses, and mount bars) for serviceability and readiness for use. FRIES equipment is inspected and maintained IAW TM 10-1670-262-12&P, *Operator and Unit Maintenance Manual Including Repair Parts and Special Tools List Personnel Insertion/Extraction Systems for STABO..., Fast Rope Insertion/Extraction System..., and Anchoring Device.*

10-17. The FRM must always store FRIES ropes and harnesses in a clean, cool, dry space, out of direct sunlight and free of chemicals or chemical vapors. Equipment that becomes wet with fresh water should be hung on hardwood pegs indoors and out of direct sunlight. Equipment that is exposed to salt water or becomes imbedded with dirt or mud should be washed and rinsed in fresh water within 72 hours and then dried as described above. FRIES equipment should be maintained in the same manner as a parachute. The unit rigger section can provide detailed guidance on appropriate inspection, care, and maintenance of FRIES equipment. Ropers complete DA Form 5752-R (Rope Log).

10-18. Before conducting a fast-rope operation, the FRM—
- Inspects the fast rope thoroughly and carefully. Checks the eyelets on the end for excessive wear. Checks the rope along its entire length for fraying. (However, snags in the rope from normal use do not weaken the rope. Also, a rope with several frayed strands in one spot should not be used.)
- Checks the rope length to make sure it is the correct rope for the operation planned.
- Checks the woven loop on the mount end for excessive wear or chemical contamination. Checks the rope along its entire length for fraying, cuts, and chemical contamination. The FRM avoids using—
  - A severely frayed rope. Light fraying on the rope from normal use does not weaken the rope.
  - A rope when any single strand is cut halfway through.
  - A rope with two or more cuts that penetrate one-third or more through the thickness of any strand within 1 foot of the running length of the FRIES.
- Inspects the rope for signs of contamination by acid, alkaline compounds, salt water, fire-extinguishing solutions, and petroleum-based solvents. Although used ropes gradually change color, such changes do not indicate a decrease in strength unless contact with strong chemicals has caused the change. Changes in color caused by chemicals are usually blotchy and have an unusual odor. Changes caused by use are usually uniform throughout the length of the rope.
- Inspects the extraction loops to the same standard as the main rope. Makes sure the woven attachment loops are secure.

> **WARNING**
>
> FRM must take out of service, mark for destruction, and turn in to the rigger section any rope known or suspected to be chemically contaminated, cut beyond allowable limits, excessively frayed, or sun-damaged.

- Inspects the harness to make sure—
  - Ropers are wearing the harness under all load-carrying equipment.
  - Ropers have properly fastened all connectors.
  - Harness material and stitching are not cut, torn, or contaminated, and all hardware is free of corrosion and is in operable condition.

Chapter 10

Note. Rigger-qualified personnel are responsible for serviceability, inspection, and maintenance of the extraction harness.

10-19. The aviation unit is responsible for the installation, removal, storage, and maintenance of FRIES mount bars. The 160th SOAR, Fort Campbell, Kentucky, prepares and provides checklists for these procedures. Aviation units participating in FRIES operations must possess and comply with current 160th SOAR guidance. Aviation units—

- Install and maintain the FRIES mount bar and check the torque and witness marks on the mount bolt.
- Check the FRIES mount bar for security and operability.
- Check the condition of the mount bar; that is, inspect for bends, dents, distortion, cracks, corrosion, and contaminants.
- Make sure the fast rope quick-release pins are installed and operable (two on each side).
- Check the security and condition of the rope-release system, as follows:
  - The mounting bolts are installed and properly torqued.
  - The safety pins are installed, and lanyards are present.
  - The system locks and unlocks as designed.
- Make sure the release cables are attached at both ends and are not broken or kinked.
- Check the release handles to make sure they are securely mounted and can move freely.
- Ensure the spring ball is installed and operational in the latch.
- Check the latch as follows:
  - Latch or close the release mechanism, and apply a downward pressure on the release arm. Try to pull the release handles—the release should not open.
  - Relax pressure on the release mechanism, and pull the handles—the release should open.
  - With no pressure on the release mechanism and with the safety pin installed, try to pull the release handles—the release should not open.

FRIES HARDWARE KITS

10-20. Each type of helicopter, and at times different versions of the same helicopter, has different FRIES hardware kits or mounting bars. The different FRIES hardware kits or mounting bars are discussed below.

UH/MH-60

10-21. All MH-60 helicopters are equipped with the FRIES special mission hardware kit I bar. Originally, UH/MH-60 helicopters were equipped with the H bar. The H bar was limited to 750 pounds. The H bar was not designed for extractions. The I bar is lighter than the H bar, can support up to 1,500 pounds, and was designed for the mission of insertion and extraction. The FRIES I bar consists of three main parts—the center beam, fast-rope attachment points, and two sliding bars (Figure 10-4, page 10-9). The center beam is fixed and supports two sliding bars. The two sliding bars are attached on opposite sides of the center beam and slide out in opposite directions. The fast-rope attachment points (Figure 10-5, page 10-9) are located at the ends of each sliding bar. The I bar is bolted through the helicopter cabin ceiling to the airframe. The sliding bars are retracted inside the helicopter during flight and normal operations. Once the doors are opened, the bars are extended out-side the aircraft to allow the fast ropes to be deployed.

MH-47

The FRIES hardware kit for the MH-47 is designed to support insertion and extraction missions. There are up to three FRIES bars or mounts that allow three fast ropes to be attached to the aircraft at the same time. The FRIES hardware kit for the MH-47 helicopter consists of two FRIES bars mounted over the ramp (Figure 10-6, page 10-9) and one fixed external mount (Figure 10-7, page 10-10) located outside and over the forward right door. If the aircraft is equipped with a hoist, the FRIES mount and hoist are connected. The forward FRIES mount and hoist cannot be used at the same time. The two aft bars attach to mounts in the ceiling of the helicopter over the cargo ramp and can be retracted inside the aircraft or extended beyond the edge of the ramp. Each bar has one fast-rope attachment point that supports one rope (Figure 10-8, page 10-10). Two fast ropes can be deployed off the ramp (Figure 10-9, page 10-10). The forward mount supports one fast rope (Figure 10-10, page 10-11). CH-47 aircraft are not normally equipped for FRIES operations, but some of them are equipped with the two aft FRIES bars.

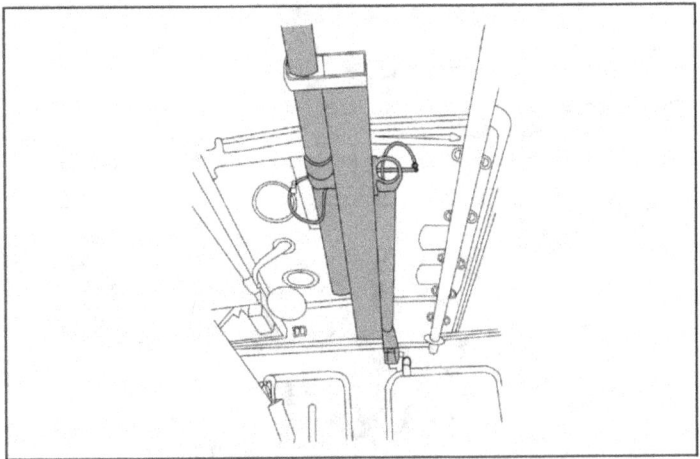

Figure 10-4. UH/MH-60 mounting bracket for FRIES (inside aircraft)

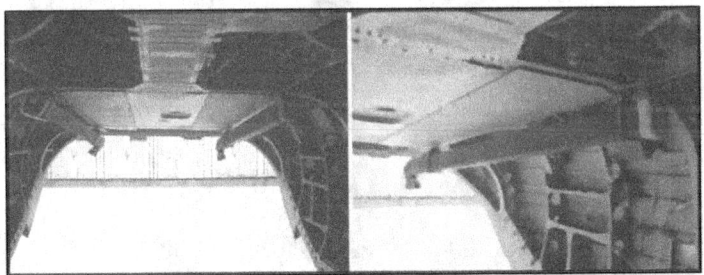

Figure 10-5. Fast-rope attachment point extended (outside aircraft)

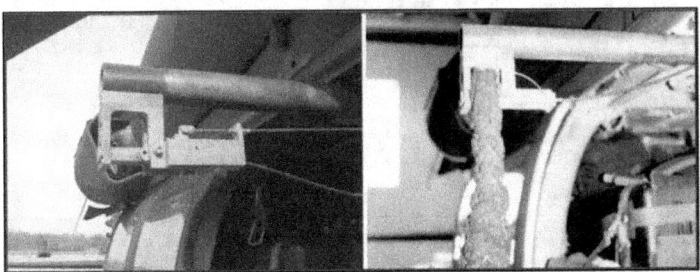

Figure 10-6. MH-47 aft-mounted FRIES bars (inside aircraft)

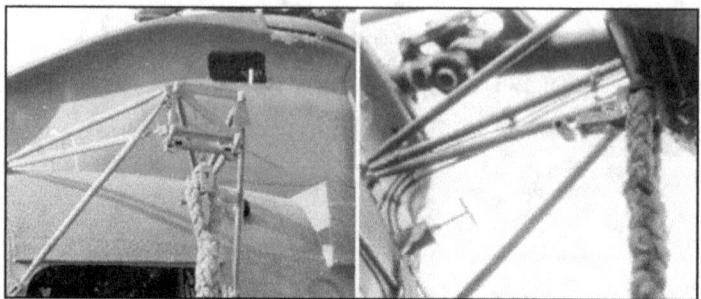

Figure 10-7. FRIES mount over the forward door (with and without hoist)

Figure 10-8. Fast rope attached to aft FRIES bar (bar retracted)

Figure 10-9. MH-47 with two aft fast ropes

Figure 10-10. Fast rope attached to forward mount

MH-6 EXTERNAL FAST-ROPE SYSTEM

10-22. The design of the external fast-rope system on the MH-6 allows up to four Soldiers to insert by using two fast ropes. This system is designed only for insertion of personnel. The external fast-rope system consists of two externally mounted fast-rope mounts, mounted over each rear door opening. The fast-rope attachment points are controlled by cables that route from the attachment points, through the cargo compartment, and to the release handles mounted on the upper center console in the cockpit. The pilots pull down the handles to jettison the ropes. The MH-6 uses 20- to 40-foot insertion fast ropes. The FRM attaches the ropes to the fast-rope attachment point and coil and places the rope on the cargo compartment floor until ready for deployment. The Soldiers ride on the external seats, and they are responsible for deploying the ropes.

Note. The external fast-rope system is not authorized for use as an extraction system.

OPERATIONAL REQUIREMENTS AND LIMITATIONS

10-23. FRIES operations require medical and communications support, including during adverse weather conditions and night operations. Personnel must use sound judgment to determine what action to take depending on the nature and severity of the condition.

MEDICAL COVERAGE

10-24. A qualified and equipped 18D, 91B medic, or a Service-qualified EMT will be at all training sites. Medics must know CASEVAC procedures and have coordinated requirements necessary to expedite evacuation and treatment of personnel on and off military installations. All medics will have as a minimum an M-5 aid bag or equivalent packed IAW unit standards. Medical transportation must also be available. The vehicle must be covered and large enough to carry an open stretcher. Units cannot conduct training without a medic, medical equipment, and transportation. If the situation warrants and the installation cannot support a MEDEVAC mission, the installation may use a FRIES aircraft as a last-resort CASEVAC. The medics will develop an evacuation plan. This plan should include, but not be limited to, the following:

- Medical facilities—location and capabilities.
- Emergency telephone numbers.
- Routes to medical facilities.

Communications Requirements

10-25. During FRIES training, the FRM or training safety officer will have radio communications with the aircraft. Voice communications are required before commencing FRIES operations. Additionally, the FRM or training safety officer will inform the pilot in command to stop operations if an unsafe condition develops. During extractions, the FRM or training safety officer will inform the pilot in command that all personnel are ready for extraction. During tactical missions, mission and aircrew personnel will use prearranged signals to communicate; for example, flashing light or chemical light signals.

Adverse Weather and Terrain Conditions

10-26. During the risk assessment for FRIES training, the S-3 considers the following conditions:
- Wind chill factors caused by the rotor wash of the helicopter or extraction cruise air speeds that could cause cold weather injuries.
- Water or ice on the fast rope, inhibiting the ability of the ropers to control their descent.
- The rope is exposed to the elements long enough for it to freeze, thereby reducing its tensile strength.
- Blowing particles produced by rotor downwash cause the aircrew or the AFRM to lose visual contact with the ground.

Night Operation Requirements

10-27. Night operations require the following additional procedures:
- Four chemical lights will be attached to the rope—two at the bottom and two 5 feet higher to aid in determining the relationship of the fast rope to the ground. The FRM places the chemical lights on opposite sides of the rope.
- The FRM secures one chemical light to the attachment point of the rope to aid in the exit.
- Soldiers do not wear night vision devices during descent.

Training Altitude Maximums

10-28. Units should conduct FRIES training at the lowest altitude possible. The length of the rope should not dictate proficiency training from aircraft.

10-29. There is no additional training value to higher altitudes, only increased chance of injury (40 feet is the recommended altitude). Training altitudes are as follows:
- Tower training: 15 to 60 feet.
- Helicopter: Initial training—not to exceed 30 feet during first two descents; conduct additional descents from maximum altitudes allowed by length of FRIES (not to exceed 90 feet). Advanced training—90 feet (descents exceeding 90 feet will be made only in a closely supervised environment).

FRIES QUALIFICATION TRAINING

10-30. Each unit (commanded by a lieutenant colonel or higher) is responsible for conducting its own FRIES qualification and sustainment program. FRIES trainers will be current FRM-qualified. Units should tailor training to unit and situation needs. However, units cannot reduce proponent requisites without written approval of the Commanding General, USASOC.

Training Prerequisites

10-31. All personnel must successfully complete initial FRIES training before they are FRIES-qualified. Participants in FRIES training must—
- Be authorized by the commander (lieutenant colonel) to participate in FRIES training.
- Have passed the Service fitness test.

- Have a current medical examination and be free of any injury or physical condition that could cause a potential safety hazard during FRIES training.
- Have demonstrated a controlled descent from a height of 10 to 15 feet.
- Have demonstrated the ability to execute a static position on a FRIES rope for 5 seconds, using hands and feet to lock-in.

> **CAUTION**
>
> Trainers must ensure the FRM has tested all personnel preparing to participate in FRIES training. The testing determines if participating personnel with equipment possess enough upper body strength to safely perform the full scope of roping duties.

10-32. Aircrews must be qualified and current to fly FRIES activities. They must be qualified IAW HQ, USASOC, the 160th SOAR Aircrew Training Program, and applicable policies.

Note. In peacetime, the maximum Soldier load will not exceed 60 pounds. This weight includes helmet, weapon, vest, web gear, and rucksack. Rucksack weight will not exceed 35 pounds.

INITIAL FRIES QUALIFICATION TRAINING

10-33. Before participating in fast-rope operations, personnel are briefed on the FRIES and its purpose, capabilities, limitations, and emergency procedures. The briefing also covers the duties and responsibilities of the pilot in command, safety, FRM, the AFRM, and any of the ground assistants. Once the FRIES master has conducted his FRIES briefing, the remainder of the initial training is hands-on practice of the proper FRIES operational techniques.

10-34. FRIES infiltration training procedures are as follows:
- Soldiers are shown the proper techniques for boarding the aircraft, moving to the door, grasping the fast rope, exiting the aircraft, and locking-in, descending, and clearing the rope. After the demonstration, all Soldiers participate in a practice exercise. During this exercise, the Soldiers properly perform the above tasks on a tower and, subsequently, from an aircraft.
- Training should be progressive, starting from a tower without equipment and then with equipment. All Soldiers must complete a successful lock-in at this level. Soldiers will not progress beyond the tower until they demonstrate the ability to stop descent, lock-in, and hold a stationary position for 20 seconds with equipment.
- Soldiers must execute two successful FRIES descents from a tower 15 to 60 feet (one descent without and one with equipment).
- Soldiers must conduct three insertions without equipment (two day and one night) and two with equipment (one day and one night) to be qualified for infiltration operations.

10-35. FRIES extraction training procedures are as follows:
- FRM shows Soldiers the proper wear of the SPIES extraction harness IAW the guidance in Chapters 12 and 13. The only modification is that Soldiers wear the same safety line with the SPIES harness as they do with the STABO harness.
- FRM shows Soldiers how to hook up to the extraction rope so that each Soldier is connected to the extraction rope by a primary and secondary attachment point. Soldiers connecting to the primary and secondary attachment points will not share the same loops. The FRM ensures the primary attachment point supports the Soldier's weight.
- FRM shows Soldiers how to correctly hook up equipment to the extraction rope.

10-36. Once trainers have demonstrated the proper FRIES procedures, Soldiers participate in practical exercises with each Soldier performing the above tasks to the satisfaction of the FRM from a static FRIES. To qualify in extraction operations, Soldiers must execute two successful FRIES extractions without equipment using a helicopter (one day and one night) and two with equipment (one day and one night).

WARNING

During initial qualification training, students will not perform rapid exits nor will there be more than three Soldiers on a rope at any one time. During advanced training, there will be no more than five Soldiers with equipment on any rope at any one time. In all cases (insertion or extraction), the load on the fast rope will not exceed the maximum rating for the rope and mount/frame being used

---

CAUTION

All Soldiers should be limited to no more than 10 roping events in a 24-hour period. No more than 6 events should be with equipment.

SUSTAINMENT TRAINING

10-37. Commanders must make sure Soldiers participating in FRIES operations receive sustainment training on equipment and procedures within 24 hours before the FRIES operation. Soldiers who do not attend FRIES sustainment training will not participate in FRIES operations. Appendix G describes the fast-rope troop briefing in detail. As a minimum, training will include a review of the following:

- Hand-and-arm signals.
- Individual equipment riggings.
- Aircraft familiarization.
- Safety procedures and emergency procedures.
- Any rehearsals the FRM or commander deems necessary.

REFRESHER TRAINING

10-38. Units provide refresher training for Soldiers who have not participated in FRIES training or operations during the past year. These Soldiers undergo refresher training before being included in an operation. FRIES refresher training consists of a complete review of FRIES and its purpose, capabilities, and limitations and FRIES emergency procedures. The refresher training also includes execution of one daylight and one night FRIES descent from a fast-rope tower and aircraft. FRM refresher training under the observation of a current FRM.

FRM SELECTION AND QUALIFICATION TRAINING

10-39. Selection and qualification of FRMs is a unit prerogative. USSOCOM M 350-6 determines the selection criteria and qualification training required.

SELECTION

10-40. Selection criteria for Soldiers to be qualified as FRM should be based on the individual's demonstrated leadership capabilities, maturity, decisiveness, and knowledge of FRIES operations. Candidates for FRM must have completed the initial FRIES training.

QUALIFICATION TRAINING

10-41. Soldiers will be qualified to perform the duties of FRM upon successful completion of the FRM training course, during which they will receive instructions on and demonstrate proficiency in—
- Mounting the fast rope to the fast-rope bar and inspecting and preparing the aircraft for FRIES operations (for example, tape those items and areas that might be an obstacle or hazard to the fast ropers exiting the aircraft).
- Performing the following FRM duties:
    - Coordination responsibilities.
    - Troop briefings (Appendix G).
    - Organization of the stick.
    - Time warnings and commands.
    - Throwing and retrieving ropes.
    - Releasing and stopping the stick.
    - Hand-and-arm signals.

10-42. During FRM qualification training, each Soldier will participate in seven operations (four daylight and three night) from a fast-rope tower (two daylight and one night) and aircraft (two daylight and two night). Specifically, personnel undergoing FRM qualification training will serve as FRM on at least two daylight operations and one night operation from an aircraft.

Note. Soldiers performing the duties of AFRM must be current FRMs.

FAST-ROPE MASTER REFRESHER TRAINING

10-43. FRMs who have not participated in FRIES operations during the past 1 year will receive refresher training by a current FRM and serve as an AFRM before performing FRM duties. Refresher training consists of an FRM briefing and participation in FRIES training.

KEY PERSONNEL DUTIES AND RESPONSIBILITIES

10-44. The following personnel duties and responsibilities provide baseline requirements for the safe conduct of FRIES operations. Unit SOPs may increase (but will not reduce) training safety requirements.

UNIT COMMANDER

10-45. The unit commander will ensure that medical personnel screen all Soldiers. The medical screening ensures Soldiers are physically and professionally able to participate in FRIES training and meet all personnel qualification requirements as previously listed.

AIR MISSION COMMANDER OR OFFICER IN CHARGE

10-46. The employing aviation unit designates the air mission commander. He makes sure all operators, aircrew members, and support elements synchronize their actions during the conduct of the mission and the FRIES operation. An air mission commander usually is not needed during small-unit qualification or proficiency training events when only one helicopter is used.

Chapter 10

PILOT IN COMMAND

10-47. The aviation unit providing helicopter support for the FRIES training and operation appoints the pilot in command. The pilot in command oversees all aspects of the flight and ensures—
- Aircrew members are current and qualified to conduct FRIES operations.
- Aircrew members know and understand their responsibilities in fast-rope operations.
- Procedures for planning, preparation, and execution are adhered to IAW this manual, USSOCOM M 350-6, and applicable SOPs and policies.
- All personnel are briefed on fire support to be provided by the aircraft, including:
  - The nature of fire support the aircraft can provide.
  - The time the fire support starts, shifts, and stops.
  - The primary and alternate commands and signals that start, stop, lift, or shift fires.
- All personnel are briefed on in-flight emergencies and safety procedures.
- The aircraft is properly configured to perform the mission.
- The aircraft is at the proper altitude, airspeed, and location as briefed.
- The command to deploy ropes is not given until the aircraft is at a stabilized hover.
- The aircraft position is maintained to keep ropes in contact with the surface until all descending ropers are on the ground (or extracting ropers are securely attached for extraction).
- Ropes are fully recovered inside the aircraft (or jettisoned) before the aircraft departs the stabilized hover position at the infiltration site.
- Ropes are never deployed with anything other than night illumination attached to the free end.

FAST-ROPE MASTER OR TRAINING SAFETY OFFICER

10-48. Units conducting FRIES operations will designate one overall FRM to organize, coordinate, and supervise the activities of the day and any additional FRMs and AFRMs, as needed. Each aircraft will have one designated FRM and a designated AFRM for each roping point. FRMs or AFRMs prepare, inspect, and control all roping activities on their points. FRMs must also—
- Coordinate all aspects of troop and unit preparation, to include procurement of enough FRIES equipment for the operation.
- Coordinate all support activities.
- Ensure the FRIES equipment is properly rigged.
- Adhere to the published time schedule and sequence of events of the operation.
- Assign qualified personnel to the duties of FRM, AFRM, and other key positions as required by the operation. Ensure they understand the proper roping procedures for FRIES operations.
- Strictly adhere to procedures for the planning, preparation, and execution of the operation as outlined in this manual, training circulars, unit SOPs, and local directives related to the specific training.
- Ensure the mission commander is briefed on the training being conducted.
- Ensure FRIES operations are conducted over terrain that permits the aircrews and FRM or AFRMs to have visual contact with the ground or vegetation.
- Relay time warnings.

Note. Primary FRM will maintain positive communication with aircrew using aircraft communication when available or prearranged hand-and-arm signals.

- Ensure all personnel understand the techniques and responsibilities for FRIES operations.
- Ensure one AFRM for each rope being used is on board and at the designated time deploys the ropes.
- Perform safety and serviceability checks on all FRIES and rigging equipment.

- Check the rope to make sure it is the correct type and length for the operation—smooth rope for infiltration or looped for exfiltration. Inspect the rope to make sure it has no contamination, damage, or defects that could make it unsafe.

> **WARNING**
>
> The FRM must make sure that nothing, other than rope illumination, is attached to the free end of the FRIES rope during deployment and that equipment is never attached and dropped.

- Ensure the rope chemical lights are correctly rigged and illuminated when needed (two at the mount, two at the end, and two 5 feet from the end).
- Ensure the attachment bar or points are serviceable and free of any defects or contamination. Ensure the quick-release mechanisms and safety pins are present, serviceable, and operate correctly. Make sure the rope is properly attached with a safety pin in place and that the rope is back-coiled.
- Ensure the proper seating arrangement for all fast ropers, to include personnel restraints, and procedures in case of an emergency landing.
- Coordinate with the pilot in command, and brief the aircrew, safeties, and ropers on the correct method of deploying the fast ropes based on the following:
    - The aircraft flight and approach and the position of the flare over the target site.
    - The height of the aircraft above the target.
    - A description of the FRIES site (rooftop or small clearing) and size and the actions of the ropers upon arrival at the site.
    - The signal to deploy the rope, who will deploy it, and how it will be done.
    - Time warnings and actions at each time warning.
    - Hand-and-arm signals.
    - The use of radios and intercoms.
- Brief the pilot in command and crew chief on verbal commands and final adjustments over the target.
- Ensure safeties know the proper procedures to—
    - Keep the aircraft on target.
    - Clear the ropes.
    - Recover the ropes.
    - Jettison the ropes.
- Observe the target and pass to the pilot in command, through the crew chief, final aircraft position adjustments over the target to make sure conditions are safe. Verify the aircraft is at a stabilized hover and remains so.
- Direct the aircrew or safety to inform the pilot when the aircraft is on target. (The pilot in command replies with the command DEPLOY ROPES.)
- Deploy the ropes when the pilot in command gives the command DEPLOY ROPES. (FRM may delegate to the safety.)
- Ensure the rope is clearly free and is touching the ground.
- When conditions are safe, command GO to the ropers and point to the rope or lead the stick out.

Chapter 10

> **WARNING**
>
> FRM will not deploy ropes until the aircraft is at a stabilized hover directly over the designated target and must fully recover ropes inside the aircraft or jettison them before the aircraft departs.

> **CAUTION**
>
> Personnel should lower equipment before the FRIES is deployed or personnel begin roping. When equipment and ropers are to be lowered simultaneously, personnel will lower the equipment from and ropers will exit from opposite doors of the aircraft.

ASSISTANT FAST-ROPE MASTER

10-49. The AFRM will be a current FRM and help the FRM conduct FRIES operations. He will become the FRM should the FRM be unable to perform his duties. The AFRM—

- Throws the rope after the aircrew gives the signal for ropes.
- Ensures the ropes are checked after deployment and before anyone descends, ensuring the FRIES attachment loops (chemical lights for night operations) are on the ground.

Notes.

1. If the FRM leads the stick out, the AFRM automatically assumes control and marshals the remaining ropers. The AFRM would then be the last man to exit.

2. During night operations, upon completing descent, the AFRM observes the other ropes of his aircraft. When he sees that all ropes are clear, he issues the prearranged all-clear hand signal to the aircrew safety. This hand signal tells the aircrew that they are free to jettison or to recover the ropes and fly away.

FRIES AIRCREW MEMBER OR SAFETY

10-50. FRIES safeties are aircrew members who perform the additional duties of safety during helicopter roping operations. Safeties help FRMs and pilots, as needed. Safeties remain in the aircraft and operate the doors, extend and retract the rope mount bars, and wear NVGs, when appropriate. During roping operations, the safety's four main duties are as follows:

- Observe all activities in the aircraft and help the pilot and FRM or AFRMs, as needed.
- Serve as the communications link between the pilot in command and the FRM. The safety monitors aircraft communications and relays information between the pilot in control and the FRM, constantly keeping both informed.
- Help the FRMs and AFRMs lower cargo and clear the aircraft of ropers.
- Clear and jettison the ropes, when appropriate.

10-51. The safety's other duties are to—

- Conduct aircraft equipment rigging and serviceability inspections, making sure mounts, doors, and ropes are ready and safe for operations both preflight and en route.
- Relay time warnings, aircraft positioning commands, and other information to help the FRM.
- Signal the AFRM to deploy ropes after verifying the aircraft is at a stabilized hover.
- Check the rope to make sure 15 feet of rope is on the target surface.

- Tell the pilot in command ROPES DEPLOYED, when appropriate.
- Help the AFRM in the control and spacing of each roper's exit.
- Monitor the rope and ropers to make sure they do not get hung up, dragged, or dropped off site.
- Observe the exit of the last roper until the roper reaches the surface, moves clear of the rope, and gives the all-clear signal.
- Tell the pilot in command ROPERS DOWN (AFT, FORWARD, LEFT, or RIGHT) when all ropers are on the surface and conditions are safe to jettison the ropes.
- Pass the signals, as required, to position the aircraft, including hold and move left, right, forward, back, up, or down.
- Act as belay man during cargo lowering, when needed.
- Jettison ropes upon receiving the command JETTISON ROPES from the pilot in command. Jettison rope, and tell the pilot in command ROPES ARE JETTISONED.
- Deploy the rope for extraction upon receiving the command DEPLOY ROPES from the pilot in command.
- Observe extracting rope and personnel, and inform or guide the pilot in command through aircraft positioning and throughout the recovery flight.

Note. MH-53 operations require only one crew member on the ramp per two ropes.

FRIES ROPER

10-52. Ropers notify the FRM, AFRM, safety, or pilot if they observe any unsafe act or condition. Ropers may halt or call for a halt of roping for safety at any time. During FRIES insertion training, the number of ropers on the fast rope at one time is limited to six. During extraction training, the total weight per extraction bar and rope must not exceed a total of 1,500 pounds. Total roper equipment weight will not exceed 60 pounds. Ropers will—

- Maintain hands at head level.
- Maintain visual contact with lower ropers during their descent.
- Maintain a minimum 1-second interval on exit to avoid collisions.
- Keep at least two points of contact on the rope (both hands) at all times.
- Use their feet for additional breaking anytime needed.
- Execute descents at a safe speed.
- Slow the rate of descent halfway down the rope to avoid landing on each other.
- Move quickly away from the rope upon arrival on the surface.
- Know correct wear of extraction (SPIES) harness and extraction procedures.

WARNING

The FRM, AFRM, and safety must make sure all personnel are on the surface and have moved clear of the ropes before aircraft fly away or ropes are jettisoned.

RIGGING OF AIRCRAFT

10-53. The aviation unit performs preattachment inspection and installation of FRIES anchor point mount bars and associated equipment. All aircraft used to conduct FRIES operations with U.S. Army personnel must be rigged and equipped IAW current parent service directives. For Army aircraft, USASOC, 160th

Chapter 10

SOAR, and U.S. Army Technology Application Program Office (TAPO) directives apply. The aviation aircrew or safety—

- Inspects the FRIES mount bar assembly, attachment hardware, and associated equipment to ensure they are serviceable and ready to use. The aviation aircrew or safety ensures—
  - The bar assembly is correctly mounted to the aircraft.
  - The mount is complete, functional, and free of any defect.
  - The mount and the aircraft floor are clean and free of any petroleum, oil, lubricants, or other contaminants that could weaken the rope or cause personnel to slip or fall.
  - The mount and aircraft are free of sharp edges or burrs that could cut the rope.

> **CAUTION**
>
> Army aircraft that have the FRIES mounts installed must possess an airworthiness release for FRIES operations from TAPO.

- Makes sure no metal fatigue or structural weakness exists that could cause the system to fail.
- Tapes and pads any sharp edges that can touch ropes or personnel.
- Extends the locking and retracting mount bars, as needed.
- Lowers and belays equipment, as needed.
- Opens, locks, and closes cargo doors, if needed.

Note. For cold-weather operations or during flights of long duration, the aircrew will operate doors or directly supervise those who do. In good weather, doors should remain locked open.

- Removes or reconfigures troop seats, if needed.
- Provides seat belts or floor-mounted personnel restraints for ropers' use during flight.
- Attaches and detaches ropes, as needed.

Note. Clusters of rucksacks or small door bundles (weight limited to 250 pounds) may be lowered (belayed) on rappelling ropes. Cargo lowered this way must never be free-dropped (the belay must start at the time the bundle is pushed [tipped] out the door).

RIGGING OF UH/MH-60

10-54. The aviation unit rigs the aircraft. The FRM and selected personnel may rig or assist in the rigging under the supervision of the aircrew. Personnel—

- Remove both of the storage pins, and allow the bars to rotate down.
- Extend the fast-rope bars out to their desired length, fully extended for insertions and halfway out for extractions, and insert the storage pin into the correct hole.
- Inspect the bar for cracks and for security of nuts and bolts.
- Rig the fast rope to the fast-rope attachment point, as follows:
  - Remove safety pin from the fast-rope release system and apply upward pressure to cabin wall-mounted release handle, releasing the gate.
  - Insert woven loop of the fast rope into the attachment point.
  - Insert the gate through the woven loop of the fast rope and into the receptacle.
  - Apply a downward pressure to cabin wall-mounted release handle while pushing the gate out until the gate is fully seated in the receptacle (locking position).

Fast-Rope Insertion and Extraction System

- Back-coil fast rope, and secure it to cabin floor, or insert the fast-rope retention strap through the coil and suspend the fast rope from the ceiling or fuel tank. Secure quick-release mechanism with the safety pin (Figure 10-11).

Figure 10-11. MH-60 rigged for fast roping

RIGGING OF CH/MH-47

10-55. The aviation unit rigs the aircraft. The FRM and selected personnel may rig or assist in the rigging under the supervision of the aircrew. Personnel—

- Inspect the FRIES hardware for cracks, rust, and security of nuts, bolts, and quick-release pins.
- Ensure the aft FRIESs are extended for insertion (not required for extraction).
- Remove the quick-release pin from the release bar assembly, then push up on the aft FRIES bar release handle (Figure 10-12), or pull out on the forward FRIES mount release handle.
- Insert the woven end loop of the fast rope into the attachment point.
- Support the rope, and insert the release bar through the woven loop and into the receptacle. Pull down on aft FRIES bar release handle or push in on the forward FRIES release handle to fully seat release bar assembly (Figure 10-13, page 10-22).
- Install quick-release pin in release bar assembly (Figure 10-14, page 10-22).

Figure 10-12. Fast-rope attachment point opened

Chapter 10

Figure 10-13. Fast rope attached and release bar seated

Figure 10-14. Quick-release pin inserted

RIGGING OF FRIES ON MH-6M

10-56. The MH-6M is a single-engine, light, utility helicopter flown by one or two pilots. There are no crew chiefs or safeties on this helicopter, so the FRM is responsible for rigging the fast rope to the fast-rope attachment point. The aircrew is still responsible for attaching the mount for the external fast-rope system to the aircraft. Because the MH-6 external fast-rope system is not authorized for extracting personnel, a caving ladder may be rigged on the helicopter to extract personnel who infiltrate by the fast rope. The FRM—

- Inspects the FRIES hardware for cracks, corrosion, and security of nuts, bolts, release pins, and cables.
- Removes the release pin from the release bar assembly.
- Lifts the release handles, located in cockpit, to the full up position.
- Inserts the woven end loop of the fast rope through the release bar.
- Supports the rope and inserts the release pin through the woven loop and the receptacle.

Fast-Rope Insertion and Extraction System

- Ensures the pin is snapped into the release bar assembly and is fully seated.
- Coils the rope (Figure 10-15) and places it in the cargo compartment, ensuring the rope will not prematurely deploy during flight.

Figure 10-15. Fast rope rigged on an MH-6

COMMANDS AND SIGNALS

10-57. Ground-to-aircraft hand-and-arm signals are military standard as found in FM 3-21.38. The following standardized hand-and-arm signals are provided for FRIES operations:
- *Time Warnings (10-, 6-, and 1-minute).* The FRM gives these time warnings by holding up the appropriate number of fingers and verbally sounding off.
- *Get Ready.* FRM extends arms horizontally, with fingers extended vertically and joined.
- *Stand By.* FRM extends arms downward at a 45-degree angle from the body, with hands closed except for index fingers, which are extended, pointing to the floor.
- *Deploy Ropes.* FRM or safety form both hands into fists at mid-thigh level and sweeps arms horizontally outward. At full extension, he opens his hands as if dropping the ropes out of the helicopter, and sounds off with DEPLOY ROPES.
- *Go.* FRM raises arms to horizontal, hands closed with index fingers extended pointing to the ropes.
- *Stop or Abort.* FRM closes hands into fists, with arms raised across the forehead.
- *Sit Down and Don Seat Belts.* FRM points to the floor, followed by moving both fists around waist, as if donning a belt.
- *Okay.* FRM raises hands to eye level, with index finger and thumb forming a circle, and other fingers extended.

FRIES PROCEDURES

10-58. Detailed planning, thorough training, and strict adherence to safe procedures can reduce the risks of personal injury and equipment damage during FRIES operations. Conducting FRIES operations with heavy loads requires personnel to be proficient in these operations. The total weight of equipment carried on Soldiers during training is limited to 60 pounds. The following paragraphs provide standardized procedures for FRMs, pilots in command, and ropers. These procedures are common to all units and aircraft, except as noted. Appendix G contains a fast-rope operational checklist.

FAST-ROPE MASTER OR ASSISTANT FAST-ROPE MASTER

10-59. The FRM or AFRM performs the following series of actions:
- Briefs members of his team and aircrew.
- Inspects team members and equipment.
- Installs the FRIES rope in the aircraft and conducts safety checks.

Chapter 10

- Relays time warnings (10-, 6-, and 1-minute) to team members. Time warnings are a tool to help keep aircrew and ropers' actions synchronized and can be modified according to user needs; however, the 1-minute warning should always be used.
- Breaks chemical lights, if required, at the 6-minute warning. During night operations, the rope is marked with six chemical lights—two at the mount, two on the end, and two 15 feet from the end.

> **WARNING**
>
> Equipment must never be attached to the FRIES and dropped. As little as 10 pounds of weight can exceed the load limit of the FRIES mounts when the load stops at the end of a 60-foot drop. The only items to be attached to the fast rope during deployment are the position markers.

- Makes sure the rope is properly configured for deployment (back-coiled to prevent tangles).
- Makes sure the team members are in order of exit no later than the 1-minute warning.
- Confirms target on final approach.
- Deploys the rope and ensures it reaches the ground. During night operations, two horizontal chemical lights are seen and verified by the crew chief wearing NVGs.
- Deploys personnel.
- Accounts for personnel and signals aircrew.

PILOT IN COMMAND

10-60. The pilot in command briefs the FRM and safety and roping personnel on the following:
- Approaching, loading, unloading, and departing the aircraft.
- Actions in the aircraft.
- Flight route and checkpoints en route.
- Altitude or height for roping.
- Time warnings for FRM, as needed (recommend 10-minute, 6-minute, and 1-minute).
- Emergencies in the following situations:
  - En route (when ropers are inside the aircraft).
  - Before deployment of ropes.
  - When fast-rope personnel are on the ropes.
  - After the ropers are on the surface.

> **WARNING**
>
> Because the FRIES mount in the CH-47 helicopter is narrow, ropers must take special care not to catch weapons or equipment on the frame during exit.

ROPERS

10-61. Ropers perform the following actions:
- At the command STANDBY (given at 1-minute warning), ropers make final check of themselves and prepare to exit position.

- At the command GO, the first roper exits the aircraft, rotates his body 90 to 180 degrees to make sure his equipment clears the aircraft, places the fast rope between the arches of his feet, and commences descent. The roper uses his hands and feet to slow his descent about two-thirds of the way down.

> **CAUTION**
>
> Ropers will not place the rope between their groin or knees because this will cause severe burns and discomfort.

- Subsequent ropers exit at 1-second intervals using the same procedure but begin slow descent about halfway down to avoid landing on each other.
- During descent, ropers keep a lookout and break as necessary to avoid landing on obstructions or on fellow ropers.
- Ropers prepare to land just before reaching the ground by spreading their legs about shoulder-width apart with their knees slightly bent.
- At landing, ropers quickly move clear of the ropes to avoid collisions with descending ropers.

WARNINGS AND COMMANDS

10-62. The following warnings and commands apply to all fast-rope operations except for some aircraft-specific procedures covered in more detail later. Fast-rope warnings and expected aircrew coordination calls are as follows:
- When the pilot in command is over the intended objective, he calls DEPLOY ROPES.
- The aircrew or FRM provides any final position adjustments of the aircraft with the commands of MOVE LEFT, RIGHT, FORWARD, BACK, UP, or DOWN as needed to place the ropes on target. Aircrew or FRM announces ON TARGET when ropes are on target.
- The aircrew or FRM commands HOLD when the pilot is to hold the position and provides an explanation as soon as practicable.
- When the FRM deploys the ropes, the aircrew member calls ROPES DEPLOYED.
- When the first roper exits the aircraft, the aircrew member calls ROPERS OUT.
- The safety watches the last roper until he is on the ground and gives the prearranged signal. The safety then tells the pilot in command ALL ROPERS AWAY and identifies his station. This alerts the pilot in command that the safety is going to either jettison or recover the ropes.
- The safety will then either recover or jettison the rope as planned, clear the aircraft all around for obstructions, and call ROPES CLEAR (station). This informs the pilot in command that the aircraft is clear for flight.

PROCEDURES SPECIFIC TO EACH AIRCRAFT

10-63. Because of differences between the various types of helicopters, units, and services, each aircraft has slight variations of the procedures covered above. Units conducting fast-rope operations for the first time together will ensure that their pre-operations briefings cover both units' SOPs so that any conflicts can be worked out in advance.

> **CAUTION**
>
> No more than six ropers will be on a rope at any one time.

Chapter 10

> **CAUTION**
>
> IAW AR 95-1 and as identified in FM 57-220 and USASOC Reg 350-2, any time operations occur over water, all personnel will wear a Service-approved personal flotation device.

UH/MH-60

10-64. The crew members will notify the FRM of each checkpoint the helicopter passes. The crew members will pass on the 10-, 6-, and 1-minute warnings to the fast-rope personnel. If the cargo doors are closed, fast-rope personnel open the doors. The helicopter will decelerate to 80 knots at the checkpoint before the target. The aircrew members will signal the fast-rope personnel to open the cargo doors. The fast-rope personnel must open the doors and ensure they are in the locked position under positive control at all times. With the doors open, the crew chief extends the fast-rope bar and inserts the pins in the bar. At the 1-minute call, the closest Soldier to the door removes the cargo strap across each cargo door opening and passes the strap back to the AFRM. He will attach the strap to the D ring on the left aft portion of the cargo doorframe.

> **WARNING**
>
> If the fast-rope personnel open the cargo doors and allow them to slide back, the doors may shear the doorstops and leave the aircraft, with potentially catastrophic results. For example, the doors might strike the tail rotor, go through the rotor system, or hit the helicopter behind. Fast-rope personnel will take extreme care to remain securely anchored to the aircraft. With the door straps removed, personnel may be knocked out of the aircraft during end-phase maneuvering of the approach.

10-65. The crew members will confirm that the helicopter is at a stable hover over the target area and give the pilot commands needed to maneuver over the precise area briefed. When the crew member is satisfied that the helicopter is in position, he will signal to the AFRM ropes. AFRM will remove the safety pin from the fast-rope restraint strap, pull the quick-release tab, and throw the rope. The crew member will confirm that the required length of rope is on the ground (fast-rope attachment loops or two chemical lights in a horizontal position). The crew member will confirm the last man is free and clear of the ropes. He will then jettison the ropes or retrieve them back inside the helicopter. Crew member(s) will use hand-and-arm signals to advise pilots left or right ropes clear, and the helicopter will depart.

MH-6

10-66. At the 1-minute warning, ropers will release their personnel restraints, break chemical lights as required, and position themselves for roping. When the aircraft is at a stabilized hover over the target area, the pilot will give the FRM the command ROPES. The FRM will relay the command and deploy the ropes. Roper 1 must ensure the attachment loops or chemical lights are on the ground and horizontal before descending. The pilot(s) will verify the ropers are free and clear of the ropes before jettison of the ropes. The maximum number of personnel on a single rope is two.

> *Note.* The FRM can be positioned on either side of the aircraft. If the aircraft has one pilot, the FRM will be on the right side.

MH-47

10-67. At the 6-minute warning, team members move to the front of the aircraft if deploying from the crew entrance door and to the forward edge of the ramp if deploying from the rear of the aircraft. At this time, the FRMs (one at the crew entrance door and one on the ramp) disconnect the fast rope from its storage point and prepare it for deployment. This may entail handing it to the first man out of each stick or setting it up on the edge of the ramp, ensuring it is back-coiled. Before conducting deployments off the ramp, the FRM briefs the deploying team members on the importance of maintaining separation between members (about 24 to 27 inches). This separation helps to maintain the aircraft in the center of gravity limits. When using the MH-47, the aircrew member located at the forward door will relay all commands to the pilot.

> **CAUTION**
>
> There is risk of hang-up on the ammunition can or weapon itself when deploying personnel from the ramp with weapons installed. If a hang-up on a weapon occurs, the aft crew member must be prepared to assist, to include pulling the tangled roper into the aircraft to untangle him from the weapon or ammunition can.

EQUIPMENT-LOWERING PROCEDURES

10-68. During FRIES insertions, equipment, and/or rucksacks may need to be lowered due to weight. The rucksacks can be lowered by one of several belay devices in the Army inventory. These devices (Figure 10-3, page 10-6) include the Sky Genie, figure eight, Seattle Manufacturing Company (SMC) ladder, and other approved devices as per FM 3-97.6, *Mountain Operations*. Personnel must ensure the weight limits are not exceeded for the particular belay device being used. The preferred method is for all ropers to exit the aircraft from one door or fast rope, and deploy the equipment from the other door or fast rope. If using the ramp of the MH-47 for both personnel and equipment, personnel will belay the equipment and then fast-rope.

10-69. The aircrew will attach the belay device to the FRIES bar, and the using unit will provide the lowering rope. The using unit will rig the rucksacks, and the crew chief will help position the equipment in the aircraft. Rigging equipment to be lowered requires rigging rucksacks in clusters. Clusters of up to five rucksacks may be lowered but smaller clusters work better. FRMs will ensure the total weight of any cluster does not exceed the weight limits of the belay device or the crew member who will belay the rucksacks. To rig the lowering rope to belay the rucksacks, the FRM—

- Places the rucksacks on the cargo deck of the helicopter, sitting up (Figure 10-16, page 10-28).
- Lays a sling rope across the frames between the shoulder strap attachment points.
- Ties a bowline knot at each end of the sling and a figure-eight knot at each rucksack.
- Attaches a snap link to the frame of each rucksack to the loops in the sling rope and the end to the lowering rope.
- For Lowe rucksacks, installs the issue H harness on the rucksacks, and join them into clusters with a sling rope tied to the quick-release attachment buckles.
- Back-coils a rappel rope on the helicopter floor at the gunner's position.

10-70. When using the SMC ladder, the FRM attaches it to the fast-rope attachment by snap links. He routes the belay rope from the back-coil, through the belay device (Figure 10-17, page 10-28), and down to the cluster snap link. He ties the end of the belay rope into a bowline with a half hitch and snaps the rope into the cluster snap link. The crew member will deploy the equipment and/or rucksacks and control the descent via the SMC ladder. The SMC ladder system can handle up to 500 pounds per rope.

Chapter 10

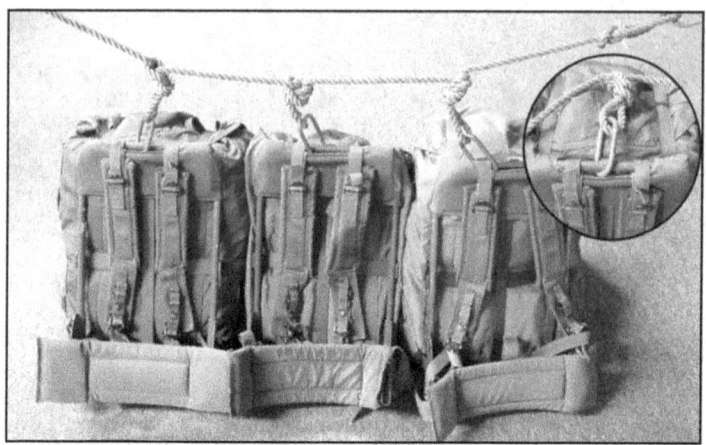

Figure 10-16. Rucksacks upright on cargo deck

10-71. When using the figure-eight device, the FRM attaches it to the fast-rope attachment by snap links. He routes the belay rope from the back-coil, through the belay device (Figure 10-18), and down to the cluster snap link. He ties the end of the belay rope into a bowline with a half hitch and snaps the rope into the cluster snap link. The crew member will deploy the equipment and/or rucksacks and control the descent via the figure-eight.

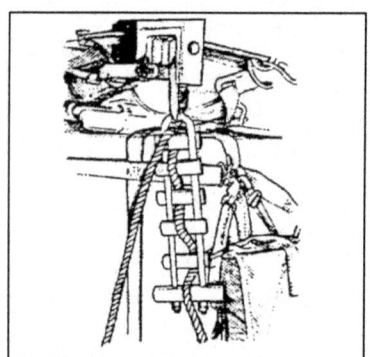 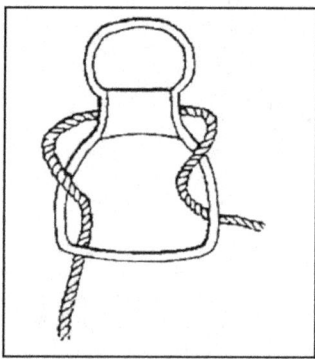

Figure 10-17. Belay device SMC ladder

Figure 10-18. Belay device figure eight

10-72. If using other types of figure eights, the rucksacks and equipment are rigged the same way and the figure eight is attached to the fast-rope attachment point in the same way. However, the lowering rope is doubled and inserted through the lowering device as shown in Figure 10-19, page 10-29.

10-73. When using the Sky Genie, the FRM it to the fast-rope attachment by snap links or the rear ceiling-mounted rappel ring. He routes the belay rope from the back-coil, through the belay device, and down, to the cluster snap link. He ties the end of the belay rope into a bowline with half hitch and snaps the rope into the cluster snap link. The crew member will deploy the equipment and/or rucksacks and control the descent via the Sky Genie.

Fast-Rope Insertion and Extraction System

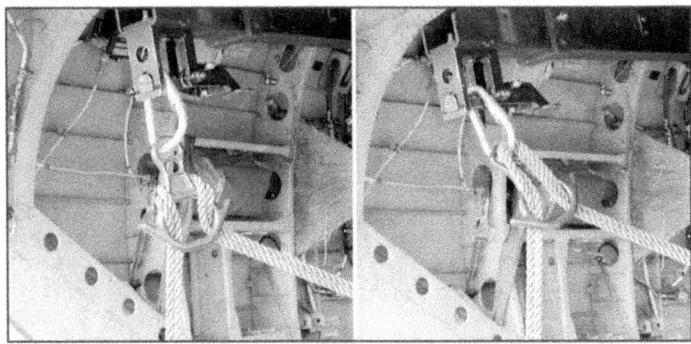

Figure 10-19. Double lowering line routing using figure eight

10-74. When using the Sky Genie, the FRM it to the fast-rope attachment by snap links or the rear ceiling-mounted rappel ring. He routes the belay rope from the back-coil, through the belay device (Figure 10-20), and down, to the cluster snap link. He ties the end of the belay rope into a bowline with half hitch and snaps the rope into the cluster snap link. The crew member will deploy the equipment and/or rucksacks and control the descent via the Sky Genie.

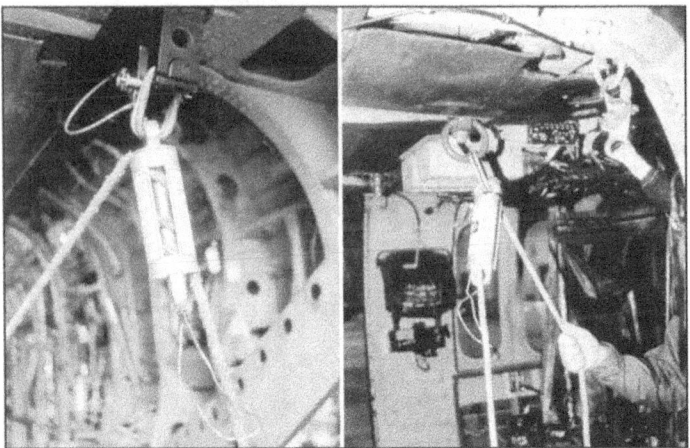

Figure 10-20. Lowering line routing using the Sky Genie

10-75. The crew chief or safety will lower the equipment during most operations while the deploying unit should delegate two Soldiers to push each cluster out of the helicopter. However, in case both door guns are being used, the using unit should be prepared to provide one FRM-qualified individual to control the equipment during lowering. The crew chief or safety will use gloves in the lowering process. This equipment-lowering system can be used to resupply items such as ammunition, water, fuel, and so on to ground units where helicopters cannot land. Additionally, this method of resupply may be configured for deployment on each side of the helicopter.

10-76. In performing his belaying duties, the AFRM or safety must—
- Provide enough slack to allow the pushers to eject the cluster from the aircraft.
- Control the descent until the cluster reaches the surface.

Chapter 10

- Unsnap the upper and lower snap links, and drop the remainder of the rope from the aircraft when the cluster is on the surface.

10-77. If lowering equipment before personnel, the FRM performs the following:
- After the crew chief or safety releases the snap hooks, immediately looks out to make sure conditions are safe to deploy the FRIES.
- Informs the FRM and pilot when conditions are safe.
- Deploys FRIES on command.

Note. Expect to reposition aircraft before ropers exit to avoid ropers landing on equipment.

EMERGENCY ACTIONS

10-78. Procedures for correcting an emergency condition that personnel could reasonably encounter are as follows. Multiple emergencies, adverse weather, or other unusual conditions may require modification of these procedures. The nature and severity of the emergency dictate the response necessary; therefore, personnel must use sound judgment in determining the correct action to take.

EMERGENCIES BEFORE ROPING STARTS

10-79. All personnel sit down, don their seat belts, and take further instructions from the pilot or crew chief. If a crash occurs, all personnel follow emergency procedures the FRM or pilot briefed before the operation.

EMERGENCIES AFTER ROPING STARTS

10-80. In an emergency, personnel follow specific procedures. The following paragraphs explain these procedures.

Unsafe Drift or Premature Liftoff

10-81. Anyone who detects the aircraft having drifted off the site must immediately stop training and inform the FRM, AFRM, safety, or pilot in command. The procedures for unsafe drift or premature liftoff are as follows:
- FRM, AFRM, safety, or roper stops stick.
- Ropers stop descent and lock in.
- FRM or crewman informs the pilot in command and guides him in moving the aircraft back on target.
- Unit continues operations.

Rope Hung or Snagged

10-82. Anyone who detects the rope being snagged or hung up must immediately stop training and inform the FRM, AFRM, safety, crew chief, or pilot in command. The procedures for a snagged or hung rope are as follows:
- Safety makes sure ropers are off the rope and are clear.
- Aircraft descends or lands, as needed.
- Ground personnel free the rope.
- Unit resumes the operation.

Premature or Unintentional Deployment of the Fast Rope

10-83. If the fast rope is released prematurely or accidentally falls from the helicopter, the FRM—
- Notifies the pilot in command.
- Follows the aircrew's instructions.

Lost Communications

10-84. All training and operations must include the use of the intercom between the pilot in command or crewmen and the FRM. If the intercom fails, the following hand-and-arm signals can be used until the rope can be cleared and the intercom is restored:

- Stop stick: A clenched fist touching the chest.
- Ropes: Open palm toward the door in a horizontal motion.
- Aircraft movement: An open palm moving and facing in the direction required.
- Stop aircraft movement: A clenched fist.

This page intentionally left blank.

Chapter 11
Air-Water Operations

Air-water operations encompass several operations. A team uses an aircraft to travel the majority of the distance to their objective or BLS and travels by boat, surface swimming, or subsurface operations the remainder of the distance. Air-water operations include water landing, water jumps, helocasting, ERDS, rolled or tethered duck, hard duck, and recovery operations. Although aircraft provide the most practical and rapid means of transporting infiltration swimmers to the vicinity of the BLS, air operations can be more complicated than other means, such as surface craft or submarine infiltrations. There is a variety of assault-type aircraft, as well as tactical and utility types, that a team can use to infiltrate with or without a CRRC or Zodiac. Air-water exfiltrations can be conducted using airland, FRIES, SPIES, or ladder operations. Because USASOC Reg 350-2 covers water jumps in detail, this chapter will concentrate on the other air-water operations.

SAFETY

11-1. The objectives of air-water operations training are to safely conduct and to maintain maximum proficiency in the execution of air-water operations. Safety is everyone's responsibility. All personnel involved in air-water operations are responsible for identifying hazardous situations and preventing injuries. Anyone who observes an unsafe condition or act is authorized to halt the operation and inform the castmaster (CM), JM, NCOIC, officer in command, or pilot in command. Refer to USASOC Reg 350-2 and 350-6 for the most current safety requirements.

SAFETY BRIEFING

11-2. A safety briefing must precede any air-water operation. The briefing should consist of, but is not limited to, a review of the following:
- Area hazards.
- General aircraft safety.
- Characteristics of equipment associated with air and water operations.
- Equipment inspection.
- Method of infiltration and/or recovery operation to be used.
- Hand-and-arm signals and emergency signals.
- Location of safety boats and marking procedures.
- Medical coverage.
- Primary and alternate communications requirements.
- Night operation requirements.

SAFETY CONSIDERATIONS

11-3. Because of the hazards involved with air-water operations, all aspects of planning and execution will emphasize safety. USASOC Reg 350-2 contains the safety requirements for conducting a water jump. Personnel planning a water jump must review USASOC Reg 350-2 before conducting the water jump. The water LZ requirements and procedures for fixed-wing aircraft appear in Chapter 5 of this manual. USSOCOM M 350-6 and this manual contain the safety considerations for water landing, helocasting,

ERDS, rolled or tethered duck, hard duck, and recovery operations. The following safety rules and limitations will be followed:
- Immediately before an air-water operation, the cast area or DZ will be physically reconnoitered by a minimum of two safety swimmers to verify water depth (no less than 10 feet deep) and the absence of obstacles and debris. Marking the cast area or DZ will be accomplished by positioning the support craft or anchored floats to indicate the safe area.
- Safety boats with motors running must be present and in the water before conducting helocasting operations. A minimum of one boat per 20 swimmers is recommended for air-water operations other than airborne operations. The boat operators must have the required training and licensing to operate the equipment they are using. All boat crew personnel must wear Service-approved personal flotation devices. Each safety boat must contain the equipment listed below:
  - Medical kit and backboard.
  - Primary and back-up radios.
  - Buoys with weights and enough line to mark suspected areas of lost equipment.
  - Appropriate lights for night operations.
- A minimum of one safety swimmer will be aboard each safety boat. The swimmer will be a graduate of the Combat Diver Qualification Course or a USSOCOM-approved waterborne infiltration course, scout swimmer course, or current Red Cross lifesaver or water safety instructor course. The safety swimmer must have swim fins, a face mask, and a Service-approved personal flotation device to help personnel, as needed. The swimmer cannot be the boat driver.
- During night operations, swimmers deploy in no less than two-man teams.
- An emergency evacuation vehicle must be stationed at the nearest boat-landing site.

Note. Absence of any safety equipment or personnel requires a cancellation of the helocasting operation.

- Personnel in the safety boats and aircraft establish radio communications with one another. If personnel cannot communicate by radio, the casting area safety officer (CASO) must use prearranged visual signals (for example, the safety boats circling in the water at the infiltration site) to convey conditions are safe to continue the operation. Lack of air-to-ground radio communications and/or prearranged visual signals requires a no drop.
- The CM must have voice communications with the pilots.
- A qualified medic must be in one of the safety boats.
- Helocasting drop altitude will not exceed 10 feet above the surface of the water (5 feet when launching CRRC).
- Helocasting drop speed will not exceed 10 knots GS.
- Casting operations are done into the wind. In rivers or strong currents, casting is into the current regardless of the wind conditions.
- The pilot ensures the recovery speed does not exceed 10 knots GS.
- If a swimmer is injured, the operation will cease until the cause and extent of the injury are determined.

AIR-WATER OPERATIONS QUALIFICATION TRAINING

11-4. Before participating in helocasting, rolled, tethered, ERDS, or recovery operations, all personnel must meet the following prerequisites:
- Be assigned or attached to a USSOCOM unit.
- Have passed the Service fitness test.

- Have a current medical examination and be free of any injury or physical condition that could cause a potential safety hazard during training.
- Have conducted drown proofing and swim test IAW USASOC Reg 350-1, *Training ARSOF Active Component and Reserve Component*, within the last 6 months.
- Successfully completed initial, sustainment, or refresher training for the specified operation.
- Have been thoroughly briefed IAW the CM briefing listed in Appendix H.
- If applicable, identified themselves as weak swimmers and swimmers with no confidence to the CM before the training.

INITIAL TRAINING

11-5. Soldiers will be considered qualified when they—
- Know the procedures, techniques, and equipment needed to conduct the type of operation by demonstrating confidence and proficiency.
- Complete a minimum of one daylight and one night operation with combat equipment and weapon.

SUSTAINMENT TRAINING

11-6. Units routinely conduct sustainment training to maintain the acquired skills. Within 24 hours of conducting air-water operations, units will receive formalized training in the procedures to be used during the operation. At a minimum, this training will include the following:
- Rigging and inspecting individual equipment.
- Rigging and inspecting accompanying equipment in the CRRC (if applicable).
- Hand-and-arm signals and emergency signals.
- Water entries.
- A dry-land rehearsal.

REFRESHER TRAINING

11-7. Soldiers who have not participated in the air-water operation during the past year will undergo refresher training before being included in an operation. Refresher training consists of sustainment training and conducting at least one operation under the observation of a current CM.

PERSONNEL QUALIFICATION REQUIREMENTS

11-8. Before conducting helocasting, ERDS, rolled or tethered duck operations, the CM and CASO or NCO must meet the criteria discussed below. The CM and CASO must be qualified for each type of operation. Just because a person meets the qualifications for being a helocast CM does not mean they can conduct rolled duck operations.

CASTMASTER

11-9. Units should select personnel to become qualified as CM based on the individual's demonstrated leadership capabilities and knowledge of air-water operations. The individuals selected will be qualified officers or NCOs who will be assigned to each mission. The individuals selected must have previously participated in that specific operation within the past year. Individuals selected will be qualified to perform the duties of CM when they have—
- Completed the initial training for that operation (candidates).
- Received instructions on and demonstrated proficiency in equipment rigging, inspecting and preparing equipment to be used in the operation, and conducting the operation.

Chapter 11

- Received instructions and demonstrated proficiency in the performance of the following CM duties:
  - Coordination responsibilities.
  - Troop and aircrew briefings.
  - Organization of the personnel involved in the operation.
  - Emergency procedures.
  - Instruction to pilots for maintaining the aircraft in position over the target.
  - Hand-and-arm signals.

CASTING AREA SAFETY OFFICER OR NCO

11-10. The CASO or NCO must—
- Be designated by the unit commander (or designated representative).
- Be qualified in helocast.
- Have previously participated in the type of operation being planned.

PERSONNEL DUTIES AND RESPONSIBILITIES

11-11. Air-water training and operations require the designation of key personnel to perform assigned tasks. Depending on the exact operation being planned, there may be some differences in the duties and responsibilities of the various positions. The positions are unit commander, air mission commander, pilot in command, CM, safety boat NCO, and safety swimmers.

UNIT COMMANDER

11-12. Before a unit participates in training, the unit commander or designated representative—
- Screens of Soldiers to ensure they meet the prerequisites for conducting air-water operations.
- Ensures required safety personnel are tasked and qualified to perform their duties.
- Ensures all equipment and support personnel required for the operation are available, including the training area.
- Ensures unit personnel follow all provisions of USASOC Reg 350-2 or USSOCOM M 350-6.
- Terminates the operation at any time because of any unsafe condition, safety requirement, weather, or lack of training requirement.

AIR MISSION COMMANDER

11-13. The employing aviation unit designates the air mission commander when more than one aircraft is involved in the operation. The air mission commander ensures—
- All aircraft and aircrews are at the appropriate locations for training, rehearsals, and the operation.
- All aircraft are properly configured for the appropriate operations.
- The aircrews and all personnel who are not part of the aircrews are briefed and understand their responsibilities during the operation, including aircraft safety and emergency procedures.
- All aircraft deploy the personnel and boats at the designated target.

PILOT IN COMMAND

11-14. The pilot in command assumes the duties of the air mission commander on single ship missions and also—
- Ensures all crew members are current and qualified to perform the appropriate operation.
- Ensures all participants onboard his aircraft are thoroughly briefed on aircraft safety and the appropriate operation and understand their responsibilities concerning the conduct of the operation.
- Ensures the CM or crew chief has inspected and safely prepared personnel and equipment.

- Ensures all aircraft requirements have been met; for example, weight and balance, flight planning, fuel requirements, and so on.
- Terminates the operation at any time because of any unsafe condition, unmet safety or training requirement, or weather.
- Orders the emergency release of the CRRC.

CASTMASTER

11-15. The CM directly oversees the conduct of the operation. He must have previously participated in an operation within the past year. He primarily—

- Assists the CASO, when possible, in conducting a reconnaissance of the proposed drop area to ensure all safety and obstruction criteria have been adhered to.
- Conducts the CM briefing IAW Appendix H.
- Conducts a visual safety check of the aircraft to ensure the proper rigging of all equipment.
- Conducts a safety inspection and equipment check of all swimmers verifying their equipment is properly positioned and functional. Inflatable flotation devices will be inspected IAW FM 20-11, *U.S. Navy Dive Manual, Volumes I and II*. During helocasting operations, inflatable flotation devices will be inflated to at least 1/4 of their total capacity (this requirement does not apply to diving operations).
- Briefs the aircrew on all aspects of the operation, including hand-and-arm signals used for directing the aircraft into the exact position for the operation, and on no-drop conditions and situations.
- Ensures radio voice communications are functional between all elements of the operation.
- Assigns swimmer buddy teams and ensuring all swimmers sit in stick order. Verifies all swimmers understand their assigned duties and the commands to be followed. Gives command for swimmers to exit the aircraft.

Note. Swimmers exit only on the command of the CM.

- Ensures swimmers exit the aircraft as a buddy team. Casts the swimmers only if the aircraft is correctly aligned and safely within the limits of speed and altitude and the safety boats are operational and at least 50 meters left or right of the line of flight. When the swimmers surface, they signal to their buddy and the safety boat that all is okay or that they need assistance.

CAUTION

When launching the CRRC, no swimmers will exit the aircraft until the CRRC is in the water. The CM aborts the helocasting operation if any unsafe condition exists.

CASTING AREA SAFETY OFFICER OR NCO

11-16. The CASO or NCO—

- Conforms to the time schedule as closely as practicable in compliance with safety standards and conditions existing at the time of the operation.
- Gives a complete briefing to all participating personnel.
- Ensures all safety and equipment requirements are met before initiating the operation.
- Ceases operations if any unsafe condition arises.
- Conducts a reconnaissance of the cast site or DZ, verifying water depth and an obstacle-free environment.

Chapter 11

SAFETY BOAT NCO

11-17. The safety boat NCO ensures the required numbers of safety boats are on site for training. He ensures there is a minimum of one boat per 20 swimmers IAW risk assessment and swimmer proficiency. The safety boat NCO—

- Attends the operations briefing and thoroughly understands the intent of the operation.
- Maintains effective control of all support in the area of operation.
- Briefs all safety boat personnel in the conduct of their assigned duties. Ensures all personnel know the day and night emergency signals for "Pick me up now. I need help."
- Supervises the boat crews and the safety swimmers.
- Establishes and maintains boat-to-air radio voice communications.
- Inspects the casting area or DZ for safe water depth, obstacles, and any potentially hazardous debris. The casting area or DZ and obstacles will be marked as required by the unit SOP.
- Obtains and reports accurate weather, wind, surf, and sea state conditions.
- Ensures the following items are present in the safety boats:
  - Medical kit and backboard.
  - Primary and back-up radios.
  - Buoys with weights and enough line to mark suspected areas of lost equipment.
  - Appropriate lights for night operations.
- Ensures cast area or DZ is kept clear of debris, unnecessary personnel, equipment, and boats.
- Ensures the safety boats move parallel and at least 50 meters left or right of the line of flight of the aircraft.
- Accounts for and ensures each swimmer is not injured after water entry. Until all swimmers are accounted for, the aircraft maintains radio contact with the safety boat.
- Aborts the operation if any unsafe condition exists or arises.

SAFETY SWIMMERS

11-18. The safety swimmers perform duties and follow commands as directed by the CASO, and follow prescribed safety procedures. Safety swimmers—

- Assist in recovering equipment.
- Aid injured or weak individuals in the water.
- Remain with the individual until the safety boat or help arrives.

OPERATIONAL REQUIREMENTS

11-19. The operational requirements must be followed as closely as possible during training under usual conditions and unusual conditions (adverse weather or terrain conditions and night operations). Personnel must use sound judgment to determine what action to take depending on the nature and severity of the condition.

MEDICAL REQUIREMENTS

11-20. A qualified and equipped 18D, 91W, or 91B medic or Service-qualified EMT will be at all training sites. Medics must know CASEVAC procedures and have coordinated requirements necessary to expedite evacuation and treatment of personnel on and off military installations. All medics will have as a minimum an oxygen bottle and M5 aid bag or equivalent packed IAW unit standards. Medical transportation must also be available. The vehicle must be covered and large enough to carry an open stretcher. Absence of a medic, medical equipment, or transportation is cause for terminating training. If the situation warrants and the installation cannot support a MEDEVAC mission, the installation may use the

helocast aircraft as a last-resort CASEVAC vehicle. The medics will develop an evacuation plan. This plan should include, but not be limited to, the following:
- Medical facilities—location and capabilities.
- Emergency telephone numbers.
- Route(s) to medical facilities.

COMMUNICATIONS REQUIREMENTS

11-21. During training, the CASO communicates by radio with the aircraft. The CM has voice communications with the pilot. Voice communications are required between the pilot in command, CM, and the CASO before commencing air-water operations. In addition, the CM or CASO informs the pilot in command to stop operations if an unsafe condition develops. During extractions, the CM informs the pilot in command that all personnel are ready for extraction. During tactical missions, mission and aircrew personnel use prearranged signals to communicate between the mission and aircrew personnel (for example, flashing light or chemical light signals).

PREMISSION PLANNING

11-22. The type of aircraft and delivery method selected determine the planning and preparations required during premission planning. A successful waterborne operation needs detailed intelligence, planning, preparations, and precise execution. First, planners should consider personnel proficiency and ability to conduct the desired infiltration technique. The personnel may need additional training and rehearsals for specific missions. Second, planners select, issue, and prepare the equipment needed for the mission. Combat loads should be light, small, and include only equipment, weapons, and ammunition needed for the mission. Planners should develop a detailed equipment-loading plan with appropriate waterproofing and bundle rigging. Third, planners should consider the method used to transport the swimmers or divers to the debarkation point. The method may be by aircraft, surface craft, or submarine. The method depends upon the mission, time, and distance.

11-23. In all infiltrations by aircraft, the pilot and flight crew must coordinate with the operational element. The pilot briefs on in-flight communications, NAVAIDs, abort plans, and other related general flight procedures. The operational element informs the pilot and flight crew of the number of personnel to be infiltrated, the type and quantity of accompanying supplies, the DZ or LZ markings, and other mission-related information. Required rehearsals for each phase of the infiltration should include the actual pilot and flight crew for the mission.

11-24. Air-water operations require careful planning, preparation, and rehearsals. Each unit may have different SOPs for exactly how they conduct the various operations. However, they should all follow the guidelines established in this chapter.

EQUIPMENT

11-25. Operational units will determine equipment requirements necessary to accomplish their mission IAW the unit SOPs. However, all personnel involved in water operations will wear an approved flotation device.

AIRBORNE OPERATIONS

11-26. Airborne water infiltrations can be conducted either with or without a boat. USASOC Reg 350-2 covers the setup and operation of water DZs. A variety of inflatable boats can be airdropped from C-130 and C-141 aircraft. Details on platform dimensions are contained in FM 4-20.142, *Airdrop of Supplies and Equipment: Rigging Loads for Special Operations*.

BOAT PLATFORM

11-27. Rigging of the drop platform is the ultimate responsibility of the rigger section. The rigger section rigs the drop platform IAW FM 4-20.142. The section mounts the boat on a combat expendable platform designed to mate with the contour of the boat. The platform is constructed of 3/4-inch marine or exterior plywood and other wooden support members. The rigging section fills the area between the boat and the bottom of the platform with sandbags weighted to raise the total airdrop weight of the package to the specified weight.

11-28. Personnel secure the motor, fuel, equipment, and weapons (depending on unit SOP) to the inside of the boat. Knowledgeable personnel supervise the loading and securing of individual and team equipment in the rubber boat. This equipment should be secure enough that no equipment will fall out during the free-fall period.

11-29. A sling provides the attachment point for a G-12D 64-foot cargo parachute. The parachute is equipped with a quick-release device that detaches the parachute from the sling upon contact. When rigging personnel deliver the fully prepared boat to the departure airfield, ground support equipment consisting of either a K-loader or a special long-tine forklift is required to load the platform onto the aircraft.

11-30. Waterborne operations include several boat and loading combinations. They are as follows:
- One CRRC loaded with all the teams' equipment (minus load-carrying equipment and weapons). The obvious disadvantage is that, should the parachute not deploy, the team will have no backup with which to accomplish the mission. A fully loaded second platform on the infiltration aircraft is optional. Should the first platform sink, the aircraft may make a second pass and drop the remaining platform. If this option is used, the team must coordinate with the aircrew before the drop to establish a signaling device.
- Two CRRCs loaded on one platform—called a double duck operation. Two separately rigged small inflatable boats, military amphibious reconnaissance system, or Zodiacs are loaded on the aircraft. The primary disadvantage of this method is that current USAF regulations require two separate passes (boat and personnel on one pass and the same on the next). In addition to disclosing the location of the drop, multiple passes make it extremely difficult for the two boat teams to link up in the water at night.

INDIVIDUAL PARACHUTISTS

11-31. The rigging of individual parachutists should be done IAW FM 3-05.212, *Special Forces Waterborne Operations*. Because rucksacks are secured inside the boat during the airdrop, parachutists may water-proof and place in a kit bag load-carrying equipment, two or three rations, one pair of boots, one set of fatigues and socks, and any other equipment deemed necessary for minimal mission accomplishment. Parachutists then rig the kit bag with a single-point release harness without a lowering line. Parachutists then rig as for a combat swim operation, to include exposure suits where required, except the kit bag replaces the rucksack.

11-32. The rationale for the use of the kit bag described above is that the cue for parachutist exit is pilot parachute inflation only; the possibility still exists that the G-12D may not open or may malfunction. If the parachutists wait for the G-12D to open fully, because of its slow opening characteristics, they will be so far from the boat that assembly on it will be extremely difficult. If this happens and the boat sinks or becomes inoperable, the team still has the option to try to swim and or drift with tides and currents to the nearest land mass and try mission accomplishment or escape and evade.

11-33. If the team jumps with the special operations waterproof bag, a lowering line should be used. Should a parachutist fail to linkup with the boat, he can inflate the bag and use it as a flotation device, which will aid the swimmer stranded in the water.

PROCEDURES

11-34. Whatever boat and loading combination is used, the JM or safety hooks the boat to the left side anchor line cable while the parachutists use the right one. The first parachutist (JM) positions himself at the tailgate hinge, even with the leading edge of the platform. The other parachutists line up behind him. When the platform starts moving, the parachutists follow it. However, the first parachutist stops at the edge of the tailgate and does not exit until he sees the pilot parachute of the G-12D deploy. The team then tries to assemble on the boat in the air and water.

11-35. Individual parachutists should release both canopy release assemblies as soon as they feel their feet touch the water. They should immediately swim away from the canopy to preclude entanglement with the sinking parachute. Doing so is particularly critical because the exposed weapon is prone to entanglement with suspension lines. (Parachutists may place weapons in the boat to preclude entanglement.)

11-36. The parachutist should then remove the harness assembly and let it sink. During training, parachutists should attach the harness to the canopy with half of a B-7 life preserver to prevent loss of the parachute. The parachutists then swim to and assemble on the boat.

11-37. The first man to the boat moves around to the rear and smells for gas (in case the gas cans have burst or leaked). If there is no gas smell, he then feels in the area around the motor and gas cans, again looking for spilled gas. Once he has determined that there is no spilled gas, he enters the boat. If there is evidence of spilled gas, he makes a decision on the extent of the damage and makes an estimate as to the relative danger of trying to start the motor. He then disconnects the risers by taking the pins out of the clevises. He cuts the center ring, allowing the platform to separate it from the boat and sink under the weight of the sandbags. He then starts the motor and picks up the remaining personnel before heading for the BLS.

11-38. Marking of the boat and personnel with chemical lights facilitates assembly and pickup of personnel. The chemical lights have minimal tactical significance due to the "over the horizon" nature of the operation. However, personnel must be cautious as to the disposition of the chemical lights because they float and are easily seen from the air.

11-39. Trash bags with weights should be stored in the boat when rigging. Parachutists place all trash from the platform (for example, 80-pound test line, tape, and so on) in the trash bag. The parachutists secure the opening of the trash bag with the weights inside. They then allow the trash bag to sink to the bottom.

HELOCASTING

11-40. Helocasting is a very effective means of inserting combat swimmers, combat divers, and CRRCs. The speed, range, and lift capability of today's rotary-wing aircraft make them excellent waterborne delivery and recovery vehicles (Figure 11-1).

Figure 11-1. Helocasting

Helocasting Preparation

11-41. Mission planners should consider the following guidelines when planning all helocasting operations:

- When planning for the number of personnel per type of aircraft, use the standard troop loading planning figures. Adjust these figures depending on aircraft configuration, type of equipment, and casting or recovery procedures. Coordinate these items in advance with the pilot in command.
- Rehearse the operation with all the swimmers, the actual aircrew, the accompanying equipment, and support personnel. Emphasize proper body exit position, exit timing, commands, and water entry positions during live casting rehearsals.
- For effective communications, ensure all personnel use the same frequency (CM, pilots, and safety boats).
- Ensure the casting area is clear of all surface and subsurface obstacles.

11-42. The CM ensures helocasting personnel preparations include the following:

- Attaching all surface swim equipment to the swimmers by using 1/4-inch 80-pound test cotton web. This normally includes masks, fins, web belts with knives, and flares.
- Ensuring all swimmers wear life jackets or vests.
- If using side doors for casting, securing the doors in the open position and taping all edges.
- Ensuring preparations for CRRC helocasting operations include the following:
    - Tying down or securing all equipment inside the boat.
    - Securing the motor in the up position with the travel lock engaged. For engines without a travel lock, tying up the engine and reinforcing it with honeycomb.
    - Securely attaching and isolating the gas can, if possible.
    - Securing the paddles under the gunwales, out of the way of the rest of the gear.
    - Securing the rucksacks and twin-80s at the four corners of the boat. Fastening them as securely as possible.
    - Waterproofing all equipment in the CRRC as if Soldiers would take it subsurface.
    - Regardless of the type of aircraft used, tying down or securing all loose equipment.

Helocast Operations

11-43. While in flight, the loadmaster passes on whatever time warnings the CM coordinated with the aircrew. Actions of the swimmers and CM will vary depending on the type of aircraft being used, number of swimmers, and presence or absence of a CRRC. The following commands are given by the CM:

- *Get Ready.* Swimmers or divers remove their seat belts only on this command.
- *Stand Up.* Swimmers stand up and face the CM.
- *Check Equipment.* Swimmers check the equipment of the person ahead of them in the stick. CM and/or safety personnel inspect equipment of swimmers.
- *Sound Off With Equipment Check.* The swimmer at the back of the stick gives the thumbs-up signal and passes a verbal okay up the stick. The first swimmer gives the thumbs-up signal and tells the CM All Okay, Castmaster.
- *Go.* The preferred method is: The CM commands GO and follows the last swimmer out of the aircraft. The CM's exit position is based on the tactical situation.

11-44. The CM and swimmers should follow the procedures and safety considerations below:

- The CM ensures the aircraft does not exceed 10 feet of altitude at 10 knots when dropping personnel only. The CM starts the initial training drops at 5 feet at 5 knots.
- When casting from a ramp, swimmers assume a normal "prepare to land" attitude.
- When casting from a side door, cast from a seated door position or from a skid. On the CM's command, swimmers push off and face opposite the direction of flight, assuming a normal "prepare to land" attitude.

Air-Water Operations

- If casting from two side doors, swimmers (in the same position) may exit simultaneously from both doors on the CM's command.
- With a CH-46/47, the CM ensures the ramp is secured in the open or casting position. The ramp should be 10 degrees below horizontal.
- If using bundles or rucksacks, throw them before the swimmer's exit.
- Upon entering the water, give an okay signal to the CM and safety boats.
- If using a CH/MH-47 and a CRRC, just before giving the commands, the CM and crew chief move the CRRC to the end of the ramp and act as follows:
  - Ensure the helicopter has slowed to 10 knots and descended to 5 feet (or as low as possible).
  - Cast the CRRC on the CM's command. Swimmers follow the CRRC.
- Upon entering the water, swimmers give an okay signal to the CM and any safety boats.

ERDS OR K-DUCK

11-45. The ERDS or K-duck provides the capability to deliver a fully inflated CRRC and team into the water using an UH/MH-60 helicopter. The maximum allowable weight for external helicopter transport is 1,500 pounds.

PERSONNEL QUALIFICATION REQUIREMENTS

11-46. Within 24 hours of the conduct of ERDS operations, units must receive formalized training in the procedures for the operation. At a minimum, this training will include all requirements as previously discussed and the following:

- Rigging and inspection of individual equipment.
- Rigging and inspection of accompanying equipment in the CRRC.
- Rigging of the CRRC to the aircraft.

PERSONNEL DUTIES AND RESPONSIBILITIES

11-47. In addition to the duties and responsibilities already covered, ERDS operations require the following: a ground hookup crew consisting of adequate personnel to attach the CRRC to the aircraft and at least one crew chief aboard each aircraft. The crew chief and CM must—

- Before the operation, ensure the aircraft is properly rigged, conduct a visual inspection of the ERDS system and I bar (if used), and confirm the removal of aircraft belly antennas.
- Brief all personnel on the operation.
- Conduct an inspection of personnel and equipment before boarding the aircraft.
- Oversee the attaching of the raft to the aircraft, and ensure compliance with the airworthiness release.
- Observe the security of the CRRC from pickup through release, keeping the pilot in command advised of the stability.
- Direct the pilot to maneuver the aircraft into the proper position for deployment.
- Deploy the CRRC when in the drop area and cleared by the aircraft pilot in command.
- Abort any part of the operation if any unsafe condition exists.

CAUTION

If the load becomes unstable during flight, a smooth reduction of airspeed may be required. If the load is unmanageable, it may be necessary to land or for the pilot in command to order an emergency release of the CRRC.

Chapter 11

RIGGING PROCEDURES FOR ERDS OR K-DUCK

11-48. There are three methods for rigging the CRRC: the K-duck fixture, the cradle method, and the harness method. Between the cradle and harness methods, the cradle method is preferred because it more firmly secures the CRRC to the bottom of the helicopter.

K-Duck Fixture Method

11-49. The K-duck fixture (Figure 11-2) is an aluminum framework that fits into the CRRC. The fixture provides an attachment or suspension point in the boat to support the boat and a full mission equipment load (up to 1,000 pounds) on a helicopter cargo hook. When using the fixture, the CRRC is carried fully loaded and inflated and requires only the attachment of the motor in preparation for getting underway.

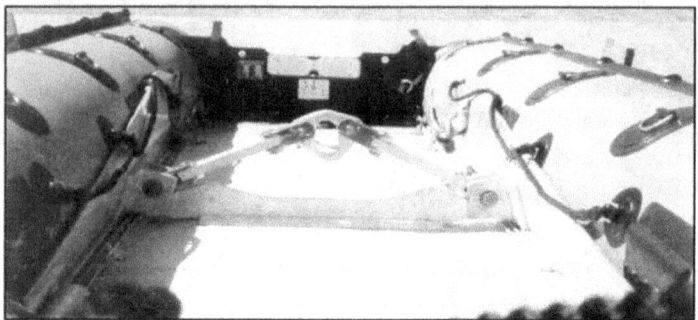

Figure 11-2. K-duck fixture

11-50. Two Soldiers can prepare and rig the CRCC with the K-duck fixture in about 20 minutes. They prepare the CRCC as follows:
- Inspect and lay out K-duck fixture components: port and starboard rails, bow frame, cross bars, and floorboards.
- Lay three forward floorboards on the floor, connected with the rounded edges facing aft.
- Snap port and starboard rails on the side of floorboards.
- Place lower (compression bar) and upper suspension link assembly in the suspension brackets on the rails. Insert pins in each end of the suspension assembly. Insert the pins through holes in rail brackets.
- Insert bow frame in sockets at front of rails.
- Lay out deflated CRRC on floor beside assembled K-duck fixture.
- Pick up K-duck fixture, insert bow frame first into CRRC, and rest the aft end of the rails on the transom of the boat.
- Place fourth floorboard, with quick-pin assemblies (Figure 11-3, page 11-13), under transom crossbar. Rest the rear edge of the fourth floorboard in its place along the rear floor of the boat. Engage the forward edge of the fourth floorboard with the rear edge of the last floorboard already in the rails to form a tepee shape.
- Ensure the front edge of the first floorboard engages the wood crossbar in the floor of the CRRC.
- Route the cargo strap through the access holes in the side rails, and tighten. This pulls the side rails snug to the aft floorboards. Line up suspension links between floorboards and side rails, and insert quick pins.
- Inflate the boat in the normal manner. Verify that, as boat inflates, the floorboards remained positioned under the boat spacers.

Air-Water Operations

Figure 11-3. Quick-pin assemblies

Cradle Method

11-51. The cradle method is certified for external helicopter transport by the U.S. Army Natick Research, Development, and Engineering Center for UH 60A and MH-60A helicopters with the fast-rope H bar or the FRIES I bar installed. Certification is granted for airspeeds up to and including 130 knots without the optional floor extension installed and 145 knots with the floor extension installed. The following equipment is required for rigging the CRRC using the cradle method:

- CRRC (empty weight = 280 pounds [fully inflated]).
- CRRC harness (Figure 11-4, page 11-14), nose strap (Figure 11-5, page 11-14), crow's feet (Figure 11-6, page 11-14, and Figure 11-7, page 11-15), and bow stiffener (optional).

Note. The CRRC harness must be maintained IAW TM 10-1670-201-23, *Organizational and Direct Support Maintenance Manual for General Maintenance of Parachutes and Other Airdrop Equipment*, especially when being used in salt water.

- Clevis assembly, medium, MS 70087-2, national stock number 4030-00-678-8562 (two each), only needed if using I bar of FRIES.
- Tape, adhesive, 2-inch wide.
- Webbing, cotton, 1/4-inch, 80-pound breaking strength.
- Cord, nylon, Type III, 550-pound breaking strength.
- Webbing, nylon, tubular, 1-inch.
- Webbing, nylon, tubular, 1/2-inch, 1,000-pound breaking strength.
- Padding, felt, cellulose, or suitable substitute.
- Energy-dissipating paper honeycomb.

27 February 2009 FM 3-05.210 11-13

Chapter 11

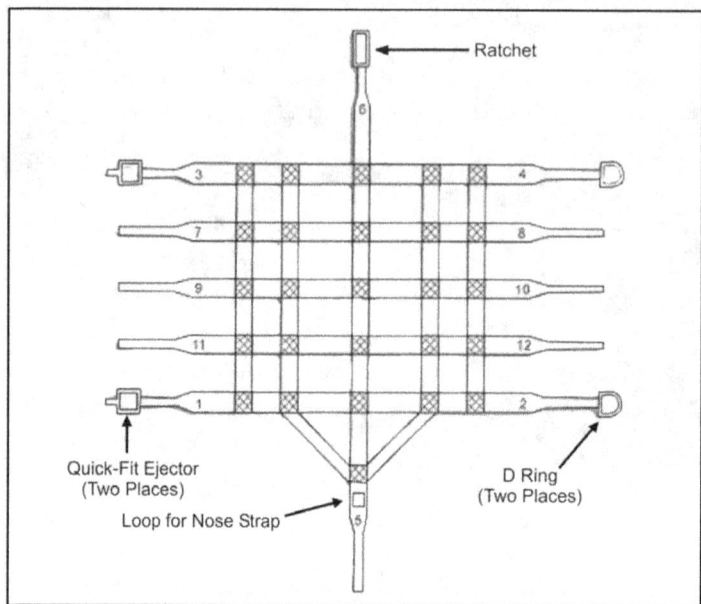

Figure 11-4. CRRC harness

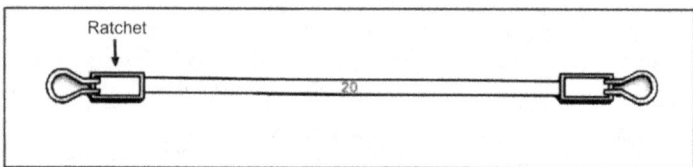

Figure 11-5. CRRC harness nose strap

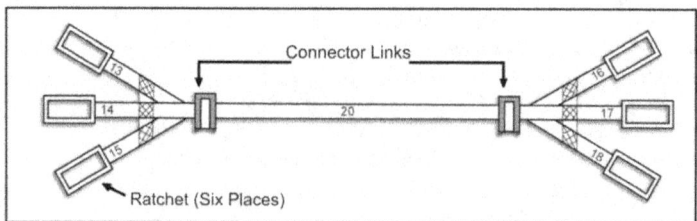

Figure 11-6. CRRC harness crow's feet

Air-Water Operations

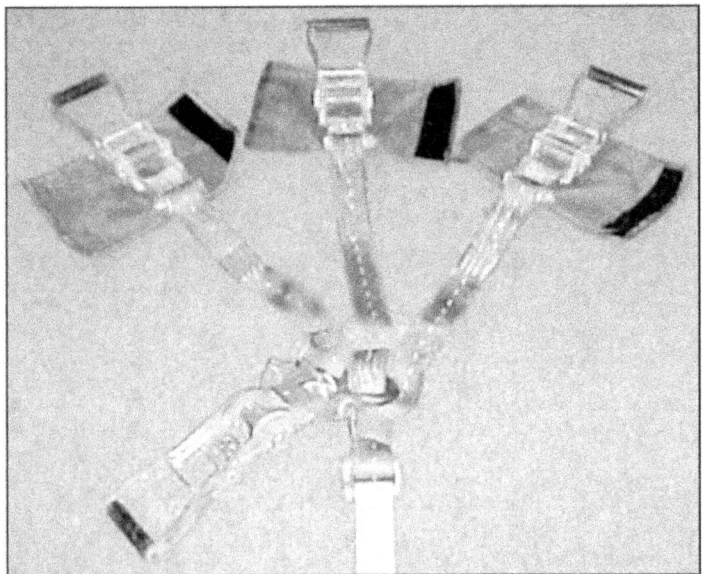

Figure 11-7. Crow's foot with bow strap ratchet attached

11-52. Two Soldiers, the hookup crew, can prepare and rig the CRRC with a cradle in 30 minutes. They and the crew chief perform the following preparations with the cradle method:
- Hookup crew installs the floor.
- Hookup crew fully inflates the CRRC.
- Hookup crew lays the cradle of the harness flat on the ground. It places the CRRC on the cradle such that the bow of the boat is at the triangular end of the cradle. Lines 1 and 2 should be at the front handles and lines 3 and 4 should be at the rear handles.

WARNING

Externally transporting the CRRC without the floor installed could cause the boat to fold up and possibly contact the helicopter rotors. Do not transport the CRRC without the floor installed. The optional floor extension is only necessary if traveling in excess of 130 knots.

- Hookup crew prepares and stows the engine, paddles, and accompanying load IAW FM 4-20.142. It must not store more than 200 pounds in the very front of the boat (beyond the end of the floor) with the bow stiffener installed. Without the wooden floor installed, they must not store more than 50 pounds in the bow.
- Hookup crew prepares and stows the motor arm up and straps the engine in place near the transom. It uses some suitable padding (honeycomb or tire) between the motor and floor, and pad the prop with honeycomb.
- Hookup crew places the paddles and gas tank in the proper positions and ties them down.

Chapter 11

- Hookup crew places the accompanying payload into the CRRC as close to the center as possible. It ensures all items are inside so only the CRRC is touching the aircraft. The hookup crew ties all large items—for example, rucksacks and water cans—with 1-inch tubular nylon line to the front towing ring. It then runs the line through each large item and ties off the line on the last item. It then ties the free end of the line to the transom. The hookup crew secures all small items to the floor by using snap links or by tying the small items to a larger item. The hookup crew pads items, as necessary.
- Hookup crew routes lines 1 and 2 over the sides of the CRRC and its payload. The hookup crew attaches lines 1 and 2 together and tightens them by adjusting the friction adapter of the quick-fit ejector. The hookup crew then folds and tapes the excess nylon to the quick-fit ejector. It repeats the process with lines 3 and 4.
- Hookup crew routes line 5 through the front towing ring and over the bow of the boat and line 6 over the transom. It attaches line 5 to line 6 and tightens the line with the ratchet. It then folds and tapes the excess nylon to the ratchet.
- Hookup crew routes the nose strap, line 20, through the loop located on line 5 at the nose of the boat and below the front towing ring. It ensures there is an even length of line 20 on each side of the loop. It then places the free ends of line 20 inside the boat in a way that allows Soldiers to easily access them once the helicopter has landed on top of the boat.
- Hookup crew removes one leg of the crow's feet section of the harness by detaching line 19 from one of the connector links.
- Crew chief inspects the helicopter after it lands near the CRRC. The crew chief dismounts after the helicopter is rigged and serves as a ground guide.
- Crew chief inflates the helicopter struts about 30 inches.
- Crew chief inspects the bottom of the helicopter and removes any antennas that may interfere with the CRRC.
- For UH/MH-60A helicopters with the fast-rope H bar installed: The hookup crew locates the two aft H bar rings at the center of the cargo doors and about 40 inches apart. It then routes line 19 through both H bar rings. It then re-attaches the removed leg of the crow's feet with the connector link. It ensures the two connector links are to the outside of the H bar rings.
- For UH/MH-60A helicopters with the FRIES I bar installed: The hookup crew ensures the I bar is in the stowed position. It locates the two release points on the I bar about 60 inches apart. It then removes the safety pins so that the release points are free to rotate fore and aft. It inserts a medium clevis assembly (G-12, MS 70087-2, national stock number 4030-00-678-8562) in each release point. It then routes line 19 through both medium clevises. It reattaches the removed leg of the crow's feet with the connector link. It ensures two connector links are to the outside of the clevises.

WARNING

The hookup crew must not route line 20 around the wheels of the helicopter. The hookup crew ensures line 20 is routed to the inside of the wheels.

- Crew chief "walks" the helicopter onto the CRRC. Helicopter pilot carefully lands on the CRRC. As the helicopter approaches, the hookup crew should lean on the CRRC to prevent it from moving because of the rotor wash. Once the helicopter is centered over the CRRC, the hookup crew may attach the CRRC to the aircraft.
- Hookup crew removes a small clevis and ratchet from the nose strap and attaches it to line 14 inside the triangular connection. It repeats this process with the other small clevis and ratchet from the nose strap and attaches line 14 to line 17.

- Hookup crew attaches lines 7, 9, and 11 of the cradle to lines 13, 14, and 15 of the crow's feet and tighten with the ratchets. It repeats with lines 8, 10, and 12 of the cradle and lines 16, 17, and 18 of the crow's feet. It inserts the free ends of line 20 into the ratchets attached to the small suspension clevises and the crow's feet (Figure 11-8).
- Hookup crew tightens ratchets until the nose of the boat is snug against the bottom of the helicopter. It folds and tapes excess nylon to the ratchets. The helicopter then lifts the boat, hovers, and carefully lands again.
- Hookup crew tightens all ratchets as much as possible. It folds and tapes excess nylon to the ratchets.

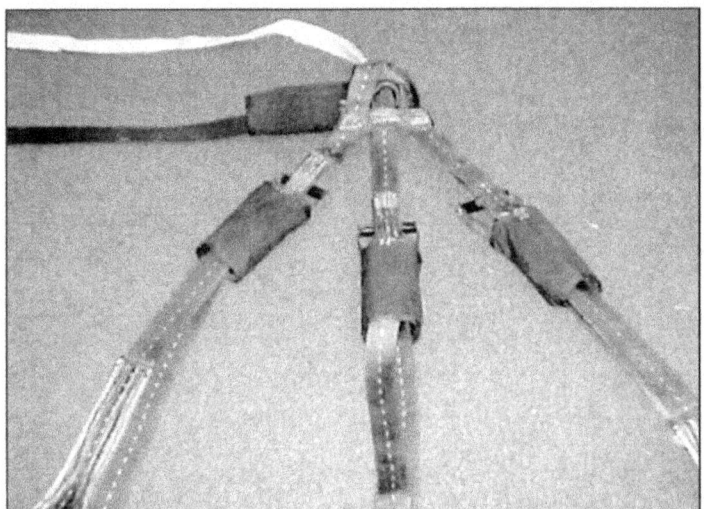

Figure 11-8. Nose strap and crow's feet connection

11-53. The hookup crew repeats the last three steps until the CRRC is rigged as tightly as possible against the bottom of the helicopter.

Harness Method

11-54. The U.S. Army Natick Research, Development, and Engineering Center developed and certified the harness method for the UH-60A helicopter. Units must obtain an airworthiness release from the U.S. Army Aviation Systems Command before externally transporting this load by air.

11-55. The hookup crew requires the following equipment and personnel for rigging the CRRC using the harness method:
- CRRC with 35-horsepower engine and two full 6-gallon fuel tanks. (CRRC with the engine and fuel weighs 475 pounds.)
- Fabricated sling assembly.
- Sling, aerial delivery, 12-foot, 3,000-pound minimum capacity, one each.
- Bow release strap, one each, fabricated.
- Belly support strap, two each, fabricated.
- Doughnut, 1-inch tubular nylon, about 1 foot in diameter. (Soldiers use doughnut to attach equipment and rucksacks into the CRRC.)
- Clevis assembly, large, MS 70087-3, one each.

Chapter 11

- Padding material, cellulose wadding or felt sheet.
- Tape, adhesive, pressure-sensitive, 2-inch-wide roll.
- Webbing, nylon, 1/2-inch tubular, 1,000-pound breaking strength.
- Tie-down strap, cargo, CGU/IB, one each.
- Energy-dissipating paper honeycomb, as required.
- Cord, nylon, Type III, 550-pound breaking strength.

Note. The personnel must maintain CRRC harness IAW TM 10-1670-201-23/NAVAIR 13-1-17, especially when being used in salt water.

- Four personnel to hook up the CRRC to the aircraft.

11-56. Two Soldiers can prepare and rig the CRRC using the harness method in 20 minutes. However, four Soldiers (hookup crew) are required to hook up this load. They perform the following steps:
- Before installing the floor panels, drill two 1/2-inch holes in both sides of the three rear floor panels. Run a 10-foot length of 1/2-inch tubular nylon webbing through both holes in each side of each floor panel. Tie an overhand knot to form a "Y".
- Install floor panels, and inflate the boat about 50 percent. Place a 12-foot sling under the boat about 6 feet from the rear of the boat with the ends of the sling coming up over the gunwales. This will be just forward of the second set of carrying handles from the rear.
- Route each end of the sling band through a loop end of the belly support straps (Figure 11-9). Attach the large clevis to the sling band. Route the belly support straps at the end of the friction adapter over the transom and under the boat. Ensure both straps are routed along the inside of the speed skegs and are between the sling band and the boat hull.

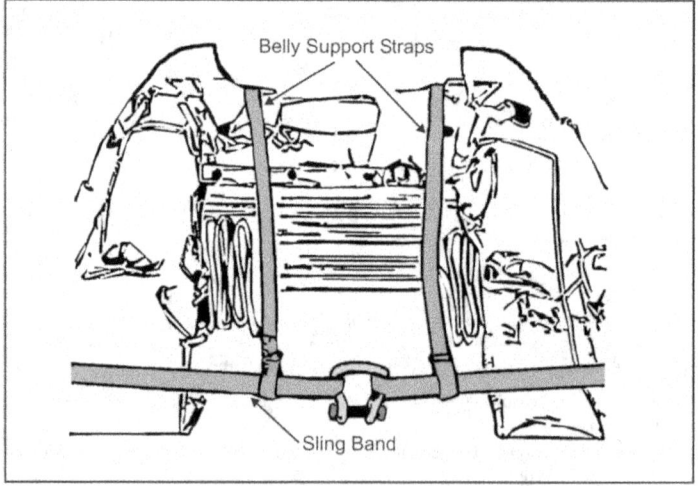

Figure 11-9. Belly support strap connected to sling band

- Tie the engine cover in place with two lengths of Type III nylon cord.
- Fold the operating handle. Tie a 4-inch by 9-inch piece of honeycomb between the handle and engine cover. Tape a 10-inch by 10-inch piece of honeycomb to the lower unit.

Air-Water Operations

- Place the fuel tanks against the transom. Set a 12-inch by 16-inch piece of honeycomb against the fuel tanks. Center the engine in the boat with its upper unit on the honeycomb (Figure 11-10).
- Secure the fuel tanks and motor with the 1/2-inch tubular nylon ties installed in the floor panels.
- Route one leg of the 1/2-inch tubular nylon around each end of the sling band to pull the sling closer to the boat. Ensure the belly support strap is between the tie and the clevis (Figure 11-11).

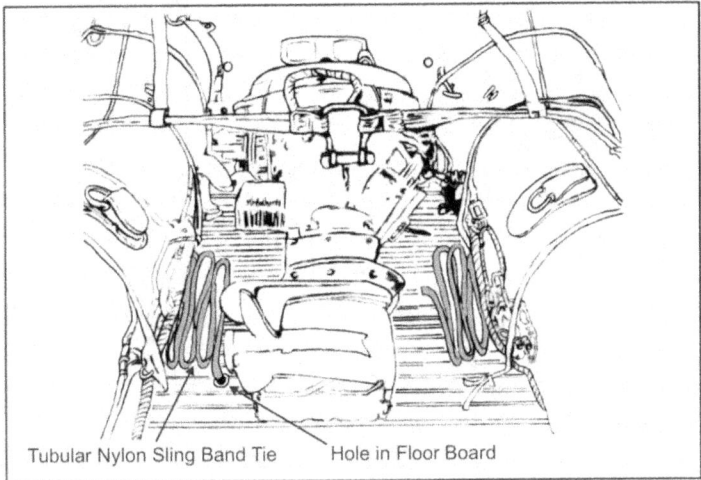

Figure 11-10. Fuel tank and engine in place

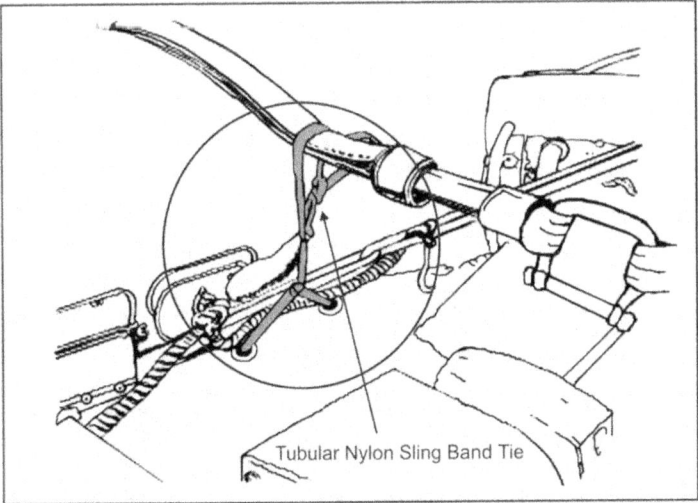

Figure 11-11. Sling band tie

- Attach the tubular nylon doughnut to the floor of the CRRC. Attach any equipment to the doughnut by using snap links and ropes.
- Wait for helicopter to land before performing hookup. Do not extend the struts of the helicopter to fit the boat underneath. The helicopter must not be flown with struts extended.
- Attach the CGU/IB tie-down strap to the forward towing loops of the aircraft, and ratchet as tightly as possible. Position the ratchet between the anchor point and the fuselage of the aircraft. Tape the ratchet, and tape the strap every 8 to 12 inches to reduce vibration.
- Remove small fin-shaped antenna from the underside of the helicopter, and cover the hole with tape.
- Position the boat under the aircraft, bow forward, and align the clevis with the cargo hook. Ensure the point of the bow is on centerline with the aircraft. Once underneath the helicopter, inflate the boat to as close to 100 percent as possible.
- Route the loop end of the bow release strap over the CGU/IB and around the clevis. Attach the clevis to the cargo hook. Route the ends of the bow release strap through the towing rings. Secure the end through the friction adapters of the belly support straps. To ensure the bow of the boat is against the belly of the helicopter, pull the bow release strap as tight as possible. Secure the running ends with tape or 1/4-inch cotton webbing.
- Once the aircraft is clear of the ground, a Soldier reaches through the hellhole and works the pump to inflate the boat to 100 percent.

Note. Upon departure, the pilot obtains a safe airspeed to determine how well the CRRC is riding before he accelerates to mission airspeed.

ERDS OPERATION

11-57. The CM or crew chief will monitor the load en route and keep the pilot informed as to the stability. The CM or crew chief will give corrections as to aircraft alignment with the drop area. The pilot not on the controls will call out the aircraft altitude at 10-foot intervals and GS at 10-knot intervals. The pilot will call out MARK when the aircraft is within the drop area and is no higher than 10 feet AGL and no faster than 10 knots. The pilot's visibility may become limited because of the spray from the water. The pilot not on the controls will turn on the wipers, if required.

11-58. Upon arrival at the drop site, the pilot initiates a progressive deceleration and descent. The CM or crew chief will release the CRRC by cutting the shear strap in the cabin area when at the proper location and call out RAFT AWAY. The pilot will maintain altitude and GS until the last swimmer has exited the aircraft. After deployment, the pilot terminates drop profile by increasing altitude and GS and attaining the appropriate mode of flight. The CASO notifies the pilot of the status of personnel before the aircraft departs the area.

CAUTION

Maximum airspeed en route is 130 knots without the optional floor extension and 145 knots with the floor extension installed.

WARNING

The CRRC will not be deployed if any personnel or safety boats are in the immediate vicinity of the drop area.

Air-Water Operations

11-59. If an emergency occurs, the pilot in command takes the appropriate action and makes the decision to jettison the CRRC. If the pilot decides to jettison, the CM or crew chief will jettison the CRRC only on command of the pilot in command.

ROLLED OR TETHERED DUCK OPERATIONS

11-60. Rolled or tethered duck operations provide the capability to deliver a team and CRRC into the water within the full fuel range of a UH/MH-60 helicopter. Because the boats are transported uninflated inside the helicopter, they do not produce drag, which would lessen the fuel range.

PERSONNEL QUALIFICATION REQUIREMENTS

11-61. Before the conduct of rolled or tethered duck operations, participants must receive formalized training in the procedures and rigging to be used during the CRRC operation. At a minimum, the training will include the following:
- Rigging and inspection of individual equipment, to include combat equipment.
- Rigging of the CRRC.
- Rigging of the aircraft (tethered duck operations only).
- Deployment of the CRRC from the aircraft.
- Employment of the CRRC in the water.

PERSONNEL DUTIES AND RESPONSIBILITIES

11-62. Rolled or tethered duck safety personnel consist of a CM and CASO. Tethered duck operations require the use of the helicopter crew chief. The CM—
- Ensures aircraft is properly rigged. Rolled duck requires no special rigging. Tethered duck rigging for aircraft is IAW Chapter 15. Fast rope is properly attached to the CRRC.
- Briefs all personnel on the operation.
- Oversees the loading of the aircraft.
- Directs the pilot to maneuver the aircraft into the proper position for deployment.
- Deploys the CRRC when in the drop area and cleared by the aircraft pilot in command.
- Aborts any part of the operation due to any unsafe condition.
- Conducts an inspection of personnel and equipment before boarding the aircraft.
- Inspects the rigging of the fast rope to the CRRC.

11-63. The crew chief disconnects the fast rope from the FRIES bar. He ensures the fast rope drops free of the aircraft.

CONDUCT OF ROLLED DUCK OPERATIONS

11-64. The rigging personnel prepare the CRRC as follows:
- Install the roll-up floor and carbon dioxide inflation system. Verify the carbon dioxide bottle is full before rigging. A full bottle weighs 48 pounds.
- Open all valves, and roll up boat to expel air in the flotation tubes. Unroll boat, and install the valve caps into the valves. Ensure the valve arrows are located in the inflate (orange) position.
- Prepare a military amphibious reconnaissance system 35-horsepower motor by wrapping cellulose wadding around the motor cover and the propeller. Secure the wadding with heavy-duty tape (100-mph tape). When necessary, substitute rags for cellulose wadding. This padding is more to protect the boat from cuts than to prevent impact damage to the motor.
- Prepare fuel bladder by wrapping it in three layers of cellulose wadding and placing the fuel bladder in an aviator's kit bag. Ensure the fuel bladder contains no more than 5 gallons of fuel. The fuel bladder could split if it is overly full.

Chapter 11

- Prepare paddles by drilling a 1/4-inch hole in the center of the grip. Route a loop of 550 cord through the grip. Tape the paddles together in a bundle.
- Stow foot pumps and hoses in pump pockets located in the CRRC.
- Fold in sides of the CRRC (in the same manner as rolling the boat for transport or storage). Place the motor crosswise on the CRRC against the transom with the motor cover on the port side tube, the lower end on the starboard side tube, and the stem brackets facing the bow.
- Locate the fuel bladder between the midsection shaft and the transom. Place the paddle bundle next to the motor. Route the motor cable through the paddle loops and the kit bag handles. Secure the paddle bundle to the transom plate with a snap link or threaded clevis.
- Fold CRRC from bow to stern (about three folds). The final fold will encase the motor between the folded boat and the transom.
- Fold the rearmost portion of the inflation tubes sideways. Tie the rolled boat with 1/4-inch tubular nylon by routing the tubular nylon around the boat and through the appropriate stem-towing ring. Repeat this procedure on the opposite side. Place the knots securing the rolled CRRC near the opening of the final fold of the CRRC. Tie the running ends of both loops together to form a "crossbar" above the opening of the final fold. (550 cord or 1/2-inch to 1-inch tubular nylon can be substituted for 1/4-inch tubular nylon.)
- For nighttime operations, mark the boat with a minimum of two chemical lights. Locate a chemical light of a different color on the tied crossbar. (Doing so helps the derigging team distinguish the top of the CRRC.)

11-65. Planning weight for a rolled duck is 400 pounds per duck. Rigging personnel prepare the helicopter for rolled duck operations as follows:

- Tape off any sharp edges that may interfere with the CRRC.
- Load the CRRC into the cargo bay; team members can use CRRC as a seat.

Note. Rigging personnel can load two rolled ducks into one UH-60 if troop seats and extended range fuel cells are removed.

11-66. Once the helicopter is located in the casting area and the CM determines the helicopter is at the proper height and speed, the team members launch the CRRC. The CM ensures the CRRC is in the water. The CM then gives the command for the cast personnel to exit the aircraft. Cast personnel assemble on the rolled CRRC. The CM conducts a headcount. Predesignated personnel then derig and inflate the rolled CRRC, attach the engine, and put CRRC into service. The team sergeant then accounts for all equipment and personnel and ensures all packing material is secured before departing the area.

Note. When launching the CRRC, no swimmers will exit the aircraft until the CRRC is in the water.

CONDUCT OF TETHERED DUCK OPERATIONS

11-67. The tethered duck is used for launching a CRRC in rough seas. The recommended height of the casting platform is 20 feet above water level. The recommended speed is 5 knots. Casters deploy into the water by sliding down the fast rope. The CRRC being tethered to the aircraft allows the cast personnel to easily assemble on the rolled CRRC. The tether also provides a positive path to the CRRC in rough seas or currents.

11-68. The rigging personnel prepare the CRRC for tethered duck operations in the same manner as described in paragraph 11-66. Rigging personnel also prepare the helicopter for tethered duck operations. They follow the steps listed in paragraph 11-67 and the steps below:

- Connect the ground end of the fast rope to the tied crossbar of the rolled duck. Use a sling rope routed through the extraction loop and then routed around the tied "crossbar." Coil the fast rope on top of the CRRC.

Air-Water Operations

- Connect the helicopter end of the fast rope to the aircraft IAW procedures outlined in Chapter 15. Do not use a safety rope with this procedure.
- Inspect the release mechanism of the FRIES bar, and ensure the crew chief is familiar with its operation.

Note. When launching the CRRC, no swimmers will exit the aircraft until the CRRC is in the water.

11-69. Cast personnel perform the following steps to launch the CRRC for tethered duck operations:
- Once in the casting area, the CM ensures the helicopter is at the proper height and speed, and team members deploy the CRRC.
- CM ensures the CRRC is in the water. CM then gives the command for the cast personnel to mount the fast rope and slide into the water. Once a caster enters the water, he follows the fast rope hand over hand to the CRRC.
- Cast personnel assemble on the rolled CRRC, and CM conducts a head count.
- Crew chief releases the fast rope once cast personnel are clear. Cast personnel disconnect the fast rope and allow it to sink (during training events safety boat picks up the fast rope).
- Predesignated personnel then derig and inflate the rolled CRRC, attach engine, and put CRRC into service. Team sergeant then accounts for all equipment, personnel, and ensures all packing material is secured before departing area.

RECOVERY OPERATIONS

11-70. Several techniques can be used to extract teams from the water. Some of these techniques—FRIES and SPIES—are almost the same techniques as previously discussed in this manual. Other techniques, such as ladder operations, are slightly different.

LADDER RECOVERY OPERATIONS

11-71. If a wire ladder is to be used for recovery, the CM secures it on the floor to a "wire doughnut" (must be 5/8-inch wire and secured in at least five points with snap links) for a UH-1H and to the floor cargo rings in UH-60s and CH-46/47s. When a single rotor aircraft is to be used for recovery, the CM lowers a wire ladder to the swimmers who are on-line at 50-meter intervals in the casting area. At night, each swimmer attaches an IR chemical light to the upper portion of his uniform or equipment. The IR chemical light is visible from above the water. As the aircraft flies over, the swimmers hook the lowest rung on the ladder with their leading arm and climb to a designated height where they hook up (with snap link and rope seat) to the ladder.

FRIES AND SPIES OPERATIONS

11-72. Recovery personnel may recover swimmers by FRIES or SPIES extraction. The swimmers don their harnesses before the arrival of the helicopter. The helicopter simply hovers over the grouped swimmers as they attach their harness to the extraction ropes. The aircraft then slowly makes a vertical ascent until the suspended personnel are about 100 feet above the surface.

WATER LANDING OPERATIONS

11-73. A CH/MH-47 helicopter must land in the water. Helicopters should try not to land in salt water during training operations due to aircraft maintenance.

11-74. When using a CRRC with motor, Soldiers should use a low silhouette to allow the CRRC coxswain the best possible view of the aircraft and to protect themselves from colliding with the aircraft ceiling. Additionally, one person on each side of the CRRC guides the boat into the aircraft. As the aircraft passes overhead to the pickup point, the CRRC will move in trail toward the ramp. When the crew chief lowers the ramp into the water and signals the CRRC, the operator will increase speed to penetrate the rotor wash and move up onto the ramp. Upon contact with the ramp, the CRRC operator cuts the engine power and raises the motor out of the water. As the boat touches the ramp, the two guides jump out, pull the boat into the aircraft, and tie the bowline of the CRRC to an interior hard point on the aircraft ASAP.

11-75. When recovery personnel do not use a motor, they lower a rope hooked to the aircraft winch that has a 10-pound padded weight attached. The aircrew lowers the rope behind the boat and drags it over the boat. The swimmers to be picked up secure the rope, and the winch pulls in the boat.

11-76. At night, recovery personnel attach a chemical light to the rope and weight. The crew chief marks the aircraft ramp door with a red chemical light on the left side and a green chemical light on the right side.

11-77. When recovery personnel recover only swimmers, the swimmers enter the aircraft by ladder or ramp. If the aircraft is in the water, the swimmers simply swim up to the ramp.

Appendix A
Weights, Measures, and Conversion Tables

Tables A-1 through A-5, pages A-1 and A-2, show metric units and their U.S. equivalents. Tables A-6 through A-15, pages A-2 through A-5, are conversion tables.

Table A-1. Linear measure

Unit	Other Metric Equivalent	U.S. Equivalent
1 centimeter	10 millimeters	0.39 inch
1 decimeter	10 centimeters	3.94 inches
1 meter	10 decimeters	39.37 inches
1 decameter	10 meters	32.81 feet
1 hectometer	10 decameters	328.08 feet
1 kilometer	10 hectometers	3,280.84 feet

Table A-2. Liquid measure

Unit	Other Metric Equivalent	U.S. Equivalent
1 centiliter	10 milliliters	0.34 fluid ounce
1 deciliter	10 centiliters	3.38 fluid ounces
1 liter	10 deciliters	33.81 fluid ounces
1 decaliter	10 liters	2.64 gallons
1 hectoliter	10 decaliters	26.42 gallons
1 kiloliter	10 hectoliters	264.17 gallons

Table A-3. Weight

Unit	Other Metric Equivalent	U.S. Equivalent
1 centigram	10 milligrams	0.15 grain
1 decigram	10 centigrams	1.54 grains
1 gram	10 decigrams	0.04 ounce
1 decagram	10 grams	0.35 ounce
1 hectogram	10 decagrams	3.53 ounces
1 kilogram	10 hectograms	2.20 pounds
1 quintal	100 kilograms	220.46 pounds
1 metric ton	10 quintals	1.10 short tons

Appendix A

Table A-4. Square measure

Unit	Other Metric Equivalent	U.S. Equivalent
1 square centimeter	100 square millimeters	0.16 square inch
1 square decimeter	100 square centimeters	15.50 square inches
1 square meter (centaur)	100 square decimeters	10.76 square feet
1 square decameter (are)	100 square meters	1,076.39 square feet
1 square hectometer (hectare)	100 square decameters	2.47 acres
1 square kilometer	100 square hectometers	0.39 square mile

Table A-5. Cubic measure

Unit	Other Metric Equivalent	U.S. Equivalent
1 cubic centimeter	1,000 cubic millimeters	0.06 cubic inch
1 cubic decimeter	1,000 cubic centimeters	61.02 cubic inches
1 cubic meter	1,000 cubic decimeters	35.31 cubic feet

Table A-6. Temperature

Conversion	Formula
Fahrenheit to Celsius	Subtract 32, multiply by 5, and divide by 9
Celsius to Fahrenheit	Multiply by 9, divide by 5, and add 32

Table A-7. Approximate conversion factors

To Change	To	Multiply By	To Change	To	Multiply By
Inches	Centimeters	2.540	Ounce-inches	Newton-meters	0.007
Feet	Meters	0.305	Centimeters	Inches	.394
Yards	Meters	0.914	Meters	Feet	3.280
Miles	Kilometers	1.609	Meters	Yards	1.094
Square inches	Square centimeters	6.451	Kilometers	Miles	0.621
Square feet	Square meters	0.093	Square centimeters	Square inches	0.155
Square yards	Square meters	0.836	Square meters	Square feet	10.764
Square miles	Square kilometers	2.590	Square meters	Square yards	1.196
Acres	Square hectometers	0.405	Square kilometers	Square miles	0.386
Cubic feet	Cubic meters	0.028	Square hectometers	Acres	2.471
Cubic yards	Cubic meters	0.765	Cubic meters	Cubic feet	35.315
Fluid ounces	Millimeters	29.573	Cubic meters	Cubic yards	1.308
Pints	Liters	0.473	Millimeters	Fluid ounces	0.034
Quarts	Liters	0.946	Liters	Pints	2.113
Gallons	Liters	3.785	Liters	Quarts	1.057
Ounces	Grams	28.349	Liters	Gallons	0.264
Pounds	Kilograms	0.454	Grams	Ounces	0.035
Short tons	Metric tons	0.907	Kilograms	Pounds	2.205
Pounds-feet	Newton-meters	1.356	Metric tons	Short tons	1.102
Pounds-inches	Newton-meters	0.113	Nautical miles	Kilometers	1.852

Weights, Measures, and Conversion Tables

Table A-8. Area

To Change	To	Multiply By	To Change	To	Multiply By
Square millimeters	Square inches	0.002	Square inches	Square millimeters	645.160
Square centimeters	Square inches	9.155	Square inches	Square centimeters	6.451
Square meters	Square inches	1,550.000	Square inches	Square meters	0.001
Square meters	Square feet	10.764	Square feet	Square meters	0.093
Square meters	Square yards	1.196	Square yards	Square meters	0.836
Square kilometers	Square miles	0.386	Square miles	Square kilometers	2.590

Table A-9. Volume

To Change	To	Multiply By	To Change	To	Multiply By
Cubic centimeters	Cubic inches	0.061	Cubic inches	Cubic centimeters	16.390
Cubic meters	Cubic feet	35.310	Cubic feet	Cubic meters	0.028
Cubic meters	Cubic yards	1.308	Cubic yards	Cubic meters	0.765
Liters	Cubic inches	61.020	Cubic inches	Liters	0.016
Liters	Cubic feet	0.035	Cubic feet	Liters	28.320

Table A-10. Capacity

To Change	To	Multiply By	To Change	To	Multiply By
Milliliters	Fluid drams	0.271	Fluid drams	Milliliters	3.697
Milliliters	Fluid ounces	0.034	Fluid ounces	Milliliters	29.573
Liters	Fluid ounces	33.814	Fluid ounces	Liters	0.030
Liters	Pints	2.113	Pints	Liters	0.473
Liters	Quarts	1.057	Quarts	Liters	0.946
Liters	Gallons	0.264	Liters	Gallons	3.785

Table A-11. Statute miles to kilometers and nautical miles

Statute Miles	Kilometers	Nautical Miles	Statute Miles	Kilometers	Nautical Miles
1	1.61	0.87	60	96.60	52.16
2	3.22	1.74	70	112.70	60.85
3	4.83	2.61	80	128.80	69.55
4	6.44	3.48	90	144.90	78.24
5	8.05	4.35	100	161.00	86.93
6	9.66	5.22	200	322.00	173.90
7	11.27	6.08	300	483.00	260.80
8	12.88	6.95	400	644.00	347.70
9	14.49	7.82	500	805.00	434.70
10	16.10	8.69	600	966.00	521.60
20	32.20	17.39	700	1127.00	608.50
30	48.30	26.08	800	1288.00	695.50
40	64.40	34.77	900	1449.00	782.40
50	80.50	43.47	1000	1610.00	869.30

Appendix A

Table A-12. Nautical miles to kilometers and statute miles

Nautical Miles	Kilometers	Statute Miles	Nautical Miles	Kilometers	Statute Miles
1	1.85	1.15	60	111.00	69.00
2	3.70	2.30	70	129.50	80.50
3	5.55	3.45	80	148.00	92.00
4	7.40	4.60	90	166.50	103.50
5	9.25	5.75	100	185.00	115.00
6	11.10	6.90	200	370.00	230.00
7	12.95	8.05	300	555.00	345.00
8	14.80	9.20	400	740.00	460.00
9	16.65	10.35	500	925.00	575.00
10	18.50	11.50	600	1110.00	690.00
20	37.00	23.00	700	1295.00	805.00
30	55.50	34.50	800	1480.00	920.00
40	74.00	46.00	900	1665.00	1035.00
50	92.50	57.50	1000	1850.00	1150.00

Table A-13. Kilometers to statute and nautical miles

Kilometers	Statute Miles	Nautical Miles	Kilometers	Statute Miles	Nautical Miles
1	0.62	0.54	60	37.28	32.40
2	1.24	1.08	70	43.50	37.80
3	1.86	1.62	80	49.71	43.20
4	2.49	2.16	90	55.93	48.60
5	3.11	2.70	100	62.14	54.00
6	3.73	3.24	200	124.28	108.00
7	4.35	3.78	300	186.42	162.00
8	4.97	4.32	400	248.56	216.00
9	5.59	4.86	500	310.70	270.00
10	6.21	5.40	600	372.84	324.00
20	12.43	10.80	700	435.00	378.00
30	18.64	16.20	800	497.12	432.00
40	24.85	21.60	900	559.26	486.00
50	31.07	27.00	1000	621.40	540.00

Table A-14. Yards to meters

Yards	Meters	Yards	Meters	Yards	Meters
100	91	1000	914	1900	1737
200	183	1100	1006	2000	1829
300	274	1200	1097	3000	2743
400	366	1300	1189	4000	3658
500	457	1400	1280	5000	4572
600	549	1500	1372	6000	5486
700	640	1600	1463	7000	6401
800	732	1700	1554	8000	7315
900	823	1800	1646	9000	8230

Table A-15. Meters to yards

Meters	Yards	Meters	Yards	Meters	Yards
100	109	1000	1094	1900	2078
200	219	1100	1203	2000	2187
300	328	1200	1312	3000	3281
400	437	1300	1422	4000	4374
500	547	1400	1531	5000	5468
600	656	1500	1640	6000	6562
700	766	1600	1750	7000	7655
800	875	1700	1859	8000	8749
900	984	1800	1969	9000	9843

This page intentionally left blank.

Appendix B
Moon Phases

In planning night air operations, knowledge of the various moon phases and the light levels during each phase is necessary. The moon phases influence the tides, which could influence planning for air-water operations.

B-1. The moon revolves eastward about the earth; however, its rotational speed is slower than that of earth and it appears to move east to west. A complete revolution around the earth requires 29 days, 12 hours, 44 minutes, and 28 seconds. Because the time required for a revolution around the earth does not vary, the same side of the moon is always exposed to earth observers. Since the orbital plane of the moon is tilted 5 degrees 9 minutes toward the orbital plane of the earth, its orbit is closer to the northern hemisphere during winter months. As a result, moonlight is brighter in the winter than in the summer.

B-2. As the moon revolves on a vertical arc, the distance from a stationary point to the moon varies as the moon moves on its easterly orbit. This distance is referred to as the altitude of the moon and is one of the most important factors influencing night illumination. The light level, light produced by natural sources of skylight combined with moonlight, is determined by the phase angle and altitude of the moon.

B-3. The constant change in the phase angle of the moon causes varying levels of light received from the moon. At low altitude, the vertical component of moonlight incident to a horizontal surface is small compared to that at high altitude. At lower altitudes light is further reduced by the relatively long distance it travels through the earth's atmosphere. As the moon ascends in the sky, the distance light travels through the atmosphere decreases and the vertical component of moonlight increases, thus providing greater illumination. The greatest light level is achieved when the moon is directly overhead.

B-4. Because the rotation of the moon never changes and follows an exact time frame, timetables for each moon phase (new moon, first quarter, full moon, and last quarter) can be accurately computed for any year. The Air Weather Service normally provides these timetables. Geographical location is not a consideration in computing moon phases.

B-5. Figure B-1, page B-2, shows the phases of the moon as seen from the earth. The following paragraphs explain each phase:

- *New Moon.* The phase angle of a new moon begins at the 180-degree position and extends to the 90-degree position. It occurs when the moon rotates to a position between the sun and the earth. This phase always begins during the day and is not visible at night. Visual observation of the new moon at night is not apparent until the moon has reached the 173-degree position, and this takes about 2 days. The time required to complete this phase is about 8 days. During the first portion (5 days) of the new phase, a low light level will exist. As the phase progresses, illumination increases and light conditions will reach mid light level. Moonrise will occur during the day, and moonset will occur before midnight. About 40 to 50 percent of the moon will be illuminated at the end of the new moon phase. Night air operations conducted during this phase can anticipate that a low light level will prevail most of the time. Best light conditions will exist shortly after darkness when the moon is at its highest observable altitude during the night.
- *First Quarter.* The phase angle of the moon at the first quarter begins at the 90-degree position. During this quarter, more than one-half of the moon surface but not all the apparent disk is illuminated. About 4 days are required to complete the first quarter phase. During the first days of the first quarters and the last days of the new moon (about 5 days), the light will be in the mid light range with increasing intensity to a high light level (about 3 days) toward the end. Moonrise occurs during daylight near the end of the day. Moonset changes from midnight to the early morning hours. The best time for conducting air operations will normally be about

Appendix B

midnight when the moon is at its highest altitude. Light intensity is becoming brighter during this moon phase. When the moon is low on the horizon, flights toward the moon should be avoided.

Orbital paths of the earth and moon are traced through one lunar month as if viewed from a point north of the plane of the solar system. The path of the moon is exaggerated for clarity. Actually the path is always concave to the sun.

Siderial month (27 1/2 days) is the time it takes the moon to travel around the earth and realign (arrow "C") with the same star aligned at arrow "A."

Synodical month (29 1/2 days) is the time it takes the moon to travel around the earth and realign with the sun (arrow "B"). This period, also called a lunation, begins with the new moon.

Arrow "A" aligns earth and moon with the sun 93,000,000 miles away. It also aligns earth and moon with a star at infinite distance.

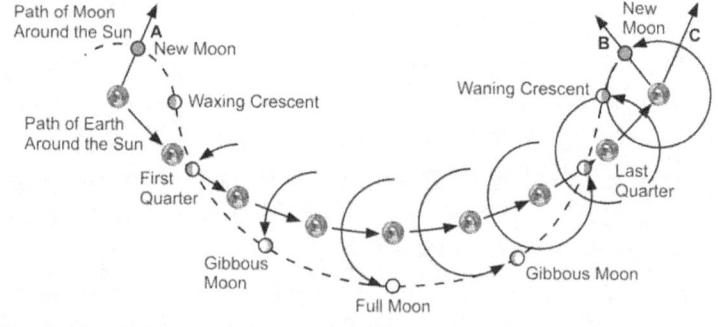

Figure B-1. Moon phases

- *Full Moon.* A full moon occurs when the sun, earth, and moon are aligned at the 0-degree position. At this time, the moon is radiating its greatest illumination. The full moon phase includes about 3 days before and 3 days after the full moon. High light conditions begin during the last days of the first quarter and extend to the first days of the last quarters (about 12 days). During the early part of the phase, moonrise occurs just before nautical twilight and progressively increases into the hours of darkness. Moonset will occur during the early morning daylight hours. Because of the intense brightness of the moon when it is at low altitude, air operations should be avoided. The optimum time for conducting flights will be the first few hours after midnight.
- *Last Quarter.* This phase of the moon is very similar to the first quarter but in reverse sequence. It begins when a portion less than the entire disk is visible and ends when only one-half of the moon is visible. The last quarter will normally last about 5 days. Light will decrease from a high light level (about 5 days). Moonrise will occur after midnight. The most desirable time for conducting night air operations is just before beginning morning nautical twilight (BMNT). Moonset occurs during the daylight hours.
- *Transition Phase.* Although there is no term that describes the period following the last quarter, there is a period of about 7 days after the end of the last quarter before the new moon phase begins. This period is similar to the new moon phase but in reverse order. Illumination of the moon decreases from half to no visual form. Moonrise occurs a few hours before BMNT, and moonset will always be during the daylight hours. Light will vary from a mid light level during the first few days (about 3 days) to a low light condition (about 4 days). To achieve any benefit from the moon illumination, air operations must be conducted 2 to 3 hours before BMNT. The longest period of low light conditions exists from the transition phase to the first quarter. During this period, there are about 16 days when the moon is less than one-half illuminated and is visible less than 50 percent of the hours of darkness.

B-6. The rhythmic rise and fall of the waters of the earth (tides) is caused mainly by the pull of the gravity of the moon (Figure B-2). The gravity of the sun also influences tides, but because of its very great distance from the earth, it only modifies the effect of the moon.

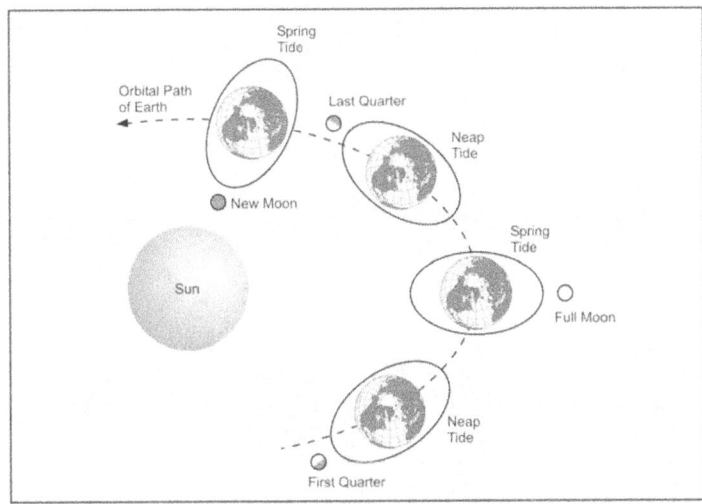

Figure B-2. Tides as related to moon phases

Appendix B

B-7. At times of new and full moon, the sun and moon align with the earth. Their combined forces produce the two high monthly tides called spring tides (not related to the season). During the first and the last quarter, the sun and moon are at right angles to the earth. This position creates the two low monthly tides, or neap tides.

B-8. While the moon pulls water to a bulge on its side of the earth, centrifugal force causes a bulge of water on the opposite side. The rotation of the earth beneath these bulges produces daily high and low tides.

B-9. The earth rotates daily about an axis through its own center. The earth-moon system rotates monthly about its center of gravity, but its center of gravity is not in the center of the earth. Thus, the earth wobbles about this eccentric point, causing a centrifugal force in the opposite direction from the moon.

Appendix C
Reports and Requests

The current SAVSERSUP governs procedures and formats for communications within SF. The formats for messages that pertain to air operations are contained in this Appendix. For additional message formats or for clarification and examples of these message formats, refer to the current SAVSERSUP. The below-listed message formats are included:
- ALVAR—Air/Maritime Mission Departure Report.
- BREAD—Operations Schedule and Exfiltration Report.
- COVER—Tactical Air Support Mission Request.
- DOUGH—Operations Schedule and Support and Resupply.
- DUMAS—Air/Maritime Mission Abort Report.
- EXCEL—Air Postmission Report.
- FABLE—LZ, DZ, recovery zone (RZ), BLS, or HLZ Postmission Report.
- GLASE—Recovery Mission Request.
- GRAIN—Operations Schedule and Infiltration Report.
- GRAZE—DZ, LZ, or RZ Survey Report.
- HELIX—HLZ Survey Report.
- JAVIS—DZ, LZ, or RZ Mission Request.
- PACER—Tactical Air Support Mission Confirmation.
- RINGO—DZ, LZ, or RZ Mission Confirmation.
- SHEAT—Beacon Bombing Request.
- SITED—LZ, DZ, RZ, BLS, or HLZ Approval and Disapproval.

ALVAR—AIR/MARITIME MISSION DEPARTURE REPORT

C-1. The air/maritime support unit will submit a mission departure report to HQ as soon as practical. If the departure location is at an AOB or launch base, the SFODA liaison officer or area specialty team (AST) should submit the report to the forward operating base (FOB) ASAP. The lines of the report contain the following information:
- Line 1. Mission designator and detachment code name.
- Line 2. Departure time.
- Line 3. Additional information.

BREAD—OPERATIONS SCHEDULE AND EXFILTRATION REPORT

C-2. The FOB transmits the operations schedule and exfiltration (OPSKED 5) to the subordinate operational bases (AOBs), command and control headquarters (SOCCE) and SF liaison cells as soon as practical following completion of the supporting unit mission planning. SOA assets require 48 hours for mission planning purposes and the USAF assets require 72 hours before the TOT. The FOB should follow the exfiltration report with a RINGO or DONOR report. If a RINGO or DONOR report is required, the subordinate base may include approved GRAZE, SMOKE, and GRANT reports to support operations. The subordinate base should receive notification from the FOB NLT 24 hours before the TOT. The FOB will send an ALVAR or DUMAS report ASAP to alert the subordinate base. The SFODA will send all

Appendix C

exfiltration postmission reports (FABLE and EXTON) ASAP (within 4 hours). The FOB will send follow-up BREAD reports concerning the same mission by using a sequential numbering system, which indicates a subsequent report (for example, BREAD 2). The SFODA code name and mission designator number (line AAA) will always be included. The lines of the report contain the following information:

- Line 1. SFODA code name and mission designator number.
- Line 2. Scheduled arrival time at airfield, intermediate staging base (ISB), or AOB.
- Line 3. Scheduled departure location (if different than above) and arrival and departure DTG.
- Line 4. Type of exfiltration platform, number of platforms, and number of personnel to be picked up for exfiltration.
- Line 5. Primary DZ, LZ, BLS, or HLZ code name.
- Line 6. Primary DZ, LZ, BLS, or HLZ location and DTG TOT.
- Line 7. Alternate DZ, LZ, BLS, or HLZ code name.
- Line 8. Alternate DZ, LZ, BLS, or HLZ location and DTG TOT.
- Line 9. Additional information.

COVER—TACTICAL AIR SUPPORT MISSION REQUEST

C-3. Units will pass requests for tactical air support through appropriate channels as quickly as possible. HQ will evaluate each request and, if approved, effect necessary coordination to provide tactical air (TACAIR) support through available AOR air resources. Should the tactical situation dictate units send emergency requests for rapid response, units may forward requests for support through the area tactical air control system. The requesting units enter the request net at the most convenient level and request immediate relay to the tactical air control center. The established format meets the requirements, to include item identifier, of the current AOR TACAIR request form. Senders may omit lines not applicable (NA). The lines of the message contain the following information:

- Line 1. Priority (emergency or routine).
- Line 2. Type mission:
  - BRAVO—Interdiction.
  - CHARLIE—CAS.
  - ECHO—Escort.
- Line 3. Results desired:
  - ALPHA—Destroy.
  - BRAVO—Neutralize.
  - CHARLIE—Harass.
- Line 4. Target location and description.
- Line 5. Requested TOT or NLT TOT.
- Line 6. Position of friendlies (if required, otherwise omit).
- Line 7. Forward air guide call sign and frequency (if appropriate).
- Line 8. Run-in heading (azimuth of desired attack angle in degrees).
- Line 9. Additional information.

DOUGH—OPERATIONS SCHEDULE AND SUPPORT AND RESUPPLY

C-4. The FOB transmits operations schedule and support and resupply (OPSKED 4) report to the subordinate operational bases (AOB), command and control headquarters (SOCCE), and SF liaison cells as soon as practical following the supporting unit's completion of the air and/or naval mission planning. SOA assets require 48 hours for mission planning purposes, and the USAF assets require 72 hours before the TOT. The subordinate AOB or SOCCE should receive notification NLT 24 hours before the TOT to synchronize and deconflict operations. The FOB should follow the resupply report with a confirmation report (RINGO or DONOR). The FOB sends an ALVAR or DUMAS report ASAP to alert the subordinate

Reports and Requests

base. If the resupply mission extends the SFODA in the AO for a follow-up mission, the FOB also sends a fragmentary order (SHORT report). The SFODA sends all resupply postmission reports (FABLE and EXTON) within 6 hours. The FOB sends follow-up DOUGH reports concerning the same mission by using a sequential numbering system, which indicates a subsequent report (for example, DOUGH 2). Senders must always include the SFODA code name and mission designator number (line AAA). The lines of the report contain the following information:

- Line 1. Mission designator and SFODA code name.
- Line 2. Scheduled arrival time at departure airfield or ISB.
- Line 3. Departure location and scheduled arrival and departure DTG.
- Line 4. Type platform, number of platforms, and number of personnel to be picked up for exfiltration.
- Line 5. Type of resupply:
  - Automatic (AUTOM).
  - On-call (ONCAL).
  - Emergency (EMERG).
  - Cache (CACHE).
- Line 6. Resupply contents. (Use MARGE/NANCY bundle codes.)
- Line 7. Primary DZ, LZ, BLS, RZ, or HLZ code name.
- Line 8. Primary DZ, LZ, BLS, RZ, or HLZ location and DTG TOT.
- Line 9. Alternate DZ, LZ, BLS, RZ, or HLZ code name.
- Line 10. Alternate DZ, LZ, BLS, RZ, or HLZ location and DTG TOT.
- Line 11. Point of no return. Based on the following mission profile criteria, use UTM coordinates and DTG:
  - Round-trip flight—Decision point at which flight commander must verify mission accomplishment based on fuel consumption for the mission.
  - One-way flight—Decision point at which flight commander cannot return to base and must continue to the objective area or next refuel point.
- Line 12. En route checkpoints or designated phase lines.
- Line 13. En route abort criteria.
- Line 14. Additional information.

DUMAS—AIR/MARITIME MISSION ABORT REPORT

C-5. The air/maritime support units will submit a mission abort report to HQ ASAP after determination that a mission will be delayed/not accomplished/cancelled due to circumstances rising during the prelaunch or launch phase. The lines of the report contain the following information:

- Line 1. Mission support designator and detachment code name.
- Line 2. Reason for abort.
- Line 3. Present status of aircraft/ship and load (if applicable).
- Line 4. Intentions.
- Line 5. Additional information.

EXCEL—AIR POSTMISSION REPORT

C-6. The air support (Air Force special operations base [AFSOB]) unit will submit a postmission report to the HQ ASAP following recovery of the mission aircraft or when mission results become known, whichever occurs first. The lines of the report contain the following information:

- Line 1. Mission designator and/or call sign. For multiple aircraft flights, list all aircraft call signs.
- Line 2. Target identification/location. (Target name, line number, and coordinates—UTM coordinates supplemented by geographic coordinates when warranted.)

Appendix C

- Line 3. TOT. If applicable, enter total time in search/recon after TOT. Report all times by DTG using Greenwich mean time ZULU.
- Line 4. Results of mission:
  - GREEN—Mission accomplished IAW plan.
  - YELLOW—Mission accomplished with limiting factors; explain.
  - RED—Mission not accomplished; explain.
- Line 5. Crew evaluation of results:
  - Number of personnel and/or pounds of cargo infiltrated or exfiltrated.
  - Degree of target destruction:
    - BLACK—Not attempted.
    - BROWN—00 to 25 percent destroyed.
    - FLESH—26 to 50 percent destroyed.
    - WHITE—51 to 75 percent destroyed.
    - AMBER—76 to 100 percent destroyed.
  - Photo target coverage.
  - Leaflets delivery coverage.
- Line 6. Additional information. Information not mentioned in the above items:
  - Significant sightings: unusual or new enemy equipment, troop concentrations, and so on.
  - Enemy aircraft: number, type, markings, armament, and tactics.
  - Antiaircraft fire.
  - Surface-to-air missiles.
  - Weather, conditions over target, and conditions along the route.
  - Essential elements of information.

FABLE—LZ, DZ, RZ, BLS, OR HLZ POSTMISSION REPORT

C-7. Outstations will submit postmission reports to the base at the first scheduled contact following an air or maritime support operation. Air or maritime support operations will be classified as successful only if all personnel and equipment were airdropped or airlanded on prescribed DZ, LZ, or BLS or all personnel and equipment to be removed were successfully extracted from prescribed RZ or BLS. Additionally, all supplies delivered to the SFOD via airdrop, airland, or maritime mission must have arrived in a usable condition. The lines of the report contain the following information:

- Line 1. Mission support designator and/or code name of LZ, DZ, RZ, BLS, or HLZ.
- Line 2. Results of mission:
  - GREEN—Mission accomplished IAW plan.
  - YELLOW—Mission accomplished with limiting factors (injured personnel, broken or lost equipment). Explain limiting factors. For LZ or DZs, identify unserviceable items or injured personnel.
  - RED—Mission not accomplished (abort, delay, and so on). Explain why mission aborted.
- Line 3. If personnel and/or equipment are not on proper DZ, LZ, or BLS, give distance and direction from intended infiltration or exfiltration point (point of impact).
- Line 4. Call sign of aircraft.
- Line 5. Line number or mission number (if different from line AAA).
- Line 6. Time of drop.
- Line 7. Type of drop (for example, CARP, GMRS, and VIRS).
- Line 8. Type of load and number of troops involved.
- Line 9. Current weather (estimated ceiling and visibility, surface winds, and new MEWs).
- Line 10. Additional information.

GLASE—RECOVERY MISSION REQUEST

C-8. The requesting party will submit the recovery mission request ASAP in advance of the requested TOT. The request will indicate a primary and alternate RZ, if an alternate RZ is available. If the RZ has not been previously reported, a GRAZE (DZ, LZ, or RZ) report must precede or accompany the mission request. When a kit (surface-to-air recovery or Fulton recovery system) is dropped, package or personnel will be recovered 20 minutes later unless otherwise specified in the mission and confirmation message. If the requesting party does not make the recovery time at the primary or alternate RZ, the aircrew will return to the staging base. Requesting party must resubmit their request if the party still desires a recovery. This format applies to message/materiel pickup missions also. The lines of the report contain the following information:

Note. The Fulton recovery system is currently not an option, since the USAF is currently not training this system.

- Line 1. Mission designator and code name (primary RZ).
- Line 2. DTG for recovery request (primary).
- Line 3. Authentication data (primary RZ). If a kit is to be dropped, a single panel or a single flashing light is positioned in the center of the RZ.
- Line 4. Total weight in pounds of recovery kits needed and type of package to be recovered. (None indicates a kit is with the outstation, or not needed. One indicates a kit must be dropped.)
- Line 5. RZ code name for alternate RZ.
- Line 6. DTG for recovery request (alternate RZ).
- Line 7. Authentication data (alternate RZ).
- Line 8. Offload airfield (spell out if USAF code is not available).
- Line 9. Special handling.
- Line 10. Additional information.

GRAIN—OPERATIONS SCHEDULE AND INFILTRATION REPORT

C-9. The FOB transmits the infiltration schedule 2 to the subordinate operational bases (AOB or ISB), C2 HQ (SOCCE), and SF liaison cells as soon as practical following completion of the air/maritime mission brief to the SFODA. SOA assets require 48 hours for mission planning purposes, and the USAF assets require 72 hours before the TOT.

C-10. For coordination, the subordinate operational base should receive notification from the FOB NLT 24 hours before the TOT. The FOB should follow the infiltration report with a RINGO, DONOR, or BRAKE report.

C-11. The FOB will send an ALVAR or DUMAS report ASAP to alert the subordinate base. The SFODA sends all infiltration postmission reports (FABLE and EXTON) within 6 hours. The FOB sends follow-up GRAIN reports concerning the same mission by using a sequential numbering system, which indicates a subsequent report (GRAIN 2). The SFODA code name and mission designator number (line AAA) will always be included. The lines of the report contain the following information:

- Line 1. Mission designator number and SFODA code name.
- Line 2. Scheduled arrival time at departure airfield or ISB.
- Line 3. Departure location, scheduled arrival, and departure DTG.
- Line 4. Type and number of platforms and number of personnel on SFODA.
- Line 5. Primary DZ, LZ, BLS, or HLZ code name.
- Line 6. Primary DZ, LZ, BLS, or HLZ location and DTG TOT.
- Line 7. Alternate DZ, LZ, BLS, or HLZ code name.
- Line 8. Alternate DZ, LZ, BLS, or HLZ location and DTG TOT.

Appendix C

- Line 9. Point of no Return. Based on the following mission profile criteria, use UTM coordinates and DTG.
  - Round-trip flight—Decision point at which flight commander must verify mission accomplishment based on fuel consumption for the mission.
  - One-way flight—Decision point at which flight commander cannot return to base and must continue to the objective area or next refuel point.
- Line 10. En route checkpoints or designated phase lines.
- Line 11. En route abort criteria.
- Line 12. Additional information.

GRAZE—DZ, LZ, OR RZ SURVEY REPORT

C-12. Elements will survey and report as soon as practical proposed DZs, LZs, and RZs during the course of operations. Elements use the format below. The FOB will approve the GRAZE report with a SITED report. If the LZ that was surveyed is to be used for rotary-wing operations only, elements use the HELIX report. The lines of the report contain the following information:

- Line 1. Code name and type. Elements include the word "RESUP" after the code name of DZ, if DZ can only be used for resupply. Elements use the following indicators for type:
  - TIGER—Personnel DZ.
  - BRAVE—Resupply DZ.
  - MOUSE—Water DZ.
  - PLANK—Fixed-wing LZ.
  - CAMEL—Rotary-wing LZ.
  - RISER—RZ.
- Line 2. Location of DZ:
  - Use complete military grid coordinates to nearest 100 meters of center of DZ and latitude or longitude to nearest 100 yards of center for ocean DZ. For inland water DZs, elements will use grid coordinates.
  - If an area DZ, use the coordinates of Points A and B, which establish the flight path.
  - Elevation.
- Line 3. Reference point:
  - Use landmarks clearly shown on issued map or chart.
  - Report reference points by magnetic azimuth, description, and distance in kilometers from the center of DZ.
  - Measure timing point in meters from PI.
  - Measure timing point in meters from centerline.
- Line 4. Width, length, and long axis of DZ:
  - Report width and length in meters and long axis by magnetic azimuth.
  - If an area DZ, omit this item.
- Line 5. Open quadrant:
  - If open 360 degrees, report open.
  - Measure open quadrant from center of zone and report as a series of azimuths in magnetic degrees. The open quadrant indicates acceptable approaches.
- Line 6. Track. The track is the recommended magnetic azimuth on which the aircraft is to fly when executing the drop. Should the circumstances dictate a required track, the symbol "RQR" will precede the azimuth (if not otherwise stated in follow-up message). The aircraft will fly the required track within 15 degrees of either side of the track.

- Line 7. Obstacles:
  - Report by description, magnetic azimuth, and distance from the center of the DZ any artificial obstacles over 90 meters in height above the level of the DZ within a radius of 5 NM that are not shown on the issued map.
  - If there are no obstacles, omit this item.
- Line 8. Additional information.

HELIX—HLZ SURVEY REPORT

C-13. The helicopter landing report is a brief, concise format to be used during insertions, extractions, MEDEVACs, and other helicopter operations. The outstation sends this report. The lines of the report contain the following information:
- Line 1. Code name (operational element selects code name).
- Line 2. Location (6-digit UTM coordinates).
- Line 3. LZ size in meters.
- Line 4. Wind direction (cardinal directional wind is blowing from) and velocity in knots.
- Line 5. Recommended approach heading (magnetic).
- Line 6. Open quadrant(s). Measure open quadrant from center of HLZ and report as a series of azimuths in magnetic degrees. The open quadrant indicates acceptable aircraft approaches.
- Line 7. Soil composition.
- Line 8. Obstacles: type, height in feet, azimuth and distance from center of HLZ.
- Line 9. Authentication used by LZ security element (smoke, strobe, panel).

Note. Subject to change upon receipt of the SITED message.

JAVIS—DZ, LZ, OR RZ MISSION REQUEST

C-14. The DZ, LZ, or RZ mission request will include primary and alternate (if available) DZ, LZ, or RZ within the support capabilities of the requestor. If mission planners designate an alternate in the request, it will be manned. If the DZ, LZ, or RZ on which a mission is requested has not been reported previously, the requestor must submit a GRAZE report in addition to the air mission request. The requesting party will submit the recovery mission request as far in advance of the requested TOT as possible. When the aircrew drops a recovery kit (STAR), they will recover the package or personnel 20 minutes later unless otherwise specified in the mission confirmation (RINGO) message. If the aircrew does not accomplish the recovery on time, the aircrew will return to the staging base. If the requesting party still desires recovery, it must then submit another recovery mission request. The lines of the report contain the following information:
- Line 1. Code name and type using the following indicators:
  - TIGER—Personnel DZ.
  - BRAVE—Resupply DZ.
  - MOUSE—Water DZ.
  - PLANK—Fixed-wing LZ.
  - CAMEL—Rotary-wing LZ.
  - RISER—RZ.
- Line 2. DTG request (primary).
- Line 3. Authentication data:
  - Authentication procedures will be IAW SOI (day letter code) and indicated by inserting the word "STANDARD" in this paragraph. If an authentication light or any other deviation from this procedure is used, it will be so indicated in this paragraph.
  - When homing beacons are used, indicate type and frequency. State placement if positioning is nonstandard.

Appendix C

- Authentication letter for air/sea rendezvous must be a mixture of dots and dashes.
- For RZ only—if none desired, state "none."
● Line 4. Personnel and supplies requested:
  - Indicate unit number of personnel and amount of supplies to be infiltrated.
  - When supplies are requested, extract key from catalog supply system (MARGE) or specify items when not listed in bundle code.
  - Identify hazardous cargo and/or nonstandard loads, if appropriate.
  - For RZ only, indicate total weight in pounds of recovery kit and package to be recovered.
● Line 5. Contact procedures (primary) for DZ, LZ, or RZ, as appropriate.
● Line 6. DZ, LZ, or RZ code name (alternate).
● Line 7. Authentication procedures (alternate) if different than primary.
● Line 8. Contact procedure (alternate) if different than primary.
● Line 9. Additional information:
  - Indicate light pattern.
  - Indicate type of light if other than flashlight.
  - Latest DTG mission confirmation can be received.
  - On-load or off-load airfield for troops and supplies.
  - For DZ only, state desired drop altitude and method; for example, static line or HALO. (Alternate DZ altitude and method if different than primary.)
  - For RZ only, indicate any special handling requirements.

PACER—TACTICAL AIR SUPPORT MISSION CONFIRMATION

C-15. The JFACC will transmit a tactical air support mission confirmation message to the operational base for further relay to the requestor. If the mission is approved, the confirmation will include essential planning factors as coordinated by the HQ with the supporting TACAIR resources. The lines of the report contain the following information:
● Line 1. Mission designator and mission number.
● Line 2. Target locations.
● Line 3. Approval or disapproval (explain disapprovals in remarks).
● Line 4. Time on target (DTG).
● Line 5. Attack forces call signs and frequency (if appropriate).
● Line 6. Additional information.

RINGO—DZ, LZ, OR RZ MISSION CONFIRMATION

C-16. The FOB transmits the DZ, LZ, or RZ mission confirmation to the outstation ASAP after the mission request has been approved. The outstation will acknowledge receipt of mission confirmation on the next scheduled contact with the base. If the outstation does not positively acknowledge receipt of the mission confirmation, the component HQ should determine whether the mission is to be executed or canceled. Final decision authority lies with the controlling HQ. Even if the outstation does not receive a mission confirmation message from the FOB, the primary LZ, DZ, or RZ and the alternate, if requested, will be manned as per the request. The lines of the report contain the following information:
● Line 1. Mission designator and LZ, DZ, or RZ code name:
  - TIGER—Personnel DZ.
  - BRAVE—Resupply DZ.
  - MOUSE—Water DZ.
  - PLANK—Fixed-wing LZ.
  - CAMEL—Rotary-wing LZ.
  - RISER—RZ.

- Line 2. Track (or run in quadrant for RZ) on primary.
- Line 3. DTG of landing, drop, or recovery. If requesting party asks for recovery kit, aircrew will drop it on TOT and the actual recovery will be 20 minutes later.
- Line 4. DZ, LZ, or RZ code name (alternate).
- Line 5. Track (or run in quadrant for RZ) on alternate.
- Line 6. DTG of landing, drop, or recovery on alternate.
- Line 7. Type of aircraft to be used.
- Line 8. For DZ, indicate—
  - Number of personnel and containers.
  - Drop altitude in feet for primary and alternate.
  - Position of flashing-light signal in DZ marking.
  - On-load airfield (for infiltration missions, include final team briefings and load times).
- Line 9. For LZ, indicate—
  - On-load airfield and load date-time for infiltration missions. List off-load airfield and date-time of estimated time of arrival (ETA) for exfiltration missions.
  - Required LZ length in feet beyond "C" light and number of lights to be used.
- Line 10. For RZ, indicate—
  - The time that lift line lights will be turned on if recovery is to be accomplished during darkness and recovery kit has been pre-positioned.
  - Off-load airfield ETA.
- Line 11. Additional information. Confirm the type, frequency, and operation time of homing beacon, and explain any changes to original mission request.

SHEAT—BEACON BOMBING REQUEST

C-17. The SFODA uses the beacon bombing request message to provide necessary information to conduct beacon offset and on-target bombing. Different aircraft require different information for a beacon bombing mission. Refer to the SAV SER SUP 6, Item 114, for more information. The lines of the message contain the following information:

- Line 1. Target description.
- Line 2. Target location.
- Line 3. TOT (eight-digit coordinates).
- Line 4. Beacon location (no earlier than, NLT, or ASAP).
- Line 5. Beacon type code.
- Line 6. Beacon-to-target range (meters).
- Line 7. Beacon elevation (meters).
- Line 8. Target elevation (meters).
- Line 9. Additional information. Run-in heading, if applicable.

SITED—LZ, DZ, RZ, BLS, OR HLZ APPROVAL AND DISAPPROVAL

C-18. The FOB evaluates the GRAZE reports ASAP after receipt and approval or disapproval by the base using the SITED report. The report will include any limitations for approval or reason for disapproval. The lines of the report contain the following information:

- Line 1. LZ, DZ, RZ, BLS, HLZ code name and type:
  - TIGER—Personnel DZ.
  - BRAVE—Resupply DZ.
  - MOUSE—Water DZ.
  - PLANK—Fixed-wing LZ.

Appendix C

- CAMEL—Rotary-wing LZ.
- RISER—RZ.
* Line 2. Approval or disapproval.
* Line 3. Remarks (for approved sites, list any limitations; for disapproved sites, list reasons).

Appendix D

PIBAL System

Soldiers can use the PIBAL system to compute the MEW. The following text and tables discuss the equipment used to compute the MEW by the PIBAL method and the procedure for computing MEW by the PIBAL method.

EQUIPMENT

D-1. Soldiers use the following equipment to compute the MEW by the PIBAL method:
- Helium source.
- PIBAL, 10 or 30 grams.
- Drift scale or other device for measuring from 0 to 90 degrees.
- Balloon measuring tape (to measure balloon circumference). Circumference for 10-gram PIBAL is 57 inches during daylight hours and 74 inches at night. Circumference for 30-gram PIBAL is 78 inches during daylight hours and 94 inches at night.
- PIBAL lighting units (Type 5) for night use (liquid-activated lights).
- Compass.
- Conversion charts or ascension table (10- and 30-gram PIBAL) (Tables D-1 and D-2, pages D-1 and D-2).
- Watch with a second hand.

Table D-1. Wind speed in knots for a 10-gram helium balloon

Elevation Angle	Drop Altitude in Feet											Ascension Table	
	500	750	1000	1250	1500	1750	2000	2500	3000	3500	4000	4500	Time	Altitude in Feet
70	02	02	01	01	01	01	01	01	01	01	01	01		
60	03	02	02	02	02	02	02	02	02	02	02	02		
55	03	03	03	03	03	03	03	03	03	03	03	03		
50	04	04	03	03	03	03	03	03	03	03	03	03	0:10	80
45	05	04	04	04	04	04	04	04	04	04	04	04	0:20	170
40	06	05	05	05	05	05	05	04	04	04	04	04	0:30	250
35	07	06	06	06	06	05	05	05	05	05	05	05	0:40	330
30	08	07	07	07	07	07	07	07	06	06	06	06	0:50	400
25	10	09	09	09	08	08	08	08	08	08	08	08	1:02	500
24	10	10	09	09	09	09	08	08	08	08	08	08	1:15	540
23	10	10	10	09	09	09	09	08	08	08	08	08	1:20	610
22	12	11	10	10	10	10	09	09	09	09	09	09	1:30	670
21	12	11	11	10	10	10	10	10	10	10	10	10	1:43	750
20	13	12	11	11	11	11	10	10	10	10	10	10	1:50	790
19	14	13	12	12	11	11	11	11	11	11	11	11	2:25	1000
18	15	13	13	12	12	12	12	11	11	11	11	11	2:44	1100
17	15	14	13	13	13	13	12	12	12	12	12	12	3:00	1250
16	17	15	14	14	14	13	13	13	13	13	13	13	3:49	1500

Appendix D

Table D-1. Wind speed in knots for a 10-gram helium balloon (continued)

Elevation Angle	Drop Altitude in Feet											Ascension Table	
	500	750	1000	1250	1500	1750	2000	2500	3000	3500	4000	4500	Time	Altitude in Feet
15	18	16	15	15	14	14	14	14	14	14	14	14	4:30	1750
14	19	18	18	18	16	15	15	15	15	15	15	15	5:11	2000
13	21	19	18	17	17	17	17	16	16	16	16	16	6:34	2500
12	22	20	19	19	18	18	18	18	18	17	17	17	7:58	3000
11	24	22	21	21	20	20	20	19	19	19	19	19	9:22	3500
10	27	25	23	23	22	22	22	21	21	21	21	21	10:44	4000
09	30	27	25	25	25	24	24	24	23	23	23	23	12:08	4500

Table D-2. Wind speed in knots for a 30-gram helium balloon

Elevation Angle	Drop Altitude in Feet											Ascension Table	
	500	750	1000	1250	1500	1750	2000	2500	3000	3500	4000	4500	Time	Altitude in Feet
80	01	01	01	01	01	01	01	01	01	01	01	01		
70	03	03	03	02	02	02	02	02	02	02	02	02		
60	04	04	04	04	04	04	04	04	04	04	04	04		
55	05	05	05	05	05	05	05	05	05	05	04	04	0:10	120
50	06	06	06	06	06	06	06	06	05	05	05	05	0:20	240
45	07	07	07	07	07	07	07	07	06	06	06	06	0:30	360
40	09	08	08	08	08	08	08	08	08	08	08	08	0:42	400
35	10	10	10	10	10	10	10	09	09	09	09	09	0:50	500
30	12	12	12	12	12	12	12	11	11	11	11	11	1:02	600
25	15	15	15	15	15	15	14	14	14	14	14	14	1:10	830
24	16	16	15	15	15	15	15	15	15	15	15	15	1:17	1000
23	17	17	16	16	15	15	15	15	15	15	15	15	1:46	1250
22	18	18	17	17	17	17	17	16	16	16	16	16	2:10	1500
21	19	19	18	18	18	17	17	17	17	17	17	17	2:34	1750
20	20	20	19	19	19	19	18	18	18	18	18	17	2:56	2000
19	21	20	20	20	20	20	19	19	19	19	19	18	3:43	2500
18	22	22	21	21	21	21	20	20	20	20	20	20	4:31	3000
17	23	23	23	22	22	22	22	22	21	21	21	21	5:21	3500
16	25	25	24	24	24	24	23	23	23	23	22	22	6:09	4000
15	27	27	26	26	25	25	25	24	24	24	24	24	7:00	4500
14	29	29	28	27	27	27	27	26	26	26	25			
13	31	30	30	30	30	29	29	29	28	28	28	27		

PROCEDURE

D-2. Soldiers follow the procedures below to measure MEW using the PIBAL:

- Fill the 10-gram or 30-gram balloon with helium to the required size. Inflate the 10-gram balloon with helium to a circumference of 57 inches during daylight hours or 74 inches at night. Inflate the 30-gram balloon with helium to a circumference of 78 inches during daylight hours or 94 inches at night. This increase in size compensates for the weight of a small marking light attached to the balloon used for night observations.
- At night, attach a marking light. Attach a 6-inch chemical light or another type of light activated by immersing it in water to the balloon.

PIBAL System

- Release the balloon, and begin timing.
- Using Tables D-1 and D-2, determine the ascent times required for the balloon to reach various altitudes. This method is also used to estimate the base altitude of cloud layers by determining the ascension time for the balloon until obscured by the cloud base.
- During ascent, look for unusual movement by the balloon, which indicates erratic wind conditions. The altitude of these occurrences, if significant, should be included in the MEW report to the aircraft.
- When the balloon reaches drop altitude, measure the elevation angle with a pocket transit theodolite, clinometer, or any other accurate means available.
- Measure the magnetic azimuth to the balloon and note the reciprocal heading. This factor will give the MEW wind direction.
- Referring to the scale on the left side of Table D-1 or D-2, locate the angle that corresponds to the angle measured. Move horizontally across the table to the vertical column that corresponds to the drop altitude. The value at the intersection of these two lines is the MEW wind speed in knots.

D-3. When transmitting the MEW, the Soldier ensures it is identified as the "MEW" and the altitude to which it was taken is included. He also reports any indication of erratic winds or wind shear at this time. (Phraseology: LIFTER ONE SIX, MEAN EFFECTIVE WIND TO ONE THOUSAND FEET, THREE FIVE ZERO AT ONE NINE.)

This page intentionally left blank.

Appendix E
Malfunction Report

The post air mission forms in this appendix are the required reports for training operations. These reports are self-explanatory and require little time to complete. The post air mission reports for training operations are as follows:
- USASOC Form 1051-R-E (United States Army Special Operations Command Airborne Operations Flash Report) (Figure E-1, pages E-2 and E-3).
- USASOC Malfunction Officer/NCO Checklist (Figure E-2, page E-4).

Appendix E

UNITED STATES ARMY SPECIAL OPERATIONS COMMAND
AIRBORNE OPERATIONS FLASH REPORT
(USASOC REG 350-2)

DATE:

1. Unit Designation:	2. Air mission/Line No:
3. Type/Number Aircraft:	4. Supporting MAC Unit:
5. Date/Time of Drop:	6. Number of Jumpers:
7. Number/Type Chutes:	8. Drop Altitude: Ft. (AGL)
9. Drop Zone:	10. Track of Aircraft: Degrees
11. DZ Code Letter:	12. Winds at Drop Altitude: Knts
13. Winds at Surface: Knts	14. Weather Conditions: Temperature:

15. Type of Airborne Operations:

Static Line - CARP	GRS	Blind Drop/AWADS	VIRS	WDI	Other
MFF - HALO	HAHO	JM Directed	HARPS	WDI	Other

16. Number/Type Containers/Equipment Dropped:

Door Bundles ☐ CDS ☐ HE ☐ Other ☐

17. Number of Jumpers evacuated from DZ:
(Specify injuries in Section A on page 2 of this form.)

18. Remarks (include unusual incidents such as Towed Jumpers, Aborts, Jumpers dropped off DZ, etc.): (Specify malfunction/entanglement in Section B, page 2 of this form.)

19. Headquarters Report Submitted to:

a. Battalion	Received By:	DTG:	(Zulu)
b. Group/Regt	Received By:	DTG:	(Zulu)
c. MSC	Received By:	DTG:	(Zulu)
d. Installation	Received By:	DTG:	(Zulu)
e. USASOC EOC	Received By:	DTG:	(Zulu)

ote: USASOC EOC phone number is toll free commercial within CONS 1-800-833-3462 (within NC 1-800-642-6428).

20. DZSO/Individual Initiating Report:

Name: _____ Rank: _____

Duty Title: _____ Report Initialized: _____ (Zulu)

USASOC FORM 1051-R-E
May 98 (AOOP) Previous editions are obsolete

Figure E-1. USASOC airborne operations flash report

Malfunction Report

UNITED STATES ARMY SPECIAL OPERATIONS COMMAND AIRBORNE OPERATIONS FLASH REPORT (USASOC REG 350-2)

Section A. PERSONNEL EVACUATED

1. Name: SSN: Rank: Unit:

   Type of Injury:

   Cause of Injury-Malfunction: Entanglement: PLF: Missed DZ   DZ Obstacle
                                  Other:
   Method of Evacuation: FLA   HELO   Other:
   Medical Facility Evacuated to:

2. Name: SSN: Rank: Unit:

   Type of Injury:

   Cause of Injury-Malfunction: Entanglement: PLF: Missed DZ   DZ Obstacle
                                  Other:
   Method of Evacuation: FLA   HELO   Other:
   Medical Facility Evacuated to:

3. Name: SSN: Rank: Unit:

   Type of Injury:

   Cause of Injury-Malfunction: Entanglement: PLF: Missed DZ   DZ Obstacle
                                  Other:
   Method of Evacuation: FLA   HELO   Other:
   Medical Facility Evacuated to:

4. Name: SSN: Rank: Unit:

   Type of Injury:

   Cause of Injury-Malfunction: Entanglement: PLF: Missed DZ   DZ Obstacle
                                  Other:
   Method of Evacuation: FLA   HELO   Other:
   Medical Facility Evacuated to:

Section B. MALFUNCTION/ENTANGLEMENTS

1. Name: SSN: Rank: Unit:
   Describe Mishap:

   Action Taken:

2. Name: SSN: Rank: Unit:
   Describe Mishap:

   Action Taken:

USASOC FORM 1051-R-E
May 98 (AOOP) Previous editions are obsolete

Figure E-1. USASOC airborne operations flash report (continued)

Appendix E

USASOC MALFUNCTION OFFICER/NCO CHECKLIST

PRINT CLEARLY ALL INFORMATION BELOW. BE THOROUGH! IF MORE SPACE IS NEEDED, PRINT ON THE BACK OF THIS SHEET.

Malfunction Officer/NCO Name and Rank _____ Date _____

Unit _____ Phone Number _____

Drop Zone _____ Check In Time: _____ Check Out Time: _____

DZSTL Name/Rank/Unit/Phone Number: _____

Drop Time: _____ Type and Number of Aircraft: _____

Unit Jumping: _____ Type of Jump C/E: _____ A/NT: _____ Other: _____

Number and Type of Parachute: MC1-1C_____ T-10C_____ G-14_____ Other _____

Reserve Activation: Yes ☐   No ☐   If Yes, Reason: _____

Malfunction/Entanglement/Serious Injury/Incident

   a. Type of Incident (*Circle One*)

      1. Entanglement Until Impact

      2. Serious Injury

      3. Partial Malfunction

      4. Total Malfunction

   b. Secure the area, obtain statements, fill out DD Form 1748-2, and notify the NCOIC/OIC of the supporting rigger facility.

REPORT ANY PROBLEMS OR CHANGES (NAMES, DATES, TIMES/LOCATIONS)

Signature of the DZSTL: _____

Signature of the Malfunction Officer/NCO: _____

Figure E-2. USASOC malfunction officer/NCO checklist

Appendix F
Jump Procedures and Jumpmaster Checklists

This appendix provides tables of the standard time warnings and jump commands and the special jump procedures (time warnings and jump commands) peculiar to certain aircraft used in airborne operations. In addition, this appendix provides checklists for various aircraft.

STANDARD AND MODIFIED TIME WARNINGS

F-1. The standard time warnings for aircraft used in airborne operations are provided in Table F-1, pages F-1 and F-2. Modified time warnings are provided in Table F-2, pages F-2 and F-3.

Table F-1. Standard time warnings

	20-Minute Time Warning
	NOTE: At the 20-minute warning, all in-flight rigging should be completed.
JM	Inspect jump packs.
Parachutists	Be alert and fasten ballistic helmets. Attach A-series containers, weapons, individual equipment, Dragon missile jump packs, SMJPs, and AT-4 jump packs, as applicable.
Loadmaster	Position door bundles near jump door, and hook them up to outboard anchor line cable.
	10-Minute Time Warning
	NOTE: At the 10-minute warning, all inspections should be completed.
JM	Hook up to the inboard anchor line cable. Following hookup, begin issuing jump commands.
	NOTE: Depending on the type of jump, number of parachutists, and type of aircraft, JM may give jump commands at the 20-minute warning.
Parachutists	Stand by.
Loadmaster	Stand by.
	6-Minute Time Warning
JM	Perform jump door safety checks. Perform outside air safety checks. Perform initial clear-to-the-rear and parachutist safety checks. Check for checkpoints en route to the DZ.
Loadmaster	Once aircrew is finished with 6-minute slowdown check, take position between both jump doors and give control of the doors to the JMs.
	1-Minute Time Warning
JM	Issue time warning with the lead hand. Complete an outside safety check. Once safety check is completed, spot for checkpoints.

Appendix F

Table F-1. Standard time warnings (continued)

	30-Second Time Warning
JM	Perform final outside safety check. Obtain eye-to-eye contact with the other JM. Once eye-to-eye contact is obtained, give each other thumbs-up signal.
	NOTE: Thumbs-up signal indicates all is well outside the aircraft, the JMs have spotted the DZ, and the JMs have identified DZ-appropriate markings (smoke, VS-17 panels, or RAM). Face parachutists, and issue the command **STAND BY**.
	10-Second Time Warning
JM	After green light, JMs do the following:
	• PJM commands **GO**, PJM's Parachutist 1 exits the aircraft.
	• AJM taps his first parachutist out 1/2 second after the PJM's Parachutist 1 exits.
	NOTE: A green light is the final time warning on a USAF aircraft.
	Green Light (CARP)
JM	Command **GO**.
Parachutists	One-half second after PJM's Parachutist 1 exits, AJM taps out his Parachutist 1.
	Green Light (GMRS)
	NOTE: Green light signals only that the aircraft is over the leading edge of the DZ and all conditions are cleared to drop.
JM	Once aircraft is over RP, command **GO**.
Parachutists	One-half second after PJM's Parachutist 1 exits, AJM taps out his Parachutist 1.

Table F-2. Modified time warnings

	2 Hours and 20 (or 30) Minutes
	NOTE: All rigging and inspection should be done before 20- or 30- minute warning.
JM	Perform jumpmaster personnel inspection.
Parachutists	Begin rigging.
Loadmaster	Stand by.
Safety	Perform safety checks. Rig equipment containers.
	20-Minute Time Warning
	NOTE: The 20-minute warning may be increased to 30 minutes to provide enough time for static safety personnel to complete safety checks and to rig equipment containers.
JM	Follow standard procedures.
Parachutists	Follow standard procedures.
Loadmaster	Stand by.
Safety	Complete safety checks. Complete rigging of equipment containers.
	NOTE: Green light indicates the aircraft is at the leading edge of the DZ and all clear to drop.

Jump Procedures and Jumpmaster Checklists

Table F-2. Modified time warnings (continued)

	10-Minute Time Warning
	NOTE: The 10-minute warning may be increased to 15 minutes to allow parachutists to stow troop seats and for static personnel to complete safety checks before the 1-minute warning.
JM	Before loadmaster gives 10-minute warning, hook up to the inboard anchor line cable. Hand static line to safety, and command **SAFETY CONTROL MY STATIC LINE**. After receiving 10-minute warning, PJM issues jump commands.
Parachutists	Release and stow seats.
Loadmaster	Notify JMs of 10-minute warning.
Safety	Perform safety checks. Rig equipment containers.

STANDARD JUMP COMMANDS

F-2. Table F-3, pages F-3 and F-4, provides the standard jump commands for fixed-wing aircraft. Table F-4, page F-5, provides the standard jump commands for rotary-wing aircraft.

Table F-3. Standard jump commands

	GET READY
Parachutists	Undo safety belts.
	OUTBOARD PERSONNEL, STAND UP
Parachutists	If seated along outside wall of aircraft, stand up. Place safety belt in seat. Raise folding seats to create more room and prevent tripping over safety belts.
Safety	Ensure all parachutists place their safety belts in their seats before raising it.
	INBOARD PERSONNEL, STAND UP
Parachutists	If not seated against wall of aircraft, stand up. Place safety belt in seat. Raise folding seats.
	NOTE: On smaller aircraft with only two rows of seats, the JM may replace the commands with **PORTSIDE** (or **STARBOARD**) **PERSONNEL, STAND UP**.
Safety	Ensure all parachutists place their safety belts in their seats before raising it.
	HOOK UP
Parachutists	Hook up to the appropriate anchor line cable.
Safety	**Caution: Do not exceed the maximum number of parachutists per anchor line cable for the aircraft being used.**
	NOTE: The opening gate portion of the static line snap hook faces toward the outboard side (skin) of the aircraft. The safety wire is inserted away from the parachutist and bent down.
	CHECK STATIC LINES
Parachutists	Check own static line and that of the parachutist in front of you. Ensure line has not become unstowed or misrouted.
Safety	Ensure the last two parachutists turn toward the skin of the aircraft and the second-to-last parachutist checks the last parachutist's static line. Move from the forward portion of the aircraft to the aft end while checking all the parachutists' static lines and inspecting for the proper routing of the parachutists' leg tie-down straps for their ALICE packs. Remind parachutists to gain eye-to-eye contact while handing (arm straight and elbow locked) their static line to the safety.

Appendix F

Table F-3. Standard jump commands (continued)

	CHECK EQUIPMENT
Parachutists	Check own equipment, starting with the helmet, and ensure no sharp edges are on the rim of the ballistic helmet and the chin strap and parachutist retention straps are properly routed and secured. Then physically seat the activating lever of the chest strap ejector snap and the leg strap ejector snaps. If jumping combat equipment, ensure the ejector snap of the hook-pile tape lowering line is properly attached and seated. Use free hand to complete these actions while maintaining a firm grip on the static line bight with the other hand.
	SOUND OFF FOR EQUIPMENT CHECK
JM	Once Parachutist 1 says **"All okay, jumpmaster,"** give each other a thumbs-up that everything is okay. Make a proper bight on their static lines, turn toward skin of aircraft, and command **PARACHUTIST 1, CHECK MY STATIC LINE**.
Parachutists	If applicable, pass the **"All okay."** In case of a problem, place hand over the anchor line cable to attract the attention of the safety.
	STAND BY
JM	DOOR: Take position facing jump door to control flow of parachutists to the platform.
Parachutists	RAMP: Maintain a reverse bight in the static line with the elbow held as high as possible (elbow facing top of fuselage) to avoid a static-line mishap. Parachutist 1 moves to ramp hinge. DOOR: Parachutist 1 assumes a position by the jump door.
Safety	DOOR: Stand directly under the aft end anchor line cable support bracket. Face the parachutists. Control the static lines by accepting the static line snap hook with the outboard hand, vigorously pushing static line snap hook with the inboard hand up and into the upper trailing corner of the jump door, and not allowing any slack in the static lines to fall down into the jump door.
	GO
JM	AJM: Follow last parachutist out of the aircraft.
	PJM: Wait until AJM clears the jump platform, and then exit the aircraft.
Parachutists	RAMP: With a normal walking or shuffling pace, move onto the ramp. Exit the aircraft at an angle of about 30 degrees toward the side of the aircraft and without vigor. Place hands on the ends of reserve parachute before exiting plane.
	DOOR: First parachutist exits the aircraft. All subsequent parachutists begin moving toward the door by using a shuffle. Once the parachutists begin to move in a shuffle, they assume an elbow-locked position with the arm that is controlling their static line. The parachutists place their static line control hand so that it is nearly touching the back of the pack tray of the parachutist in front of them. This establishes the proper jump interval. Parachutists do **not** place their static line control hand in a position so that it extends past the pack tray of the parachutist in front of them. As each parachutist approaches the door, he establishes eye-to-eye contact with the safety and hands his static line to the safety. Once the safety has control of the parachutist's static line, the jumper returns his hand to the end of the reserve parachute with his fingers spread. After handing his static line to the safety (vicinity of the lead edge of the door), the parachutist executes a left or right turn (as appropriate) and faces directly toward and centered on the door with both hands over the ends of the reserve parachute, fingers spread. He continues the momentum of his movement by walking toward the door, focusing on the horizon, and stepping on the jump platform. He pushes off with either foot and **vigorously** jumps up and **out** away from the aircraft. He immediately snaps into a good tight body position and exaggerates the bend into his hips to form an "L" shape.
Safety	Immediately following PJM's exit, conduct a towed parachutist inspection. If there is no problem, turn the jump door over to the loadmaster with the thumbs-up signal.

Jump Procedures and Jumpmaster Checklists

Table F-4. Rotary-wing standard jump commands

	GET READY
JM	Command **GET READY** 4 minutes or less from the drop time. Inspect each safety belt to ensure it is clear of the parachutist and equipment.
Parachutists	Undo safety belts, and place them to the rear.
	CHECK STATIC LINES
JM	Command **CHECK STATIC LINES**. Rise and check the routing of static lines from the point of attachment to the pack tray. Ensure the static lines are properly routed and hooked up.
Parachutists	Check static lines.
	CHECK EQUIPMENT
JM	Command **CHECK EQUIPMENT**.
Parachutists	Check equipment.
	SOUND OFF FOR EQUIPMENT CHECK
JM	Command **SOUND OFF FOR EQUIPMENT CHECK**.
Parachutists	Parachutist 1 indicates equipment is okay by telling JM **"Okay jumpmaster."** Remaining parachutists do likewise in the order in which they are seated.
	SIT IN THE DOOR
JM	Command **SIT IN THE DOOR**.
Parachutists	Parachutists 1 and 2 swing their legs to the right and take sitting positions in the jump door with their feet together outside the cargo compartment. Parachutists 3 and 4 extend their legs outside and move to a sitting position. Parachutists 1 and 2 place both hands (palms down) on the floor alongside their thighs, turn their heads toward the JM, and wait. Parachutists 5 and 6 swing their legs to the left, take sitting positions in the left jump door, and follow the same procedures as Parachutists 1 and 2. Parachutists 7 and 8 extend their legs outside and move to a sitting position.
	STAND BY
JM	Command **STAND BY** 8 to 10 seconds before **GO** command.
Parachutists	Stand by, and prepare to exit.
	GO
JM	Command **GO**, and tap back of parachutist's helmet. Control parachutists' exits by ensuring 1-second interval between parachutists.
	NOTE: When aircraft is carrying cargo and parachutes, pilot releases cargo before JM gives **GO** command.
Parachutists	Exit when given the **GO** command. Exit in numerical order.

AIRCRAFT-SPECIFIC JUMP PROCEDURES

F-3. The standard jump procedures (time warning and jump commands) may differ for certain aircraft. Aircraft-specific jump procedures that differ from the standard jump procedures are listed in Tables F-5 and F-6, pages F-6 through F-8.

CASA-212

F-4. The JM issues the jump commands for the CASA-212. The CASA-212 jump procedures are provided in Table F-5, pages F-6 and F-7.

Appendix F

Table F-5. CASA-212 jump procedures

	**NOTE:** Follow standard jump procedures, and include the following:	
	**20-Minute Time Warning**	
JM	**NOTE:** The JM may give the 20-minute and 10-minute warnings on the ground depending on the time of flight. Notify parachutists of 20-minute time warning.	
	**10-Minute Time Warning**	
JM	Notify parachutists of 10-minute time warning.	
	**6-Minute Warning**	
JM	Notify parachutists of 6-minute warning.	
Parachutists	Remove seat belts or safety strap, and move them out of the way.	
	**GET READY**	
Parachutists	Focus on JM, and await jump commands.	
	**STARBOARD PERSONNEL, STAND UP**	
Parachutists	Even-numbered parachutists stand.	
	**PORTSIDE PERSONNEL, STAND UP**	
Parachutists	Odd-numbered parachutists stand.	
	**HOOK UP**	
Parachutists	Parachutists hook up to the anchor line cable with the opening gate facing port side. Even-numbered parachutists hook up between odd-numbered parachutists. Assume a reverse bite in the left hand while protecting rip cord grip with right hand.	
	**CHECK STATIC LINES**	
Parachutists	Each parachutist checks his own static line and that of parachutist in front of him.	
Safety	Inspect each parachutist's static line.	
	**CHECK EQUIPMENT**	
JM	Look for any sign of a problem. If closer than safety, attempt to correct any problems with parachutist's equipment. If parachutist's equipment cannot be corrected, unhook him, move him to front of aircraft, have him sit, and do not allow him to jump.	
Parachutists	Parachutists check their equipment. If parachutist finds a problem with his equipment, he immediately raises his left hand while keeping his right hand over the reserve rip cord grip.	
Safety	Look for any indication of a problem. If closer than the JM, attempt to correct any problems with parachutists' equipment. If the problem with a parachutist's equipment cannot be corrected, move parachutist to front of aircraft, have parachutist sit, and do not allow parachutist to jump.	
	**SOUND OFF FOR EQUIPMENT CHECK**	
JM	After receiving okay and thumbs-up signal from Parachutist 1, visually conduct a 360-degree safety check and observe for the ground reference points and DZ. **NOTE:** The JM may spot from the open aft portside jump door or from the ramp. The JM can communicate with the pilot in one of two ways: by passing information through the safety or by using the toggle by the starboard window. One click on the toggle indicates a 5-degree flat rudder left or right turn to line up the aircraft. The JM should coordinate this with the pilot before takeoff. These instructions are also located on the toggle inside the aircraft.	
Parachutists	Parachutists indicate equipment is ready by passing up the stick a thumbs-up signal. Parachutist 1 indicates to JM the parachutists' equipment is ready by stating "**All okay, jumpmaster**" and giving JM the thumbs-up signal.	

Jump Procedures and Jumpmaster Checklists

Table F-5. CASA-212 jump procedures (continued)

	1-Minute Time Warning
JM	Notify parachutists of 1-minute warning.
	STAND BY
JM	About 30 seconds from DZ, command **STAND BY**.
	GO
JM	Ensure jump light is green, and observe panel markers. Tap Parachutist 1 on back of helmet, and command **GO**. Control flow of parachutists and each parachutist's static line. After parachutists have exited, check for towed parachutist. If none, give thumbs-up signal to safety. Retrieve deployment bags. Close ramp. Unhook, store (in aviator's kit bag), and secure deployment bags until landing.
Parachutists	After JM commands **GO** and taps the back of Parachutist 1's helmet, Parachutist 1 exits the aircraft by walking off the ramp at a 45-degree angle toward the port side of the aircraft, assuming a tight body position and counting in thousands to 6,000. The remaining parachutists follow with a 1-second interval between parachutists.
Safety	After receiving thumbs-up signal from JM, notify pilot **"All parachutists clear."** Help JM retrieve deployment bags. Unhook, store (in aviator's kit bag), and secure deployment bags until landing.

C-208B

F-5. The JM issues the jump commands for the C-208B. The C-208B jump procedures are provided in Table F-6, pages F-7 and F-8.

Table F-6. C-208B jump procedures

	NOTE: Follow standard jump procedures, and include the following:
	NOTE: The aircraft is loaded one parachutist at a time in reverse order. As each parachutist enters the aircraft, he hands his static line snap hook to the JM. The JM hooks up each static line snap hook to the anchor line cable, ensuring the opening gate of the static line snap hook faces up. The JM may give the 20-minute and 10-minute time warnings on the ground depending on time of flight.
	6-Minute Time Warning
JM	Give parachutists the 6-minute warning.
	GET READY
JM	Command **GET READY**.
Parachutists	Parachutists focus on JM.
	CHECK STATIC LINES
JM	Command **CHECK STATIC LINES**. Check the first two parachutists on the starboard side and the first parachutist on the port side.
Parachutists	Parachutists check their own static line and that of the parachutist in front of them.
Safety	Check the equipment of the last three parachutists on the port side and the last two on the starboard side.

Appendix F

Table F-6. C-208B jump procedures (continued)

CHECK EQUIPMENT	
JM	Command **CHECK EQUIPMENT**. Look for any indication of a problem. When closest to a parachutist with a problem, attempt to correct problem. If the problem cannot be corrected, move the parachutist to the front of the aircraft, have the parachutist sit, and do not allow him to jump.
Parachutists	Parachutists check their equipment. When a parachutist finds any discrepancies with his equipment, he immediately raises his left hand and signals the safety or the JM, while keeping his right hand over the reserve cord grip.
Safety	Look for any indication of a problem. When closest to a parachutist with a problem, attempt to correct the problem. If the problem cannot be corrected, move the parachutist to the front of the aircraft, have the parachutist sit, and do not allow him to jump.
SOUND OFF FOR EQUIPMENT CHECK	
JM	After equipment check is complete, unfasten jump door safety strap from the leading edge of the jump door and secure it. Visually conduct a 360-degree safety check, and observe for the ground reference and DZ.
Parachutists	Each stick passes up the stick **"Okay."** Parachutist 1 of each stick tells JM **"All okay, jumpmaster"** and gives JM the thumbs-up signal.
1-Minute Time Warning	
JM	Give parachutists the 1-minute warning.
Parachutists	Parachutists watch JM and prepare for next command.
SIT IN THE DOOR	
JM	Command **SIT IN THE DOOR**. Take control of Parachutist 1's static line.
Parachutists	Starboard side Parachutist 1 moves to jump door, takes up a sitting position with his feet outside the aircraft and both hands on the floor by his sides, and prepares for a vigorous exit. Remaining parachutists maintain a reverse bite on their static lines and slide toward the jump door in stick order. Portside parachutists follow last starboard side parachutist.
GO	
JM	Ensure green light is lit, and observe the panel markers. Tap Parachutist 1, and command **GO**. Tap remaining parachutists, and command **GO**. Control parachutists' static lines once they are in the door. Once all parachutists have exited, make a towed load check. If clear, give safety the thumbs-up signal. Retrieve deployment bags, which are unhooked, stored in the aviator's kit bag, and secured until landing.
Parachutists	After JM taps him and commands **GO**, Parachutist 1 vigorously exits, assumes a tight body position, and counts in thousands to 6,000. Remaining parachutists slide into jump position, release reverse bite, and follow same procedures as Parachutist 1.
Safety	Notify the pilot **"All parachutists clear."** Help the JM retrieve the deployment bags, which are unhooked, stored in the aviator's kit bag, and secured until landing.

JM CHECKLISTS

F-6. The JM, and sometimes the pilot or his designated representative, performs an inspection of the aircraft before a jump. Figures F-1 and F-2, pages F-9 through F-11, provide aircraft-specific checklists.

Jump Procedures and Jumpmaster Checklists

Before Equipment is Loaded and Parachutists are Boarded

Seats:
- Aircraft has enough seats for number of troops.
- All seats have seat belts.
- Seat backs are secure.
- Seats are serviceable.
- Nothing is protruding through seats.
- Pairs of seats forward of each jump door have a strap attached to secure them in the upright position.

Floor:
- Nonskid surface covering is in good condition.
- Floor is clean and safe to walk on.
- Roller conveyors are stored.
- Loose equipment is secured in the cargo ramp area and does not interfere with troops.
- Equipment tie-down rings are depressed into their recesses.

Jump platforms:
- Nonskid surface covering is present and in good condition.
- There are no cracks or bends.
- Studs are locked in seat track receptacles
- Tie-down fitting is locked.
- All bolts and nuts are present.
- Platforms easily swing in and out.

Jump doors:
- No sharp edges or protrusions on doorframes.
- Doors open and close easily.

Miscellaneous:
- Day lighting system is operational.
- Night lighting system is operational.
- Airsickness bags are available.
- JM kit (extra equipment) is onboard.
- Earplugs are available.
- Heavy tape is available to secure platform and windscreen-locking lever. If jump platforms and windscreen are not installed in the doors, they must be secured to the upper ramp.
- All equipment and crew baggage are secured to the floor. During the jump briefing, parachutists warned to avoid striking or grabbing the door platform or windscreen-locking lever on the leading edge of the door.

> **WARNING**
> It is a serious hazard to the exiting parachutist if the windscreen-locking lever swings into the open door.

- Tailgate drops (MFF and bundles only).

Figure F-1. JM checklist for C-27A

Appendix F

- Ensure loadmaster installs the stops on both sides of the tailgate so it will be level with the aircraft floor when open.
- Disengage the support bracket near the door for bundle drops that use the retrieval system to pull the static lines.
- Secure the retrieval cable against the anchor line cable in several places with breakaway ties starting at the rear of the cable and ending at the tailgate hinge. This prevents the tailgate from cutting the retrieval cable during operation.

En Route Before JM Commands STAND UP

Platforms:
- Platforms are locked into the two "keyholes" on the floor and slid to the rear of the aircraft. The large portion of the keyhole slot should be visible.
- The platform-locking lever on the leading edge of the door should be in its locked position. The lug this level controls should be engaged to the doorframe.
- The platform-locking lever should be taped in place to help prevent any parachutists from inadvertently unlocking it.
- The flange on the trailing side of the platform must overlap the inside of the doorframe about 1/2 inch.

Jump lights:
- Check location of jump lights.
  - Rear at the forward, left door.
  - Rear of both jump doors.
  - High above, to the rear, and on both sides of the ramp.
  - Static-line anchor cable system.
- Ensure jump lights are functioning properly.

Anchor cable:
- Forward end of cable is firmly secured to bracket on bulkhead with three threads showing on turnbuckle.
- Cable is operational.
- Cable has no breaks, frays, or kinks.
- Cable is clean and free of rust.
- Static-line stop is present.
- Support bracket at the trailing edge of the door is locked in place to support the cables.

Emergency equipment:
- First aid kit is onboard (one).
- Fire extinguishers are onboard (two).
- Alarm system is operational.
- Emergency exits are operational and accessible.
- Sufficient emergency parachutes are available.

Aircraft Slowdown Warning at 3 Minutes Before Drop

- Doors are opened and locked in place.
- Air deflectors are extended.
- Jump platforms are locked in place.

Upon Opening Doors

- Ensure door bundles have 15-foot static lines with three drogue parachutes.
- Ensure jump platform is secure and will sustain parachutist's weight.

Figure F-1. JM checklist for C-27A (continued)

Seats:
- Adequate seats available for troop load.
- All seats have safety belts.
- Seat backs are secure.
- Seats are serviceable.
- No projections through seats.

Floor:
- Nonskid surface covering is in good condition.
- Floor is clean and safe to walk on.
- Loose equipment is secured and does not interfere with troops.

Ramp:
- There are no sharp or protruding edges on ramp.
- Nonskid surface covering is in good condition.
- Floor is clean and safe to walk on.
- Loose equipment is secured and does not interfere with troops.
- Door opens and closes easily.

Jump Lights:
- Check Set 1 (above port aft jump door) for operation.
- Check Set 2 (above starboard aft emergency door) for operation. Check alarm bell; it is the signal for exiting.

Static Line Anchor Cable System:
- *Forward support beam*:
    - Bolts, nuts, and safety wire are present.
    - Anchor cable is attached to centerline anchor point.
    - Cable bolt, locking bolt, nut, and safety wire are present.
    - Check anchor line tension indicator—red line indicator should not be seen.
- *Anchor cable*:
    - Cable has no breaks, frays, kinks.
    - Cable is clean and free of rust.
    - Swage is present.
- *Anchor line cable aft support*:
    - Cable, locking bolt, nut, and safety wire are present.

Emergency Equipment:
- First aid kits are onboard (2).
- Fire extinguishers are onboard (2).
- Alarm system is operational.

Miscellaneous:
- Lighting system is operational.
- Airsickness bags are available.
- JM kit (extra equipment) is onboard.
- Earplugs are available.
- Loose equipment and jump door (removed) are moved forward in the aircraft and secured.

Figure F-2. JM checklist for CASA-212

This page intentionally left blank.

Appendix G
Fast-Rope Troop Briefing and Operational Checklist

The fast-rope troop briefing covers detailed instructions concerning every aspect of the operation, such as the aircraft to be used, training area characteristics, uniform, equipment, and emergency procedures. The operational checklist ensures units perform the tasks for a fast-rope operation.

TROOP BRIEFING

G-1. All participants in the fast-rope operation must attend the entire briefing. The officer in charge or NCO in charge briefs all personnel participating in FRIES training. The format for the briefing is shown in Figure G-1, pages G-1 and G-2.

---

1. **Briefing Area.**
   a. Manifest check.
   b. Operations time sequence, radio call signs and frequencies, actions if radio fails, and visual signals or markings.
   c. Location, identification, and marking of—
      (1) PZ (day and night).
      (2) Infiltration site (day and night).
   d. Ground operations and loading.
   e. Heading, route, flight time, and predicted weather conditions.
   f. Altitude.
   g. Time warnings.
   h. Hand-and-arm signals.
   i. Emergencies:
      (1) PZ.
      (2) Fast-rope personnel on ropes.
2. **Rehearsal of Actions in Helicopter.**
   a. Seating order.
   b. Exit order.
   c. Wearing of seat belts or improvised restraints.
   d. Securing of equipment.
   e. Hand-and-arm signals or emergency signals for day and/or night operations.
   f. Movement as directed.
3. **Fast Roping.**
   a. Releasing of seat belts.
   b. Hand-and-arm signals.
   c. Movement as directed.
   d. Positioning of equipment.

Figure G-1. Format for fast-rope troop briefing

Appendix G

e. Exiting of aircraft.
f. Accountability of personnel and/or equipment.

4. **Emergencies.** Personnel will adhere to the following procedures in an emergency. Personnel use sound judgment to determine the correct action to take.

   a. Aircraft emergency:
      (1) Stop stick (cease FRIES operation).
      (2) Ensure ropers are clear.
      (3) Take appropriate action.
   b. Unsafe drift or premature liftoff:
      (1) Lock-in.
      (2) Stop stick.
      (3) Get back on target.
      (4) Continue operations.
   c. Hung rope:
      (1) Ensure ropers are clear.
      (2) Descend.
      (3) Release rope.
   d. No communications:
      (1) The signal for **STOP STICK** is a clenched fist in the chest.
      (2) The signal for **ROPERS** is pointing a finger toward the exit.
      (3) The signal for **AIRCRAFT MOVEMENT** is an open palm moving in and facing toward direction required.
      (4) The signal for **STOP AIRCRAFT** is a clenched fist.
   e. Known hazards on or around the infiltration site.

Figure G-1. Format for fast-rope troop briefing (continued)

FAST-ROPE OPERATIONS CHECKLIST

G-2. The sequence of actions and duties of individuals presented in Figure G-2, page G-3, are recommended. However, the FRM may modify the checklist based on unit procedures, mission, and type of aircraft.

1. **Preflight Actions.**
   a. Receive briefing from S-3 air.
   b. Conduct pilot and crew brief.
   c. Conduct aircraft inspection and rigging.
   d. Conduct safety brief and operations brief.
   e. Conduct static load rehearsal.
   f. FRM inspects personnel and equipment.
2. **Load Aircraft.**
   a. Position equipment and personnel.
   b. Ensure personnel are strapped or tied in to the aircraft.
3. **Actions in Flight.**
   a. Monitor command net.
   b. Monitor aircrew net.
   c. Monitor flight route.
4. **Actions at 10-Minute Warning.**
   a. Issue 10-minute time warning.
   b. Check jumper equipment.
   c. Check fast ropes, platform, and hookup.
   d. Open aircraft doors, if required.
   e. Secure fast-rope bar in position.
5. **Actions at 6-Minute Warning.**
   a. Issue time warning.
   b. Position personnel and equipment.
6. **Actions at 1-Minute Warning.**
   a. Issue time warning.
   b. Unhook personnel.
   c. Break chemical lights.
7. **Actions at GO.** Ropers exit aircraft upon FRM's signal.
8. **Postmission Actions.** Personnel account for themselves and equipment upon completion of the FRIES operation.

Figure G-2. Fast-rope operations checklist

This page intentionally left blank.

Appendix H
Castmaster Briefing

Helicopter cast and recovery operations begin with the CM briefing. This briefing covers detailed instructions concerning every aspect of the operation, to include a description of the aircraft to be used, casting area characteristics, uniform and equipment, and emergency procedures. All swimmers, the pilot in command, and the airborne mission commander must attend the entire briefing. A recommended briefing format is in Figure H-1, pages H-1 and H-2.

---

1. **Briefing Area.**
    a. Manifest check.
    b. Time sequence for the operation, to include radio call signals and frequencies, action for radio failure, and smoke codes and visual signals.
    c. Flight routes, checkpoints, and flight time.
    d. Location and identification of cast area:
        (1) Markings (day or night).
        (2) Obstacle markings (day or night).
    e. Cast altitude and speed (maximum altitude is 10 feet and maximum speed is 10 knots actual airspeed).
    f. Type aircraft (number and formation).
    g. Number of sticks, load order, seating arrangement, and exit order.
    h. Number of passes.
    i. Water depth and obstacles (minimum water depth is 10 feet).
    j. Location and marking of safety boats.
    k. Conduct of overall operation.
    l. Cast and recovery rehearsal, if applicable.
    m. Abort procedures and signals.
    n. Pilot and CM briefing.
    o. Positioning of equipment.
    p. Review of jump commands, hand-and-arm signals, and signals for swimmers to use once in water.
    q. Movement in aircraft, when permitted.
    r. CM inspection of personnel and equipment before boarding the aircraft. Duties and responsibilities of the pilot in command, aircrew, CM, safety boat NCO, and safety swimmers.

2. **In the Helicopter.**
    a. Secure seat belts or safety straps and equipment.
    b. Watch for CM signals.
    c. Move as directed.

Figure H-1. Recommended CM briefing format

Appendix H

3. **Cast.**
   a. Release seat belt or safety strap.
   b. Position equipment.
   c. Receive CM signals.
   d. Exit aircraft.
4. **After Exiting Helicopter.**
   a. Assume proper body position for water entry.
   b. Signal that you are okay.
   c. Don swimming gear.
   d. Secure equipment, and account for personnel and equipment.
   e. Execute remainder of operation.
5. **Recovery.**
   a. Assume correct swimmer alignment.
   b. Follow procedures or techniques for recovery system used.
   c. If a rope ladder is used, snare ladder with arm and stabilize ladder or follow ascending procedures for the recovery system used.
   d. Board aircraft.
   e. Secure seat belts or safety straps.
   f. Account for personnel and equipment.

Figure H-1. Recommended CM briefing format (continued)

Glossary

SECTION I – ACRONYMS AND ABBREVIATIONS

AA	antiaircraft
ACC	air component commander
ADEPT	alternating door exit procedures for training
ADF	automatic direction finder
AF	Air Force
AFB	Air Force base
AFI	Air Force Instruction
AFR	Air Force Reserve
AFRM	assistant fast-rope master
AFSOB	Air Force special operations base
AFSOC	Air Force Special Operations Command
AFSOF	Air Force special operations forces
AGL	above ground level
AHHS	altitude hold and hover stabilization
AHO	above the highest obstruction
AJM	assistant jumpmaster
ALCE	airlift control element
ALICE	all-purpose, lightweight individual carrying equipment
ALLTV	all-light-level television
AM	amplitude modulation
AMC	Air Mobility Command
AMP-2	Airfield Marking Pattern-2
ANG	Air National Guard
AO	area of operations
AOB	advanced operations base
AOD	automatic-opening device
AR	Army regulation
ARNG	Army National Guard
ASAP	as soon as possible
ASE	aircraft survivability equipment
AST	area specialty team
ATC	air traffic control
AWADS	adverse weather aerial delivery system
AZAR	assault zone availability report
BASS	ballistic armor subsystem
BC	Black Crow

Glossary

BLS	beach landing site
BMNT	beginning morning nautical twilight
BTR	beacon tracking radar
C2	command and control
CARP	computed air release point
CAS	close air support
CASEVAC	casualty evacuation
CASO	casting area safety officer
CBRN	chemical, biological, radiological, and nuclear
CCT	combat control team
CDS	container delivery system
CM	castmaster
COA	course of action
CONUS	continental United States
CRRC	combat rubber raiding craft
CRS	container retrieval system
CSAR	combat search and rescue
CT	counterterrorism
CWIE	container, weapon, individual equipment
CWS	combat weather squadron
DA	direct action; Department of the Army (form)
DAP	defensive armed penetrator
dB	decibel
DD	Department of Defense (form)
DF	direction finding
DIP	desired impact point
DME	distance measuring equipment
DOK	Director of Operations, Plans (USAF term)
DRU	direct reporting unit
DTA	dual target attack
DTG	date-time group
DZ	drop zone
DZSO	drop zone safety officer
DZST	drop zone support team
DZSTL	drop zone support team leader
ECCM	electronic counter-countermeasures
ECM	electronic countermeasures
EMT	emergency medical technician
ERDS	external raft delivery system
ERP	effective radiated power
ESSS	external stores support system

ETA	estimated time of arrival
EW	electronic warfare
EZ	extraction zone
F	Fahrenheit
FAA	Federal Aviation Administration
FAC	forward air controller
FARP	forward arming and refueling point
fax	facsimile
FCO	fire control officer
FFAR	folding fin aerial rocket
FID	foreign internal defense
FLIR	forward-looking infrared
FM	field manual
FOB	forward operating base
FOV	field of view
FRIES	fast-rope insertion and extraction system
FRM	fast-rope master
ft	feet
ft/sec	feet per second
GMRS	ground-marked release system
GPS	global positioning system
GS	ground speed
GSO	ground safety officer
GTA	ground-to-air
GUC	ground unit commander
GW	gross weight
HAARS	high-altitude airdrop resupply system
HAHO	high altitude high opening
HALO	high altitude low opening
HARP	high altitude release point
HE	heavy equipment; high explosive
HEDP	high-explosive dual-purpose
HEI	high explosives incendiary
HF	high frequency
HLZ	helicopter landing zone
HN	host nation
HQ	headquarters
hr	hour
HSLLADS	high-speed, low-level aerial delivery system
HV	high-velocity
IAW	in accordance with

ICS	internal communications system	
IDS	infrared detection set	
IFR	instrument flight rules	
ILS	instrument landing system	
IMC	instrument meteorological conditions	
INS	inertial navigation system	
IP	initial point	
IR	infrared	
IRCM	infrared countermeasures	
ISB	intermediate staging base	
JFACC	joint force air component commander	
JM	jumpmaster	
JP	joint publication	
JSOA	joint special operations area	
JSOTF	joint special operations task force	
kHz	kilohertz	
KIAS	knots indicated airspeed	
KTAS	knots true airspeed	
lb	pound	
LLLTV	low-light-level television	
LORAN	long-range navigation	
LZ	landing zone	
m	meter	
MEDEVAC	medical evacuation	
METT-TC	mission, enemy, terrain and weather, troops and support available–time available and civil considerations (Army)	
MEW	mean effective wind	
MF	medium frequency	
MFF	military free fall	
min	minute	
mm	millimeter	
MONOHUD	monocular head-up display	
mph	miles per hour	
MPI	multiple points of impact	
MPS	meters per second	
MSL	mean sea level	
NA	not applicable	
NAVAID	navigational aid	
NCO	noncommissioned officer	
NCOIC	noncommissioned officer in charge	
NLT	not later than	

Glossary

NM	nautical mile
NTO	night tactical operation
NVG	night vision goggle
OPLAN	operations plan
OPSEC	operations security
PI	point of impact
PIBAL	pilot balloon
PJM	primary jumpmaster
PLS	personnel locator system
PRTS	parachutist rough-terrain system
PSYOP	Psychological Operations
PZ	pickup zone
RAM	raised angle marker
RCL	reception committee leader
reg	regulation
RP	release point
RT	radar transponder
RWR	radar warning receiver
RZ	recovery zone
SATCOM	satellite communications
SAVSERSUP	signal audio visual service supplement
SCNS	self-contained navigation system
SF	Special Forces
SFOD	Special Forces operational detachment
SFODA	Special Forces operational detachment A
SFODB	Special Forces operational detachment B
SFODC	Special Forces operational detachment C
SKE	station-keeping equipment
SL	sea level
SLAP	Sabot-launched armor piercing
SMC	Seattle Manufacturing Company
SO	special operations
SOA	special operations aviation
SOAR	special operations aviation regiment
SOCCE	special operations command and control element
SOF	special operations forces
SOG	special operations group
SOI	signal operating instructions
SOLL	special operations low-level
SOP	standing operating procedure
SOS	special operations squadron

Glossary

SOTF	special operations task force
SOW	special operations wing
SPIES	special patrol infiltration and extraction system
SR	special reconnaissance
STANAG	standardization agreement
STAR	surface-to-air recovery
STG	special tactics group
STOL	short takeoff and landing
STS	special tactics squadron
STT	special tactics team
TACAIR	tactical air
TACAN	tactical air navigation
TAPO	Technology Application Program Office
TM	technical manual
TOT	time on target
TV	television
UARRSI	Universal Aerial Refueling Receptacle Slipway Installation
UHF	ultrahigh frequency
U.S.	United States
USAF	United States Air Force
USAJFKSWCS	United States Army John F. Kennedy Special Warfare Center and School
USAR	United States Army Reserve
USASOC	United States Army Special Operations Command
USMC	United States Marine Corps
USSOCOM	United States Special Operations Command
UTM	universal transverse mercator
UW	unconventional warfare
VHF	very high frequency
VIRS	verbally initiated release system
VMC	visual meteorological conditions
VOR	VHF omnidirectional range
WSVC	wind streamer vector count
ZM	zone marker

References

SOURCES USED

These are the sources quoted or paraphrased in this publication.

AF Instruction 11-201, *Flight Information*, 1 September 1997

AF Instruction 11-202, Volume 3, *General Flight Rules*, 5 April 2006

AF Instruction 13-217, *Drop Zone and Landing Zone Operations*, 10 May 2007

AMC Reg 55-60, *Assault Zone Procedures*, October 1992

AR 95-1, *Flight Regulations*, 3 February 2006

AR 25-2, *Information Systems Security*, 14 November 2003

Executive Order 12333, *United States Intelligence Activities*, 4 December 1981

FM 3-05.60, *Army Special Operations Forces Aviation Operation*, 30 October 2007

FM 3-05.211, *Special Forces Military Free-Fall Operations*, 6 April 2005

FM 3-21.38, *Pathfinder Operations*, 1 October 2002

FM 3-21.220, *Static Line Parachuting Techniques and Tactics*, 23 September 2003

FM 3-97.6, *Mountain Operations*, 28 November 2000

FM 4-20.142, *Airdrop of Supplies and Equipment: Rigging Loads for Special Operations*, 19 September 2007

FM 5-430-00-1, *Planning and Design of Roads, Airfields, and Heliports in the Theater of Operations—Road Design*, 26 August 1994

FM 5-430-00-2, *Planning and Design of Roads, Airfields, and Heliports in the Theater of Operations—Airfield and Heliport Design*, 29 September 1994

FM 10-500-3, *Airdrop of Supplies and Equipment Rigging Containers*, 8 December 1992

FM 10-550, *Airdrop of Supplies and Equipment: Rigging Stinger Weapon Systems and Missiles*, 29 May 1984

FM 20-11, *U.S. Navy Dive Manual, Volumes I and II*, March 2001

FM 90-26, *Airborne Operations*, 18 December 1990

JP 3-05, *Doctrine for Joint Special Operations*, 17 December 2003

JP 3-09.1, *Joint Tactics, Techniques, and Procedures for Laser Operations*, 28 May 1999

STANAG 3597, Edition 3, *Helicopter Tactical or Non-Permanent Landing Sites*, 20 February 1986

STANAG 3601, Edition 3, *Criteria for Selecting and Marking of Landing Zones for Fixed Wing Transport Aircraft*, 17 October 1995

TM 10-1670-201-23, *Organizational and Direct Support Maintenance Manual for General Maintenance of Parachutes and Other Airdrop Equipment*, 30 October 1973

TM 10-1670-262-12&P, *Operator and Unit Maintenance Manual Including Repair Parts and Special Tools List Personnel Insertion/Extraction Systems for STABO..., Fast Rope Insertion/Extraction System..., and Anchoring Device*, 25 September 1992

TM 38-250, *Preparing Hazardous Materials for Military Air Shipment*, 15 April 2007

USASOC Reg 350-1, *Training ARSOF Active Component and Reserve Component*, 1 September 2005

USASOC Reg 350-2, *Training Airborne Operations*, 4 June 2008

USSOCOM M 350-6, *Special Operations Forces Infiltration/Exfiltration Operations*, 25 August 2004

World Meteorological Organization, *Manual on Codes*, Number 306, Volume I.1, International Codes, 1995

DOCUMENTS NEEDED

These documents must be available to the intended users of this publication.

AF Form 3822 (LZ Survey)
AF Form 3823 (Drop Zone Survey)
AMC Form 168 (Airdrop/Airland/Extraction Zone Control Log)
DA Form 5752-R (Rope Log [History and Usage])
USASOC Form 1051-R-E (Airborne Operations Flash Report)

Index

A
aerial resupply, 1-9 through 1-13
air support unit, 1-12, 1-13
airborne infiltration techniques, 1-3 through 1-8
aircraft classifications, 4-1, 4-2

C
cargo bag, 8-5, 8-6
castmaster briefing, H-1
clandestine operations, 6-10, 6-11, 6-18, 6-23
combat search and rescue (CSAR), 5-16, 6-18
command and control (C2), C-1 through C-3
covert, 5-22, 5-27

D
drop zones,
    selection of, 3-1 through 3-8
    types of, 3-9 through 3-13

E
external raft delivery system (ERDS) operation, 11-20

F
fast-rope insertion and extraction system (FRIES), 5-29, 10-1 through 10-9
fast-rope troop briefing, G-1

H
helicopter rigging, 9-8 through 9-18
helocasting, 1-3, 11-9 through 11-11
host nation surveys, 3-38

I
inspection of SPIES, 9-6, 9-7

J
joint mission briefing, 2-7
joint special operations task force (JSOTF), 1-13
jump procedures, F-1

K
K-duck fixture, 11-12

L
landing zones considerations, 4-2, 4-3
level turning radius, 3-9

M
marking landing zones, 4-19 through 4-12
marking patterns, 3-24 through 3-26
mission, enemy, terrain and weather, troops and support available, time available, and civil considerations (METT-TC), 2-1, 2-2

N
nonstandard aircraft, 7-1

O
obstacle clearance, 4-22
operational requirements, 9-7, 10-11, 11-6

P
point of impact, 3-13, 3-14
poncho-expedient parachute, 8-7
premission planning, 11-7 through 11-9

R
rates of descent, 2-7, 2-8
rigging procedures, 8-1, 11-12

S
safety, 2-5, 10-2, 11-1
snow landing zone, 4-31
SPIES equipment, 9-5, 9-6
standard jump commands, F-3
static-line drop zone, 3-24
sustainment training, 9-2, 10-14

T
training objectives, 9-1

U
unconventional warfare (UW), 3-31, 4-21

This page intentionally left blank.

FM 3-05.210
27 February 2009

By Order of the Secretary of the Army:

                                GEORGE W. CASEY, JR.
                                General, United States Army
                                       Chief of Staff

Official:

JOYCE E. MORROW
*Administrative Assistant to the
Secretary of the Army*
      0904003

DISTRIBUTION:

Active Army, Army National Guard, and U.S. Army Reserve: To be distributed in accordance with the initial distribution number (IDN) 111113, requirements for FM 3-05.210.

PIN: 081798-000

www.ingramcontent.com/pod-product-compliance
Lightning Source LLC
Chambersburg PA
CBHW050047230526
45470CB00004B/1436